69.50 80K

COMBUSTION DIAGNOSTICS BY NONINTRUSIVE METHODS

Edited by
T. D. McCay
NASA Marshall Space Flight Center
Marshall Space Flight Center, Alabama

J. A. Roux
The University of Mississippi
University, Mississippi

Volume 92
PROGRESS IN
ASTRONAUTICS AND AERONAUTICS

Martin Summerfield, Series Editor-in-Chief
Princeton Combustion Research Laboratories, Inc.
Monmouth Junction, New Jersey

Technical papers selected from the AIAA 21st Aerospace Sciences
Meeting, January 1983, and the AIAA 18th Thermophysics Conference,
June 1983, and subsequently revised for this volume.

Published by the American Institute of Aeronautics and Astronautics, Inc.
1633 Broadway, New York, N.Y. 10019

American Institute of Aeronautics and Astronautics, Inc.
New York, New York

Library of Congress Cataloging in Publication Data
Main entry under title:

Combustion diagnostics by nonintrusive methods.

  (Progress in astronautics and aeronautics; v. 92)
  Technical papers selected from AIAA 21st Aerospace Sciences Meeting, January 1983, and the AIAA 18th Thermophysics Conference, June 1983, and subsequently revised for this volume.
  Includes index.
  1. Combustion – Congresses. I. McCay, T. D. II. Roux, J. A. III. American Institute of Aeronautics and Astronautics, IV. Series.
TL507.P75 vol. 92 629.1 s (621.402'3) 84-12425
(QD516)
ISBN 0-915928-86-8

Copyright © 1984 by the American Institute of Aeronautics and Astronautics, Inc. All rights reserved. Printed in the United States of America. No part of this publication may be reproduced, distributed, or transmitted, in any form or by any means, or stored in any data base or retrieval system, without the prior written permission of the publisher.

Progress in Astronautics and Aeronautics
Series Editor-in-Chief
Martin Summerfield
Princeton Combustion Research Laboratories, Inc.

Series Associate Editors

Burton I. Edelson
*National Aeronautics
and Space Administration*

Allen E. Fuhs
Naval Postgraduate School

J. Leith Potter
Vanderbilt University

———————

Norma J. Brennan
Director, Editorial Department
AIAA

Camille S. Koorey
Series Managing Editor
AIAA

Table of Contents

Preface ix

Chapter I.   CARS 1

CARS Diagnostics of High Pressure and Temperature Gases 3
   J.H. Stufflebeam, R.J. Hall, and J.F. Verdieck
   United Technologies Research Center, East Hartford, Connecticut

CARS Thermometry and N_2 Number-Density Measurments in a Turbulent Diffusion Flame 24
   L.P. Goss, D.D. Trump, G.L. Switzer, and B.G. MacDonald,
   Systems Research Laboratories, Inc., Dayton, Ohio

Comparison of CARS Combustion Temperatures with Standard Techniques 45
   R.R. Antcliff, *Systems Research Laboratories, Inc., Dayton, Ohio* and
   O. Jarrett Jr., *NASA Langley Research Center, Hampton, Virginia*

Electronically Resonant CARS Detection of OH 58
   J.F. Verdieck, R.J. Hall, and A.C. Eckbreth,
   United Technologies Research Center, East Hartford, Connecticut

Simultaneous CARS and Luminosity Measurements in a Bluff-Body Combustion 82
   G.L. Switzer, D.D. Trump, and L.P. Goss, *Systems Research Laboratories, Inc., Dayton, Ohio,* and W.M. Roquemore,
   R.P. Bradley, J.S. Stutrud, and C.M. Reeves, *Air Force Wright Aeronautical Laboratories, Wright-Patterson Air Force Base, Ohio*

Chapter II.   Laser-Induced Fluorescence 105

Nonintrusive Pressure Measurements with Laser-Induced Iodine Fluorescence 107
   J.C. McDaniel, *University of Virginia, Charlottesville, Virginia*

Laser-Induced Schlieren Effect in Sodium-Nitrogen Mixtures 132
   J.W.L. Lewis and J.D. Selman, *The University of Tennessee Space Institute, Tullahoma, Tennessee*

Use of Laser-Induced Fluorescence for Fundamental Gas-Phase
Kinetic Measurements....................................147
    A. Fontijn, *Rensselaer Polytechnic Institute, Troy, New York*

Chapter III.   Particle Diagnostics 175

Nonintrusive Laser-Based Particle Diagnostics - Invited Review.....177
    E.D. Hirleman, *Arizona State University, Tempe, Arizona*

Interpretation of Optical Measurements of Soot in Flames..........208
    R.A. Dobbins, R.J. Santoro, and H.G. Semerjian, *National Bureau
    of Standards, Washington, D.C.*

In Situ Measurement of the Complex Refractive Index of Combustion
Generated Particulates 238
    E.A. Powell and B.T. Zinn, *Georgia Institute of Technology,
    Atlanta, Georgia*

Chapter IV.   Combustion Diagnostics Applications 253

Temperature and Concentration Measurements in an Internal
Combustion Engine Using Laser Raman Spectroscopy 255
    A. zur Loye and D.A. Santavicca, *Princeton University,
    Princeton, New Jersey*

Rayleigh Thermometry with Low-Power Laser Sources 270
    D. Benhachmi, N. Younes, H. Yakout, P.E. Emmerman, and
    R. Goulard, *The George Washington University, Washington, D.C.*

Laser Tomography for Simultaneous Concentration and
Temperature Measurement in Reacting Flows 300
    S.R. Ray and H.G. Semerjian, *National Bureau of Standards,
    Washington, D.C.*

Flow Measurement in a Model Combustion Chamber.............325
    P. Magre, J. Labbé, and G. Collin, *ONERA, Châtillion, France*

Author Index for Volume 92.............................342
List of Series Volumes..................................343

Table of Contents for Companion Volume 91

Chapter I. Contamination Overview 1

Improved Methods for Characterizing Material-Induced Contamination 3
A.P.M. Glassford, R.A. Osiecki, and C.K. Liu, *Lockheed Palo Alto Research Laboratories, Palo Alto, California* and M. Hitchcock, *Air Force Materials Laboratory, Wright-Patterson Air Force Base, Ohio*

Potential for Cross Contamination for Payloads in the STS Bay 29
R.G. Moss, *Ford Aerospace and Communications Corporation, Palo Alto, California*

Chapter II. Sources and Prevention of Contamination 37

Debris from Spallation of Foam Insulation of Cryogenic Fuel Tanks in Space Launch Sytems 39
E.P. del Casal, *Energy Incorporated, Idaho Falls, Idaho*

Particle Dispersion around a Spacecraft 54
A.L. Lee, *Lockheed Missiles & Space Company, Inc., Sunnyvale, California*

Impact of the STS Ground/Launch Particle Contamination Environment on an Optical Sensor 73
L.E. Bareiss and F.J. Jarossy, *Martin Marietta Denver Aerospace, Denver, Colorado*

Analysis of Contamination Degradation of Thermal Control Surfaces on Operational Satellites 96
J.E. Ahern, R.L. Belcher, and R.D. Ruff, *Aerojet ElectroSystems Company, Azusa, California*

Abatement of Gaseous and Particulate Contamination in a Space Instrument 108
J.J. Scialdone, *NASA Goddard Space Flight Center, Greenbelt, Maryland*

Chapter III. Properties and Effects of Contamination 137

Infrared Optical Properties of Thin $CO, NO, CH_4, HCl, N_2O, O_2, N_2$, and Ar Cryofilms 139
B.E. Wood, *Calspan Field Services, Inc., Arnold Air Force Station, Tennessee*, and J.A. Roux, *University of Mississippi, Oxford, Mississippi*

Infrared Optical Properties of Solid Mixtures of Molecular Species at 20 K 162
K.F. Palmer, *Westminster College, Fulton, Missouri*, J.A. Roux *University of Mississippi, Oxford, Mississippi*, and B.E. Wood, *Calspan Field Services, Inc., Arnold Air Force Station, Tennessee*

Measurements of Infrared Optical Properties of Al_2O_3 Rocket Particles 180
W.L. Konopka, R.A. Reed, and V.S. Calia, *Grumman Aerospace Corporation, Bethpage, New York*

Improvements in Rocket Engine Nozzle and High Altitude Plume Computations 197
S.D. Smith, *Lockheed Huntsville Research & Engineering Center, Huntsville, Alabama*

α_s/ϵ_H **Measurements of Thermal Control Coatings over Four Years at Geosynchronous Altitude** 215
    D.F. Hall and A.A. Fote, *The Aerospace Corporation,*
    El Segundo, California

Calorimetric Measurements of Thermal Control Surfaces on Operational Satellites. 235
    J.E. Ahern and K. Karperos, *Aerojet ElectroSystems Company,*
    Azusa, California

Experimental Investigation of Bipropellant Exhaust Plume Flowfield, Heating, and Contamination and Comparison with the CONTAM Computer Model Predictions 261
    H. Trinks, *Technical University, Hamburg-Harburg, Hamburg,*
    Federal Republic of Germany and R.J. Hoffman, *Science*
    Applications, Inc., Los Angeles, California

Particle Sampling of Solid Rocket Motor Exhausts in High-Altitude Test Cells 293
    P.T. Girata Jr. and W.K. McGregor, *Sverdrup Technology, Inc./AEDC Group,*
    Arnold Air Force Station, Tennessee

Postfire Sampling of Solid Rocket Motors for Contamination Sources in High-Altitude Test Cells 312
    P.T. Girata Jr. and W.K. McGregor, *Sverdrup Techology,*
    Inc./AEDC Group, Arnold Air Force Station, Tennessee

Preface

This book represents a break with the tradition of previous thermophysics volumes by highlighting papers in only one fundamental subject area – combustion diagnostics by nonintrusive spectroscopic methods. Previous volumes have usually presented a broad selection of papers from AIAA thermophysics sessions, representing the spectrum of current research interests. The focus here is of current research interest in its own right, and the techniques being developed are finding broad acceptance as standard tools within the combustion and thermophysics research communities. The papers selected cover both research currently being conducted with these nonintrusive techniques and recent advances in the techniques themselves.

This volume outlines the state-of-the-art of two basic techniques – coherent antistokes Raman scattering (CARS) and laser-induced fluorescence (LIF) – and it demonstrates current diagnostic capabilities in two application areas, particle and combustion diagnostics. The ultimate goals for the techniques are to correctly diagnose gas and particle properties in the flowfields of interest. The need to develop nonintrusive techniques is apparent for all flow regimes, but it becomes of particular concern for the subsonic combustion flows so often of interest in thermophysics research. Thus, this volume contains scientific descriptions of the methods for making such measurements, primarily the measurement of gas temperature and pressure and of particle size.

The papers in this volume were drawn from the thermophysics sessions of the AIAA 21st Aerospace Sciences Meeting in Reno, Nevada in January 1983 and the AIAA 18th Thermophysics Conference in Montreal, Canada in June 1983. The papers selected were reviewed, revised, updated, and organized into four chapters covering CARS, LIF, particle diagnostics, and combustion diagnostic applications.

Chapter I contains five papers devoted to the development and application of CARS to combustion systems. *Stufflebeam, Hall,* and *Verdieck* discuss some of the practical aspects of CARS thermometry for high-pressure combustion systems and develop collisional narrowing models for proper interpretation of spectral data. *Goss, Trump, Switzer,* and *MacDonald* apply the CARS technique to the measurement of time-averaged and fluctuating temperatures within a turbulent flame. In addition, an approach to obtaining simultaneous N_2 number density measurements is demonstrated. The third paper by *Antcliff* and *Jarrett* describes a comparative study performed to verify the CARS technique as applied to a hydrogen-air flat flame burner. The CARS results for temperatures of 1000 - 2100 K are verified using the results of sodium line reversal, thin-wire thermocouples and standard heat balance techniques. *Verdieck, Hall,* and *Eckbreth* show how the temperature and concentration of a minor species can be made measurable by employing a laser frequency that is tuned to be resonant with an electronic transition. The theory for implementation of such a scheme and its application to a methane/oxygen flame are presented. The final paper in Chapter I demonstrates another CARS application where the technique is used to make temporally and spatially resolved temperature measurements in flame "turbules" behind a bluff-body combustor. *Switzer, Trump, Goss, Roquemore, Bradley, Stutrud,* and *Reeves* employ a conditional sampling approach using flame luminosity to enhance the value of the technique.

Chapter II deals with the use of laser-induced fluorescence for flow diagnostics. This first paper by *McDaniel* describes the iodine seeding of a flow to permit pressure measurements. A tunable dye laser is used to provide a sheet of laser radiation that permits a two-dimensional pressure distribution to be directly recorded. Results show that 1% pressure fluctuations in a nonsteady flow can be resolved at frequencies up to a few kilohertz while maintaining good spatial resolution. *Lewis* and *Selman* then discuss the development of a laser-induced Schlieren technique for the measurement of the trace species concentration in a gas mixture. The technique is demonstrated for low Mach number flows using a 1 W laser source with a millisecond pulse width. The third paper applies LIF to determine chemical kinetic rates. *Fontijn* gives specific examples of how LIF can and has been employed for fundamental kinetic measurements in combustion media.

Chapter III is devoted to the subject of nonintrusive particle diagnostics. The first paper (invited) by *Hirleman* provides an overview of the subject, including a discussion of the basic principles of particle light scattering and the physics of its application to particle sizing. The effects of size ensembles, multiple scattering and other experimental realities are considered and numerous sizing techniques are disussed in this excellent review. The second paper, by *Dobbins, Santoro,* and *Semerjian,* extends this line of discussion to the particular case of soot, which is strongly affected by agglomeration and the effects of particle packing. Examples of these phenomena and the consequent results are given for ethane/air diffusion flames. The final paper in Chapter III is also associated with combustion. *Powell* and *Zinn* discuss an in situ technique for measuring the complex refractive index of combustion-generated particles. The method is applied to several illustrative cases, including particles with a wide range of absorption coefficients.

Chapter IV deals with a variety of nonintrusive techniques applied to combustion environments. *Zur Loye* and *Santavicca* discuss the application of conventional Raman scattering to measure temperature and carbon monoxide concentrations in an internal combustion engine. The results show the CO concentration to be considerably different than the predicted equilibrium concentration. *Benhachmi, Younes, Yakout, Emmerman,* and *Goulard* demonstrate that even a low power laser (15 mW) can be used to perform Rayliegh thermometry of an air-methane flame. Special attention is given to filtering techniques and detector cooling to provide the necessary signal-to-noise ratio for good accuracy. *Ray* and *Semerjian* provide a discussion of a new optical diagnostic technique – laser tomography. The technique provides fast response measurement of the species concentration and temperature in nonuniform three-dimentional flows. Examples of the application of the technique to a seeded premixed flat flame are discussed and measurement accuracies within plus or minus 1.7% are demonstrated. In the only velocimeter paper included here, *Magre, Labbé,* and *Collin* employ a bicolor laser Doppler velocimeter to investigate the effects of combustion (heat release) on the time averages and fluctuations of velocity within a model combustion chamber. Although not completely successful, their work represents a promising application of nonintrusive diagnostic to substantiate otherwise unverified flow prediction methodology.

As coeditors we want to acknowledge the assistance of Mrs. Norma Brennan, Editorial Department Director, Mrs. Camille Koorey, Progress Series Managing Editor, and Dr. Martin Summerfield, Editor-in-Chief of the *AIAA Progress in Astronautics and Aeronautics* series. We also thank Dr. Fred Nelson for organizing the thermophysics sessions of the 21st AIAA Aerospace Science Meeting. Finally, we express our appreciation to the entire AIAA

Thermophysics Committee for the team effort and outstanding work that was done to solicit excellent papers for both the AIAA 21st Aerospace Sciences Meeting and the AIAA 18th Thermophysics Conference.

T.D. McCay
J.A. Roux
April 1984

Chapter I. CARS

CARS Diagnostics of High Pressure and Temperature Gases

J. H. Stufflebeam,* R. J. Hall,† and J. F. Verdieck†
United Technologies Research Center, East Hartford, Connecticut

Abstract

The practical aspects of CARS thermometry in the regime of high-pressure combustion are presented. The dominant effect of this environment is shown to be the phenomenon of collisional narrowing. Accurate account of this process is derived from theoretical foundations which involve the unique coherence properties of the CARS process. A model is developed and its parameters are specified from least-squares fits to experimental spectra obtained under controlled conditions. The experimental facility is described and appropriate data are presented to demonstrate the accuracy of the theoretical model. The data encompass the range 1 - 100 atm, 300 - 1500 K. The fitting parameters are molecule dependent and data are presented from CARS spectra of N_2, CO, and CO_2. Errors in CARS thermometry from neglect of the collisional narrowing phenomenon are illustrated and shown to be large even at moderate pressures.

Introduction

High-pressure combustion is extremely important in a variety of practical applications. Lack of suitable diagnostics has, until recently, limited the understanding of this media; physical probes may not survive and they are of questionable utility since their presence may seriously perturb the phenomena under study. Optical techniques appear ideally suited to diagnosing such environments and one in particular, coherent anti-Stokes Raman spectroscopy

---

Presented as Paper 83-1478 at the AIAA 18th Thermophysics Conference, Montreal, Canada, June 1-3, 1983. Released to AIAA to publish in all forms.
  *Research Scientist, Applied Laser Spectroscopy Laboratory.
  †Senior Research Scientist, Applied Laser Spectroscopy Laboratory.

or CARS, has already been demonstrated in several practical combustors.[1-3]

In CARS, incident laser beams at frequencies ω_1 and ω_2 (termed the pump and Stokes, respectively) interact through the third-order, nonlinear electric susceptibility $\chi^{(3)}$ to produce the coherent CARS radiation at frequency $\omega_3 = 2\omega_1 - \omega_2$. If the frequency difference $\omega_1 - \omega_2$ coincides with a Raman active vibrational mode of a certain species, then the CARS radiation is resonantly enhanced and uniquely characteristic of the species. Measurements of temperature are performed from the spectral shape of the CARS signature, i.e., the intensity distribution with frequency.[4] Figure 1 exhibits a library of spectra that demonstrates the sensitivity of N_2 CARS signatures to temperature at 1 atm. Experimental data are least-squares fitted to theoretical spectra such as those in Fig. 1 to determine the temperature of N_2.

CARS intensity distributions do however have a linewidth sensitivity, because of interference effects aris-

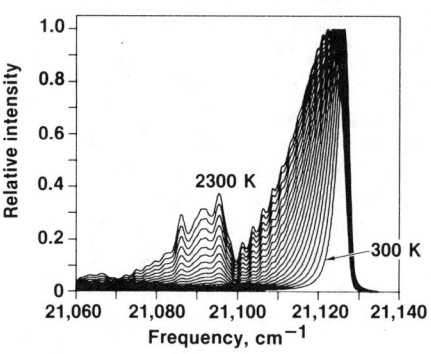

Fig. 1 Temperature dependence of N_2 CARS spectrum from 300 - 2300 in 100-K increments for 70% N_2 concentration and 2-cm^{-1} spectral resolution.

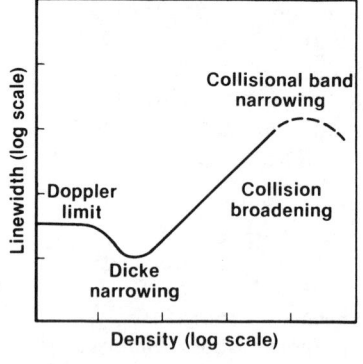

Fig. 2 Illustration of Raman linewidth dependence on density.

ing from line overlap. As the gas density increases, a variety of linewidth phenomena occur,[5] shown qualitatively by Fig. 2. As the pressure is increased the CARS signature makes a transition from the ordinary pressure broadened regime to one in which the overall bandshape is no longer adequately described by a summation of individually broadened transitions. For CARS, the collisional narrowing phenomenon is often important for pressures above 1 atm. Collisional narrowing occurs when adjacent lines have been pressure broadened to the extent that they overlap. The overlap allows communication between lines and they subsequently coalesce or collapse to a narrower bandwidth. Accurate modeling of collisional narrowing is necessary to extract temperature information from the CARS spectra and well controlled, laboratory experiments provide numerical input to the model.

The next section describes the theoretical foundations of the bandwidth narrowing phenomenon and a model is developed to predict the effect on CARS spectra at elevated pressure and temperature. This model is refined by the specification of linewidth fitting parameters which are obtained from experimental data collected under controlled conditions. The following section describes the laboratory apparatus from which the experimental spectra are obtained and then the results are presented together with the fits from the model. Results are also presented that demonstrate the errors produced when an isolated line model is used to predict temperature from collisionally narrowed spectra. The error is significant even for pressures as low as 10 atmospheres. Finally, future areas of investigation are outlined and a summary of results is provided.

Theory

The theory of pressure effects in CARS has been set forth in several literature publications,[5-8] and will not be reproduced in detail here. When the pressure has increased to the point where there is very strong overlap of adjacent transitions comprising a Raman-CARS resonance, a condition that is quite common for Q-branch vibrational signatures, it is no longer adequate to express the CARS susceptibility as independently broadened transitions. The mathematical formalism needed to describe the overall CARS bandshape for strong line overlap is very similar to that used to describe line overlap in ordinary linear absorption spectroscopy or spontaneous Raman spectroscopy.[9,10] At elevated pressure it no longer suffices to know only the linewidths of individual transitions; instead, it is necessary to specify also an off-diagonal "linewidth matrix," whose elements are related to the rates of inelastic energy transfer between vibration-rotation states in the active molecule; these elements govern the phenomenon of spectral collapse which is commonly known as collisional narrowing. In a naive view, a signature bandwidth might be expected to increase monotonically with increasing pressure; this would be the case if collisions gave rise to large, random phase shifts in radiating molecules. What is often observed however, is the collisional narrowing phenomenon[6,9,10] in which a signature is always substantially narrower than that predicted on the basis of isolated lines and which can, in many cases, actually decrease with increasing pressure. If elastic processes such as vibrational dephasing or inelastic vibrational energy transfer make important contributions to the widths of isolated lines, or if there is a strong vibrational dependence of rotationally inelastic energy tranfer rates, the off-diagonal linewidth parameters are complex and difficult to specify.[7] For most of the molecules of interest in combustion research, however, it is anticipated that the main linewidth contributions will arise from rotationally inelastic processes that will not vary strongly with vibrational state. If this is the case, then to a good approximation the off-diagonal linewidth or line mixing coefficients for Q-branch vibrational transitions ($\Delta j = 0$) are rate coefficients for inelastic energy transfer,[6,7] and the molecules jump randomly between discrete frequencies without phase changes.

The collisional narrowing effect can be regarded as the result of narrow bandwidth detection of the output of an emitter which is switching between discrete frequencies with no change of phase. When the switching frequency greatly exceeds the spacing between the discrete frequencies, an averaged signal will be recorded whose overall bandwidth will tend to narrow with further increases in the switching frequency. The recorded frequency will tend to collapse toward an averaged emission frequency. A distinction should be made here between this narrowing phenomenon which involves discrete Q-branch transitions and the phenomenon of Dicke narrowing of a Doppler profile as pressure is increased. For the molecules and pressures of interest in combustion research, Dicke narrowing is not expected to be generally important. It can be understood in terms of the same physical picture as collisional narrowing of Q-branch bands, however; instead of having discrete transition frequencies blending together through rotationally inelastic collisions, it is correct to think of Dicke narrowing arising from the blending together of discrete thermal velocity groups through elastic, momentum-transferring collisions.

A general expression for the third-order CARS susceptibility for the case in which none of the field frequencies encroaches upon electronic absorptions in the active molecule is[6,7]

$$\chi^{(3)} = (iN/\hbar)\underline{\alpha}^{\dagger} \cdot \underline{\underline{G}}^{-1} \cdot \Delta\underline{\underline{\rho}}^{(0)} \cdot \underline{\alpha} \qquad (1)$$

where N is the active molecule number density, α denotes isotropic polarizability, $\Delta\rho^{(0)}$ denotes Boltzmann population difference, and the elements of the "G matrix" are given by

$$G_{jj'} = i(\omega_p - \omega_s - \omega_j)\delta_{jj'} + \left(\frac{\Gamma_j}{2} - i\Delta_j\right)\delta_{jj'} + \gamma_{jj'}(1 - \delta_{jj'}) \qquad (2)$$

Here ω_p, ω_s, and ω_j are the frequencies of the pump, Stokes, and Raman resonance Q(j), respectively; Γ_j and Δ_j are the isolated linewidth and pressure induced line shift of Q(j), respectively; and the off-diagonal linewidth element $\gamma_{jj'}$ is equivalent to minus the first-order rate coefficient for rotational energy transfer in the direction j' - j. A perturbation expansion of Eq. 1 for small line overlap can also be useful at modest pressure.[6,11] Certain relationships involving these transfer rates follow, one of them being the principle of detailed balance

$$\rho_{jj}^{(0)} \gamma_{j'j} = \rho_{j'j'}^{(0)} \gamma_{jj'} \qquad (3)$$

and the other following from conservation of probability in inelastic collisions (scattering matrix unitarity)

$$(1 - \phi_j)(\Gamma_j/2) = -\sum_{j' \neq j} \gamma_{j'j} \qquad (4)$$

where ϕ_j is a small contribution of elastic processes to isolated linewidth. If vibrationally inelastic processes are slow, there can be no coupling between different vibrational bands, so the G-matrix will be block diagonal for the case of high-temperature gases where vibrational "hot bands" are important. At high temperatures, where large numbers of transitions must be included to accurately describe a spectrum, the analysis of line overlap effects can thus lead to rather cumbersome and time consuming calculations because, at each frequency of interest, a large, complex matrix must be inverted.

At the first level of approximation, one can employ a susceptibility expression that has been derived from an extremely basic picture of the rotational relaxation process, namely, the Gordon "rotational diffusion" model.[8,12,13] The formalism is based on the assumption that no selection rules govern the inelastic rotational energy transfer process, with a molecule losing memory of its initial rotational level and making transitions to other rotational levels at rates simply proportional to the Boltzmann population of the final state. The mixing of coherences is thus described in a simple time constant relaxation form that is not strictly as rigorous as the formalism leading to Eq. 1. The resulting susceptibility expression is extremely simple;

$$\chi_{v,v+1} = (2N/\hbar)\alpha^2_{v,v+1} \times \frac{\sum_j \Delta\rho^{(0)}_{jj}/(2\Delta\omega_j - i\Gamma_j)}{1 + i\sum_j (1-\phi_j)\Gamma_j\rho^{(0)}_{jj}/(2\Delta\omega_j - i\Gamma_j)} \quad (5)$$

with only two trivial summations required, a procedure that consumes no more computer run time than an isolated line calculation. As will be seen, calculations based on Eq. 5 give suprisingly good results, and the Gordon model result will undoubtedly play a significant role in those situations where a premium is placed on reduced computation time. Also, as will be discussed, as the pressure and line overlap increase, the predictions of various rotational relaxation models become more and more indistinguishable, with the predicted CARS spectra losing all sensitivity to model at very high pressures. In such situations, it is pointless to use the more cumbersome formalism of Eq. 1.

The problem of rigorously describing line overlap effects in CARS thus reduces, to a good approximation, to one of specifying the rotational relaxation matrix for the active molecule. This is not a trivial problem, as the rate matrix will be a function of temperature and gas composition, and can be expected to vary in a complicated way with the initial and final rotational quantum numbers. For the molecules of interest in combustion research, there are good theoretical formulations for the isolated linewidths (for example, Ref. 14), but these do not at the present time give reliable estimates of the off-diagonal elements. Even if ab-initio calculations (or detailed measurements, as in the case of molecules like CO) do eventually produce accurate relaxation rate data, it is likely that they will be cumbersome to employ in practical applications. It has been necessary to attempt to describe the relaxation rate matrix in terms of one of the semiempirical "power laws" which relate the inelastic transfer rate to simple functions of the amount of rotational energy exchanged with translation in the relaxation process,

$$N_2(V,j) + M \xrightarrow{K_{jj'}} N_2(V,j') + M + \Delta E_{jj'} \qquad (6)$$

Thus there are "exponential gap" laws,[15,16] inverse power laws,[17,18] combinations of these,[19] as well as other functional forms.[20] There is some theoretical foundation for most of these models, but they remain for the most part semiempirical fitting laws with incompletely understood physical bases. Probably the most successful of these relationships is the inverse power law; a simplified statement of this law is that the inelastic rate coefficient is assumed to be proportional to the absolute value of the rotational energy defect raised to a negative power that is the same for all initial and final quantum numbers. In these investigations the power law has been employed in the following simple form

$$-\gamma_{j'j} = K_{jj'} = K_0 \rho_{j'j'}^{(0)} |\Delta E_{jj'}|^{-\gamma} \qquad (7)$$

where K_0 is a constant independent of rotational state, and the power γ is a parameter. In this form, the model represents a kind of hybrid law involving the inverse power law and the Gordon model. Rotational symmetry should be preserved in collisions when nuclear spin statistics are a consideration ($\Delta j = \pm 2, \pm 4,...$ for N_2; in this case nuclear spin statistical weights should be divided out of the population factors $\rho_{j'j'}^{(0)}$. Detailed balance is automatically satisfied, and the parameters K_0 and γ can be determined by least-squares fitting the theoretical linewidths

$$\Gamma_j^{(th)} = 2K_0 \sum_{\ell \neq j} \rho_{\ell\ell}^{(0)} |\Delta E_{j\ell}|^{-\gamma} \qquad (8)$$

to the experimental linewidths which are assumed to be known,

$$\frac{\partial}{\partial K_0}, \frac{\partial}{\partial \gamma} \sum_j (\Gamma_j^{(exp)} - \Gamma_j^{(th)})^2 = 0 \qquad (9)$$

The linewidth fits obtained in this way are surprisingly good, as shown in Figs. 3 and 4 for N_2 and CO. In Fig.

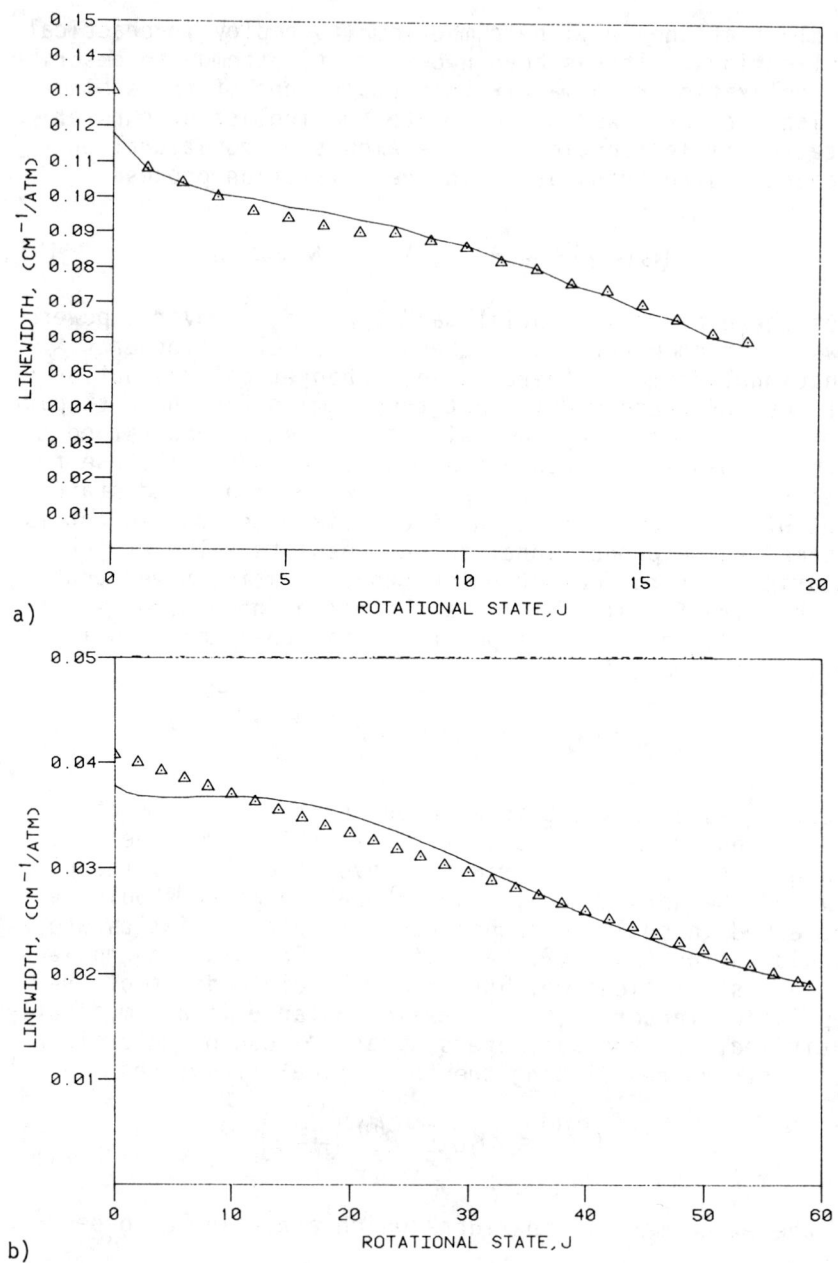

Fig. 3 Fit of simple power law model to N_2 Q-branch linewidths. a) 298 K, triangles are data of Ref. 11. b) 1730 K, triangles are data of Ref. 22.

3a the power law has been fitted to the N_2 room-temperature linewidths reported in Ref. 11. These linewidths[21] are similar to those previously reported by Rahn et al. The fit is seen to be excellent and the high-temperature fit to the theoretical and experimental N_2 linewidths reported in Ref. 22 is also quite satisfactory (Fig. 3b). In CO, the agreement with the linewidths measured in Ref.

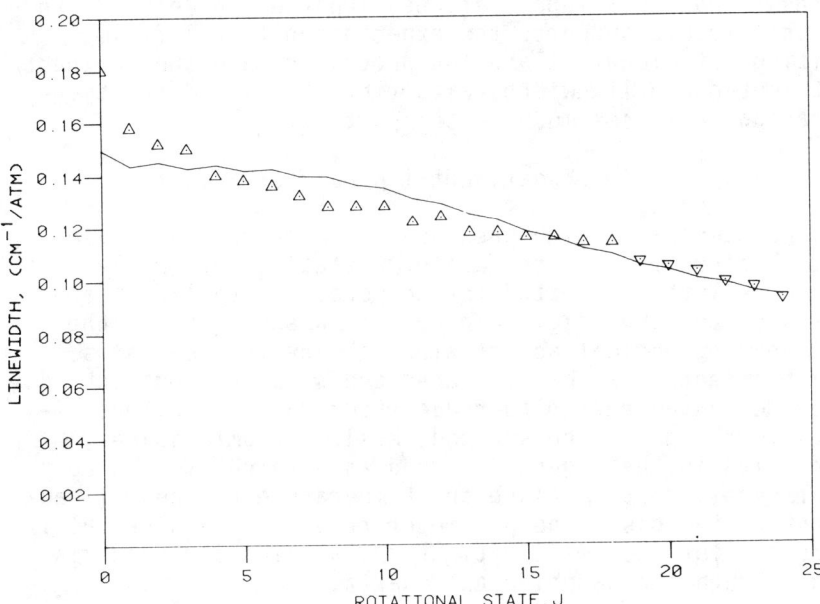

Fig. 4   Fit of simple power law model to CO Q-branch linewidths at 298 K, triangles are data of Ref. 11.

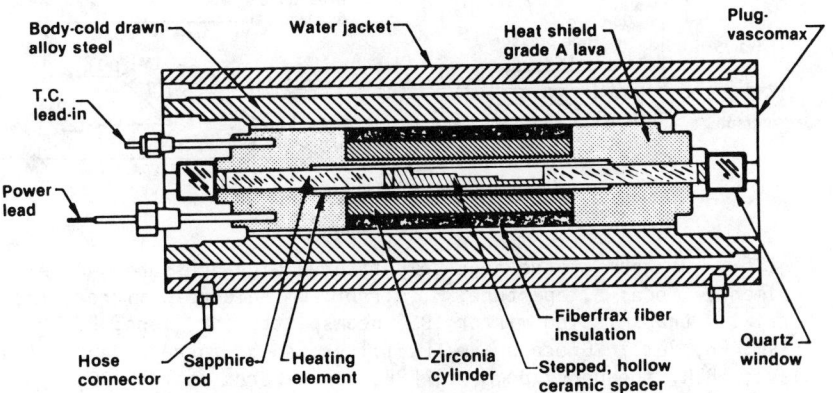

Fig. 5   Internally heated pressure vessel for CARS diagnostics.

11 is seen to be excellent for rotational quantum numbers greater than about five, with less satisfactory agreement at small values. Elastic dephasing processes may contribute to the disagreement at low $Q(j)$. The agreement is similar to the more elaborate inversions of CO P- & R-branch i.r. linewidths of Belbruno, Gelfand, and Rabitz.[23] It is noteworthy that the CO Q-branch Raman linewidths reported in Ref. 11 are very similar to the CO i.r. linewidths (fundamental and overtone) reported in Refs. 23 and 24; this result supports the expectation that lifetime limiting rotational relaxation processes make the dominant contribution to linewidth, with vibrational and rotational dephasing processes much less important.

Experimental Apparatus

The spectra of high-pressure, high-temperature gases were obtained through the use of a static, internally heated cell that is rated for temperatures to 1750 K and pressures to 5000 psig. Figure 5 is a schematic of the cell showing optical access along the cylindrical axis. Only the central 15 cm is heated and sapphire rods extend into the heated region to reduce temperature gradient effects in the gas. The stepped, hollow ceramic spacer, identified in the figure, is used as a target for an optical pyrometer to measure the temperature of the wall adjacent to the gas. The pyrometer reading is analytically corrected for the emissivity of the alumina and transmission through the sapphire and quartz.

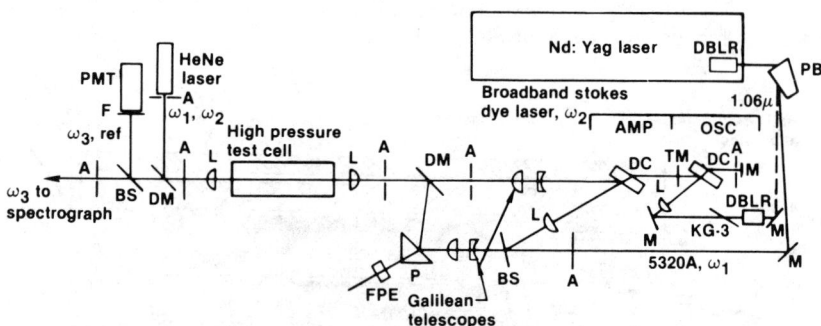

Fig. 6 Diagram of the high-temperature, high-pressure CARS experiment. Code: A, aperture; M, mirror; DM, dichroic mirror; TM, partially transmitting mirror; BS, beamsplitter; L, lens; P, prism; PB, Pellin-Broca prism; F, filter; KG-3, infrared absorbing glass; DBLR, frequency doubler; FPE, Fabry-Perot étalon; DC, dye cell; PMT, photomultiplier tube.

The limited optical access provided by the test cell dictates a collinear alignment of the pump and Stokes lasers used to generate CARS. A schematic of the apparatus used to obtain the CARS spectra is shown in Fig. 6. The output of a Quanta Ray Nd:YAG laser is frequency doubled to generate a horizontally polarized, "primary" pump beam at 5320 Å (ω_1). Residual 1.06 µ is separated from ω_1 in a Pellin-Broca prism and doubled to generate a secondary beam to pump the broadband Stokes dye laser oscillator. Part of the primary beam (30%) is split off to pump the amplifier of the dye laser ω_2. Galilean telescopes are provided to control beam waists and the focal zone locations of ω_1 and ω_2 which are combined collinearly on the dichroic mirror and focused inside the high-pressure test cell. CARS, at ω_3, is generated in the focal volume, and all three frequencies (ω_1, ω_2, ω_3) are recollimated after exiting the cell. A second dichroic mirror separates ω_3 from ω_1, ω_2 before the signal is incident on the slit of a double, 1-m monochromater with a resolution of ~0.4 cm^{-1}. The ω_1 and ω_2 beams are trapped after reflection from the second dichroic. A He-Ne laser is shown whose output is coincident with the ω_1, ω_2 path and used for

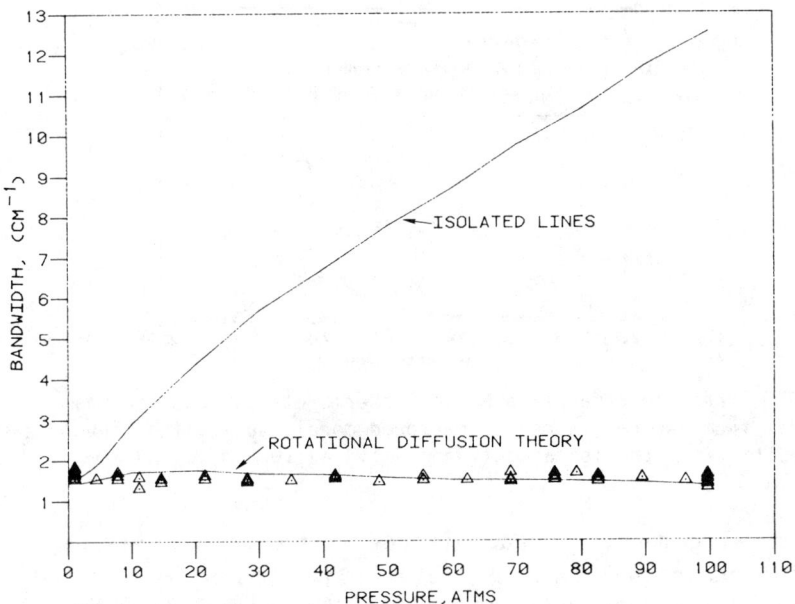

Fig. 7 N_2 bandwidth vs pressure at 300 K and 0.4-cm^{-1} resolution. Solid lines are theoretical predictions, triangles are experimental data.

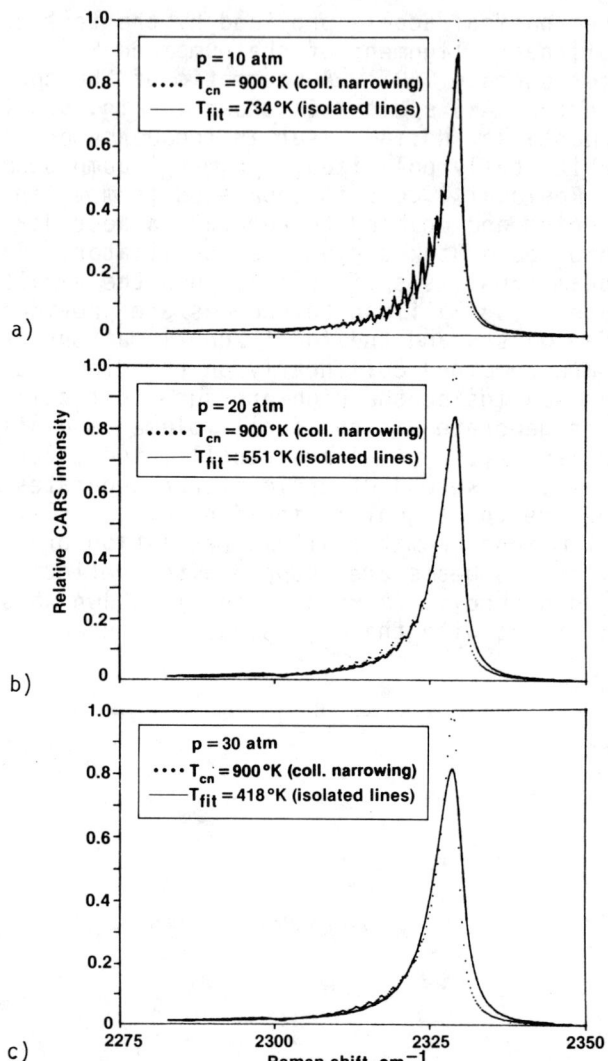

Fig. 8 Pressure effects in N_2 CARS thermometry. Dots are the spectrum from the collisionally narrowed model, solid line is the least-squares fit from the isolated line model. a) 10 atm, b) 20 atm, c) 30 atm.

optical alignment. A beamsplitter in the ω_3 leg provides a reference signal to a photomultiplier which is used to normalize the spectrally dispersed CARS signal from the monochromator. The two signals are ratioed in a boxcar averager to account for variations in laser intensity during the scan of the CARS spectrum.

At high gas pressures it is necessary to attenuate the lasers to avoid gas breakdown and stimulated Raman gain which would interfere with and perturb the CARS signal. The attenuation is achieved through a Fresnel reflection (~1.4%) from a flat surface. A prism is used for this purpose, as shown in Fig. 6, to avoid the back surface reflection that would result from a plane parallel element. Alternate methods of attenuation such as absorbing neutral density filters or reduction in the flashlamp energy input to the Nd:YAG laser are less desirable. They cause changes in the focal zone location due to thermal lensing in the filters or a change in divergence of the output of the Nd:YAG, and the focal shift can be great enough to cause damage to the sapphire rods.

A Fabry-Perot étalon is included to monitor the linewidth of the primary pump beam. The convolution of ω_1 and ω_2 determine the resolution of the CARS spectra at ω_3. An angle tuned étalon option on the Nd:YAG laser results in an ω_1 linewidth of ~0.4 cm^{-1} and the external étalon is used to detect any shift from this value which would degrade the resolution of the CARS spectra.

Experimental Results

Using the high-pressure facility, studies were performed in combustion gases from 1 - 100 atm and 300 - 1500 K temperatures. Most of the effort has been concentrated in N_2 because it is a major constituent of most combustion processes and exhibits sensitivity to temperature that makes it useful for thermometry (see Fig. 1).

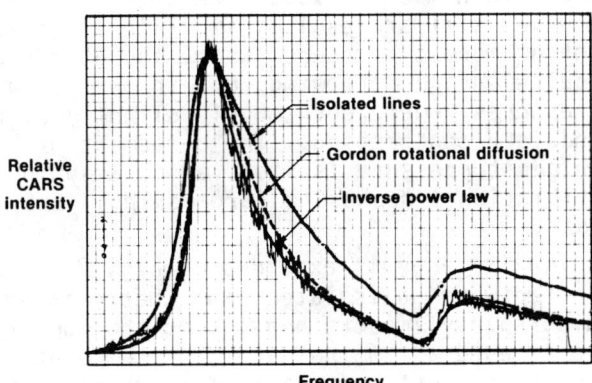

Fig. 9 N_2 CARS spectrum at 1700 K, 103 atm, 0.4-cm^{-1} resolution. The thin line is an experimental spectrum, the heavy lines are predictions of the various theoretical models.

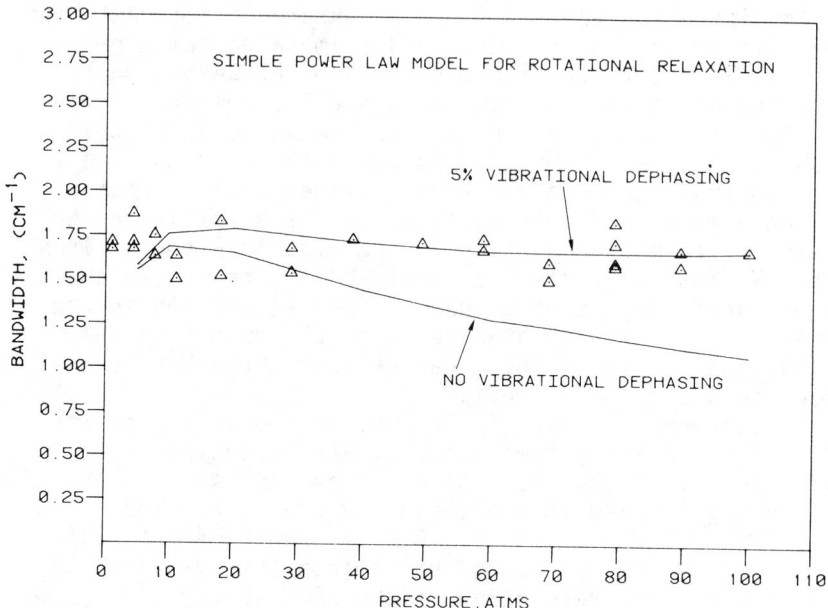

Fig. 10 CO bandwidth dependence on pressure at 300 K and 0.47-cm^{-1} resolution. Solid lines are theoretical predictions, with and without 5% vibrational dephasing, triangles are experimental data.

Figure 7 shows the variation in bandwidth of the N_2, Q-branch CARS spectrum as the pressure is increased from 1 - 100 atm at 300 K. The obvious result is that there is little variation with pressure, the effect of collisional narrowing. The theoretical calculations shown in Fig. 7 are based on the Gordon model Eq. 5. The inverse power law (Eqs. 7-9) gives similar results; it is in slightly better agreement at modest pressures where the Gordon model does not give quite enough narrowing, but in poorer agreement at high pressures, where both models slightly overpredict the bandwidth contraction. The agreement of both models at the highest pressure is improved by assuming a slight (2-4%) vibrational dephasing (broadening) contribution to linewidth.

The vertical axis range is extended in Fig. 7 to show the result expected when an isolated line theory is employed. The discrepancy impacts heavily on thermometry since bandwidth is used as an indicator for temperature; the effect of pressure must be adequately understood to accurately predict temperature from high-pressure spectra. As an example, Fig. 8 shows moderate temperature N_2 CARS spectra at 10-, 20-, and 30-atm pressure. The dots are

Table 1 Error in predicted temperatures from pressure broadened
(isolated line) fits to collisionally narrowed spectra.

Pressure, atm	T_{cn}, K	T_{fit}, K	Error, %
10	900	734	18.4
20	900	551	38.8
30	900	418	53.5

synthesized spectra from the collisional narrowing theory. These calculations were performed with the inverse power law model. The solid overlays on each spectra are the result of a least-squares fitting routine based on the isolated line (pressure broadened) model. The error in estimates of the temperature for each spectrum predicted by the isolated line model are large, in particular the error is very significant (18%) even at 10 atmospheres of pressure. For emphasis, the results are reformatted in Table 1.

From a thermometry standpoint, an important question to be answered is if the second vibrational band (hot band) disappears by collapsing or merging, at high pressure, into the ground vibrational-rotational band. The ratio of the peak height of the hot band to that of the fundamental bandhead is a parameter used to estimate temperature (see Fig.1). The spectrum of nitrogen at 1700 K and 103 atm was acquired and is shown in Fig. 9. The hot band is clearly evident, and moreover the theoretical fit using the inverse power law is excellent, demonstrating the verification of the collisional narrowing theory in this regime of simultaneous high temperature and pressure. The rotational diffusion model also gives good agreement at a savings in computer time of more than an order of magnitude.[7] The fact that the hot band does not merge into the fundamental at very high pressure is a consequence of the relative unimportance of vibrational collision processes.

CO is another diatomic molecule studied in the high-pressure facility because of its relevance to combustion chemistry. It is the direct product of the oxidation of hydrocarbon fuels. The spectroscopy of CO is quite similar to N_2, except that its vibrational Raman frequency shift is 188 cm^{-1} lower than N_2. Spectra of CO from 1 - 100 atm were obtained from the test cell and bandwidths measured from the experimental spectra. These data are presented in Fig. 10. As with nitrogen, there is little change in bandwidth as the pressure is varied, and it is

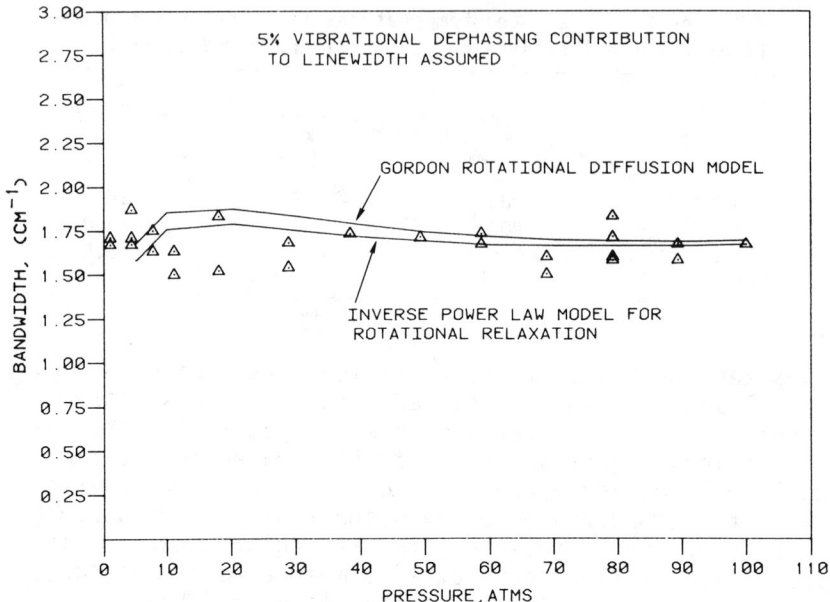

Fig. 11 Comparison of inverse power law and Gordon model of rotational relaxation. Experimental data are the same as Fig. 10.

obvious the collisional narrowing process is operative over this pressure range. The theoretical fit to the data is optimized if a small (5%) vibrational dephasing contribution is included. This dephasing is a pure broadening mechanism discussed previously that increases linearly with pressure. Including only rotational relaxation processes in the broadening model leads to an overprediction of the narrowing. Spectra of hot CO need to be obtained to further check and refine parameters of the collisional narrowing model for this molecule. The experimental data of CO have been used to demonstrate another point made the Theory section. The data are repeated in Fig. 11, together with the two popular phenomenological models that represent the rotational relaxation process. It is observed that both the rotational diffusion model and the inverse power law of rotational relaxation converge to identical results at high pressure. For very large line overlap the narrowing prediction loses sensitivity to the relaxation model, because the details of the state to state inelastic energy transfer rates become less important.

The other molecule studied in the high-pressure facility is CO_2; its importance in combustion, as the end product of fuel oxidation is well known. CO_2 is a triatomic

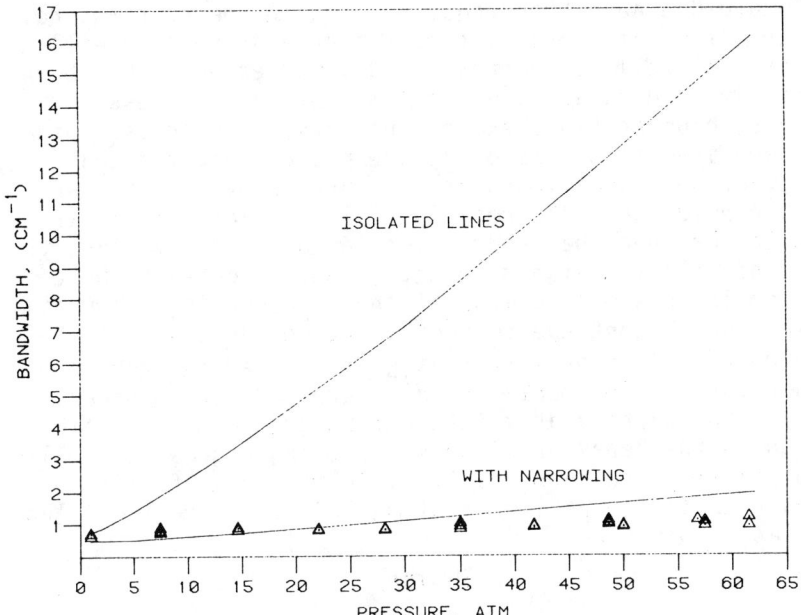

Fig. 12 CO_2 bandwidth variation with pressure at 300 K and 0.4-cm^{-1} resolution. Solid lines are theoretical predictions, triangles are experimental data.

whose Raman spectroscopy is more complicated than N_2 or CO. For Q-branch transitions of the symmetric stretch mode, bandwidths have been obtained from 1 - 65 atm and are shown in Fig. 12. Collisional narrowing is again obvious, as is the reasonable fit from the theoretical Gordon model. The power law approach might help resolve the discrepancy at the highest pressure. The importance of this data is again to refine the numerical model of the CARS spectra for this molecule so that accurate combustion parameters may be extracted from the experimental data. As with CO, it is planned to acquire hot CO_2 spectra.

Future Plans

The high-pressure CARS spectroscopy of combustion gases is an ongoing investigation at UTRC. The importance and justification for these studies has been demonstrated in this paper. It was also made clear that each combustion gas needs to be separately studied. Nitrogen data have been completed, with both hot and cold spectra obtained over the pressure range 1 - 100 atm. Room-temperature data of CO and CO_2 have been acquired and, in the near future, spectra of CO at elevated temperature and pressure will be obtained as well as spectra of CO_2. Water vapor has been initially studied at moderate temperature and pressure,[25] and hot, high-pressure experiments are in progress. High-pressure investigations of H_2 are not currently scheduled, however experiments at pressures above 100 atm are being considered. The rotational line spacing in H_2 is much larger than the other molecules and line overlap, the onset of collisional narrowing, is not expected to be a problem in this molecule until the pressure is $\geq$ 100 atm. These studies continue to shed light on the physical phenomenon of collisional narrowing and as new insights are gleaned, the experimental studies are modified appropriately. One particularly interesting avenue of investigation is the behavior of this narrowing process as collision partners are changed, i.e., in mixtures of different gases. Initial experiments along this line are scheduled for later this year.

Summary

Studies of CARS thermometry in the region of high pressure, common to gas turbine engines, rockets, and internal combustion engines, have been presented. It was shown that the effects of high pressure are accurately predicted by a theoretical model based on collisional narrowing of the rotational structure of CARS spectra. Spectra of N_2, CO, and CO_2 were acquired over the range of 1 - 100 atm and temperatures from 300 to greater than 1500 K. These data, obtained from a static high-pressure cell, were used to refine fitting parameters of the model for each gas species thus improving the validity and accuracy of temperature predictions derived from experimental spectra. Neglect of the collisional narrowing phenomenon was shown to introduce significant error in measurements of temperature even at moderate pressure ($\sim$10 atm). These errors emphasized the justification for the current work, and future investigations in the area of high-pressure CARS spectroscopy were outlined.

Acknowledgment

High-pressure CARS studies of N_2 and CO_2 were sponsored in part by the Army Research Office; the studies of CO were sponsored by the Air Force Rocket Propulsion Laboratory.

The authors are grateful to G. J. Rosasco for making available his CO and N_2 linewidth measurements prior to publication.

References

[1] Eckbreth, A. C., Dobbs, G. M., Stufflebeam, J. H., and Tellex, P. A., "CARS Temperature and Species Measurements in Augmented Jet Engine Exhausts," AIAA Paper 83-1294, 19th Joint Propulsion Conference, Seattle, Wash., June 1983.

[2] Kajiyama, J., Kayuaki, S., Kataoka, H., Maeda, S., and Hirose, C., "N_2 CARS Thermometry in Diesel Engine," International Off-Highway Meeting & Exposition, Milwaukee, Wisc., Sept. 1982; SAE Tech. Paper Series 821036.

[3] Rahn, L. A., Johnston, S. C., Farrow, R. L., and Mattern, P. L., "CARS Thermometry in an Internal Combustion Engine," *Temperature, Its Measurement and Control in Science and Industry*, American Institute of Physics, New York, 1982, p. 609.

[4] Eckbreth, A. C. and Hall, R. J., "CARS Thermometry in a Sooting Flame," *Combustion and Flame*, Vol. 36, Sept. 1979, p. 87.

[5] Hall, R. J. and Eckbreth, A. C., "Coherent Anti-Stokes Raman Spectroscopy (CARS): Application to Combustion Diagnostics," to appear in *Laser Applications* Vol. V, edited by R. K. Erf, Academic Press, New York.

[6] Hall, R. J., Verdieck, J. F., and Eckbreth, A. C., "Pressure-Induced Narrowing of the CARS Spectrum of N_2," *Optics Communications*, Vol. 35, Oct. 1980, pp. 69-75.

[7] Hall, R. J., "Coherent anti-Stokes Raman Spectroscopic Modeling for Combustion Diagnostics," *Optical Engineering*, Vol. 22, May 1983, pp. 322-329.

[8] Hall, R. J. and Greenhalgh, D. A., "Application of the Rotational Diffusion Model to Gaseous N_2 CARS Spectra, *Optics Communications*, Vol. 40, Feb. 1982, pp. 417-420.

[9] Gordon, R. G., "On the Pressure Broadening of Molecular Multiplet Spectra," *Journal of Chemical Physics*, Vol 46, Jan. 1967, pp. 448-455.

[10] Alekseyev, V. et. al., "Stimulated Raman Scattering in Gases and Gain Pressure Dependence," *IEEE Journal of Quantum Electronics*, Vol. QE-4, Oct. 1968, pp. 654-656.A.

[11] Rosasco, G. J. et. al., "Line Interference Effects in the Vibrational Q-Branch Spectra of N_2 and CO," *Chemical Physics Letters*, Vol. 97, June, 1983, p. 435.

[12] Brueck, S. R. J., "Vibrational Two-Photon Resonance Linewidths in Liquid Media," Chemical Physics Letters, Vol. 50, Sept. 1977, pp. 516-520.

[13] Temkin, S. I. and Burshtein, A. I., "On the Shape of the Q-Branch of Raman Scattering Spectra in Dense Media," Chemical Physics Letters, Vol. 66, Sept. 1979, pp. 52-56.

[14] Robert, D. and Bonamy, J., "Short Range Force Effects in Semi-classical Molecular Line Broadening Calculations," Journal de Physique, Vol. 10, Oct. 1979, pp. 923-943.

[15] Polanyi, J. C. and Woodall, K. B., "Mechanism of Rotational Relaxation," Journal of Chemical Physics, Vol. 56, Feb. 1972, pp. 1563-1572.

[16] Bernstein, R. B., "Note on the Polanyi-Woodall Equation for Rotational Relaxation," Journal of Chemical Physics, Vol. 62, June 1975, p. 4570.

[17] Pritchard, D. E. et. al., "Power Law Scaling for Rotational Energy Transfer," Journal of Chemical Physics, Vol. 70, Mar. 1979, pp. 2115-2120.

[18] Brunner, T. A., Smith, N., and Pritchard, D. E., "New Experimental Evidence for the Energy Corrected Sudden Scaling Law," Chemical Physics Letters, Vol. 71, April 1980, pp. 358-362.

[19] Dexheimer, S. L. et.al., "Dynamical Constraints on the Transfer of Angular Momentum in Rotationally Inelastic Collisions of I_2 with He and Xe," Journal of Chemical Physics, Vol. 76, May 1982, pp. 4996-5004.

[20] Whitaker, B. J. and Brechignac, P., "A New Fitting Law for Rotational Energy Transfer," Chemical Physics Letters, Vol. 95, Mar. 1983, pp. 407-412.

[21] Rahn, L. A., Owyoung, A., Coltrin, M. E., and Koszykowski, M. L., "The J Dependence of Nitrogen 'Q' Branch Linewidths," Proceedings of the VIIth International Conference on Raman Spectroscopy, Ottawa, Canada, W. R. Murphy, Editor, 1980, pp. 694-695.

[22] Hall, R. J., "Pressure-broadened Linewidths for N_2 Coherent Anti-Stokes Raman Spectroscopy Thermometry," Applied Spectroscopy, Vol. 34, Nov. 1980, pp. 700-701.

[23] BelBruno, J. J., Gelfand, J., and Rabitz, H., "Collision Dynamical Information from Pressure Broadening Measurements:

Application to Carbon Monoxide," Journal of Chemical Physics, Vol. 78, Mar. 1983, pp. 3990-3998.

[24]Varghese, P. L. and Hanson, R. K., "Tunable Infrared Diode Laser Measurement of Strengths and Collisional Widths of CO at Room Temperature," Journal of Quantitative Spectroscopy and Radiative Transfer, Vol. 24, Dec. 1980, pp. 479-489.

[25]Greenhalgh, D. A., Hall, R. J., Porter, F. M., and England, W. A., "Collisional Narrowing Effects in the CARS Spectrum of Water Vapor," in Raman Spectroscopy Linear and Nonlinear, J. Lascombe and P. V. Huong, Editors, Wiley Heyden Ltd., Chichester, England, 1982, pp. 301-302.

CARS Thermometry and N_2 Number-Density Measurements in a Turbulent Diffusion Flame

L. P. Goss,* D. D. Trump,† G. L. Switzer,‡ and B. G. MacDonald§
Systems Research Laboratories, Inc., Dayton, Ohio

Abstract

An extensive X-Y profiling of a turbulent propane diffusion flame has been conducted and the data analyzed to yield average temperature and temperature fluctuation contour maps. The average temperature contour resembles still photographs of the flame, while the temperature fluctuation contour shows the true nature of the turbulent flame as displayed by 500-frame/s ciné photographs. Included in the data analysis is a discussion of the probability distribution functions (PDF) which were obtained throughout the flame. Both single and multimodal PDF's were obtained at various locations in the flame. In an attempt to explain the origin of the various modes of the PDF's, a conditional sampling technique was used to measure the CARS temperature in the presence and absence of visible flame emission. The results indicate that in the presence of visible flame, the same high-temperature distribution is observed regardless of flame location, while the temperature distribution in the absence of flame emission varies considerably with location and mixing.

Introduction

The study of turbulent combustion on a fundamental level requires spatially resolved information on the instantaneous values of a large number of scalar quantities such as temperature, species concentration, velocity, and pres-

---

Presented as Paper 83-1480 at AIAA 18th Thermophysics Conference, Montreal, Canada, June 1-3, 1983. Copyright © American Institute of Aeronautics and Astronautics, Inc., 1983. All rights reserved.
    *Senior Chemist, Research Applications Division.
    †Associate Engineer, Research Applications Division.
    ‡Research Engineer, Research Applications Division.
    §Research Chemist, Research Applications Division.

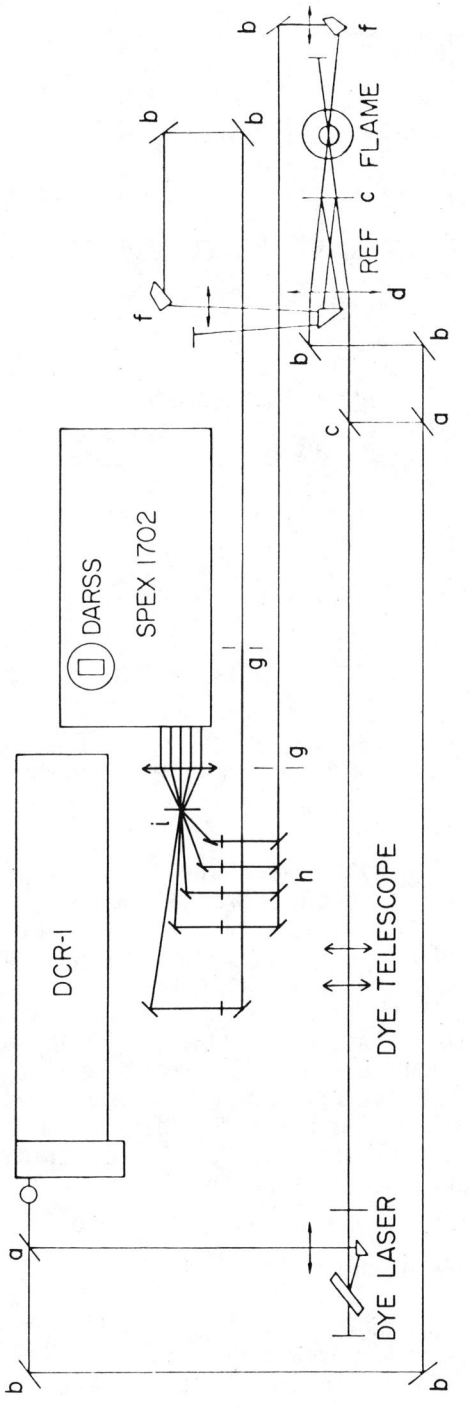

Fig. 1 CARS optical schematic. Arrows indicate lenses. a) 50% beamsplitter for 532 nm, b) total reflector, c) dichroic, green reflector, red transmitter, d) 50-cm lens, e) 20% beamsplitter, f) Pellin-Borca prism, g) iris, h) beamsplitter array (0.1%, 1%, 10%, and total reflector), i) 532-nm filter.

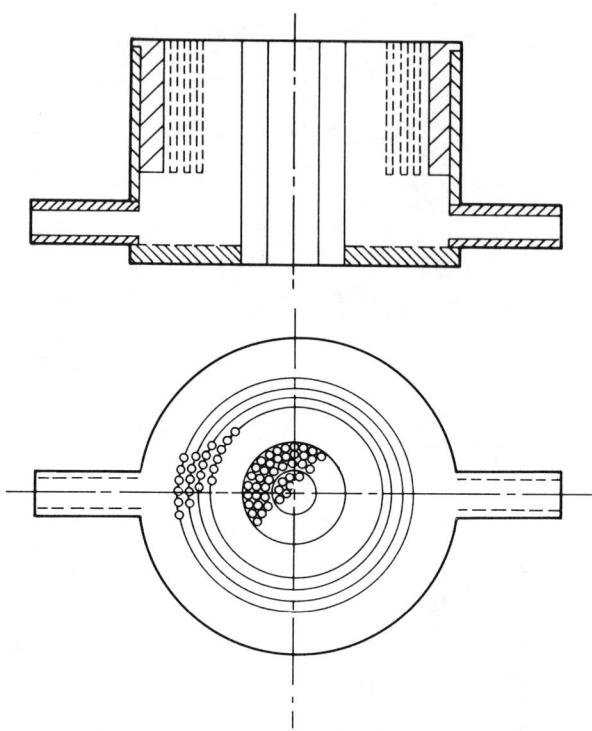

Fig. 2 Concentric tube burner assembly used for turbulent flame studies. Outer capillaries correspond to an argon sheath which was not employed in these studies.

sure. Coherent anti-Stokes Raman spectroscopy (CARS) has proven to be an attractive experimental technique for this purpose because it can provide time resolved (10 ns) and spatially resolved (0.1 mm^3) measurements of temperature and major flame species, without appreciably perturbing the flame under study. This paper presents results obtained on a propane-air turbulent diffusion flame at a Reynolds number of 6000 (based upon air flow). These results are presented in the form of average temperature and temperature fluctuation contour maps and species concentration PDF's. Conditional sampling has also been conducted on the turbulent flame in an attempt to explain the various contributions to the PDF's.

Experimental Methods

The CARS apparatus which was employed for these studies (see Fig. 1) is described in detail in Refs. 1 and 2 and, therefore, will be discussed only briefly here. The heart

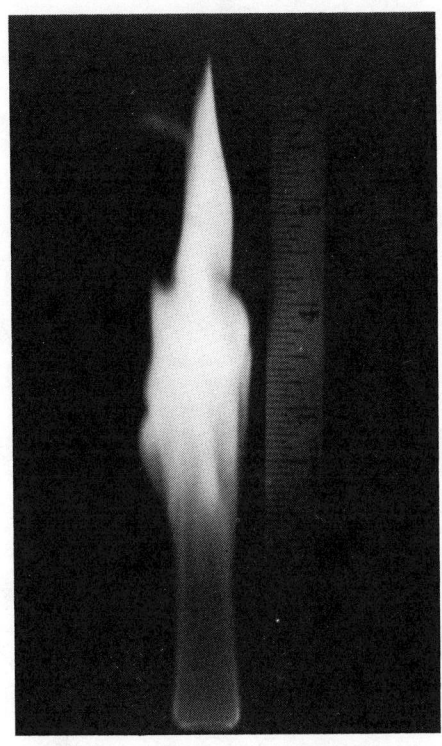

Fig. 3 Still photograph of the $R_e \sim 6000$ turbulent propane diffusion flame. Exposure time was ~ 100 ms.

of the system is a Nd:YAG laser which is employed both in the CARS process and in pumping a broadband dye laser. A BOXCARS configuration is used to achieve a spatial resolution of 1 mm along the longest beam axis. A power reference is employed for concentration measurements along with a Reticon detector for single-shot temperature measurements. The system was especially designed to acquire CARS data at a 10-Hz rate, which required interfacing to a minicomputer for control of the data acquisition. Data reduction consists of fitting the observed N_2 CARS spectral bandshape to obtain the temperature and integrating the CARS signal to obtain the concentration. The temperature fit employs a nonlinear least-squares iterative routine requiring typically three cycles of the iteration for convergence.[3] Once the temperature has been fit, the concentration can be determined from the integrated CARS signal. The N_2 number density is given by[4]

$$N_T = N_{300} \sqrt{R_T \cdot I_T/I_{300}} \qquad (1)$$

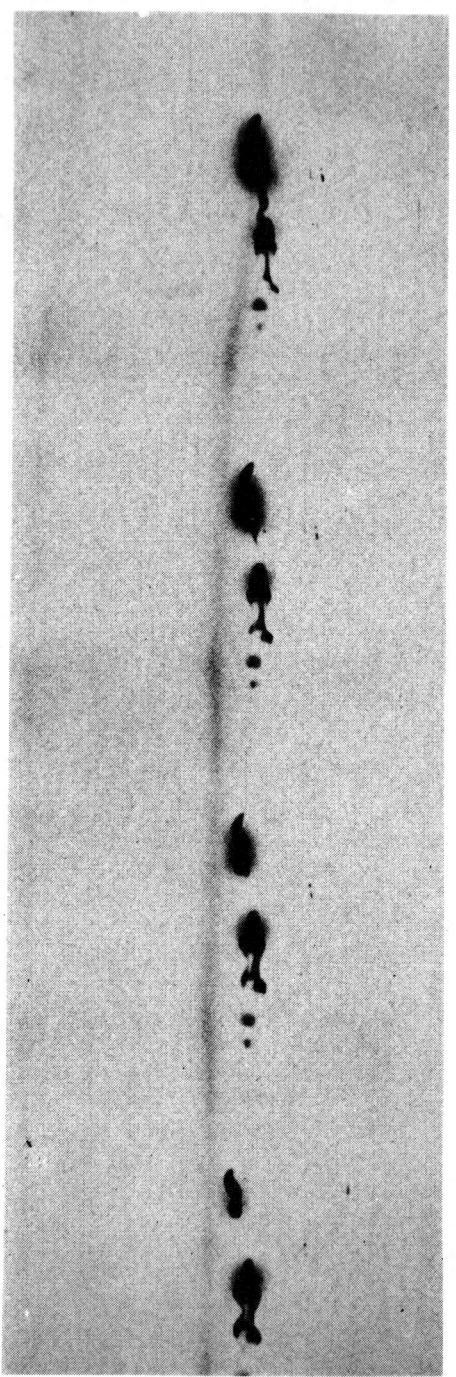

Fig. 4 Sequence of shots taken from a 500-frame/s movie of the $R_e \sim 6000$ turbulent propane flame.

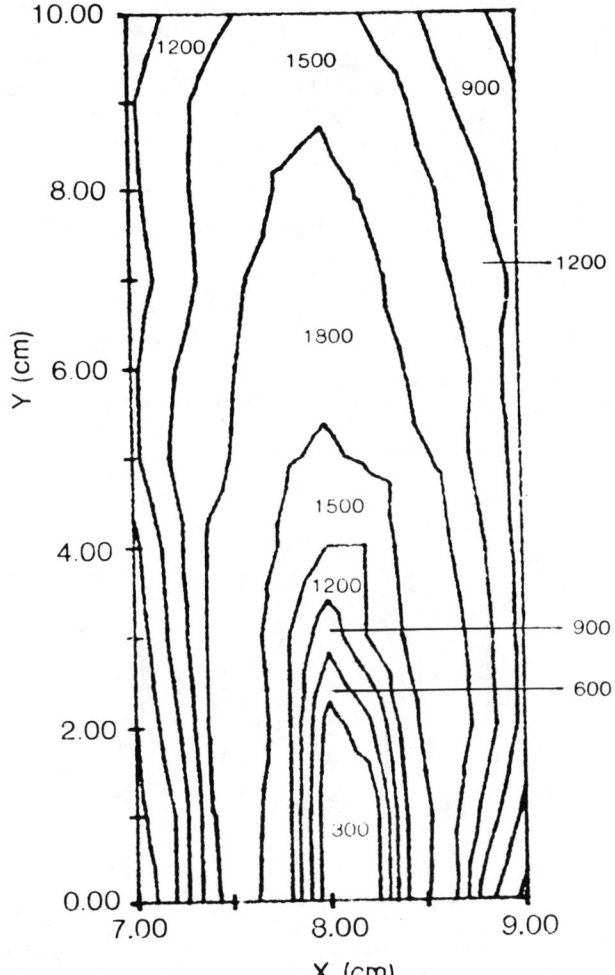

Fig. 5 Average-temperature contour of the $R_e \sim 6000$ turbulent diffusion flame. Position 0.0 is 2 cm above burner surface.

where N_T is the N_2 number density at temperature T; N_{300} is the N_2 number density at 300 K; R_T is the ratio of the bandshapes of the CARS spectrum at 300 K to that of the CARS spectrum at temperature T; I_T is the experimentally measured integrated CARS intensity at temperature T; and I_{300} is the experimentally measured integrated CARS intensity at 300 K. The ratio R_T takes into account the change in the Raman linewidths and population redistribution with temperature. Approximately 1 s of CPU time per spectrum is required for the data analysis discussed above.

The burner employed for these studies is a Perkin-Elmer assembly (No. 290-0107)[5,6] consisting of a concentric-tube arrangement which can be operated independently for either a premixed flame or a turbulent diffusion flame. The burner surface consists of a capillary plate which is used to stabilize the flame under high-flow-rate conditions. The burner design is depicted in Fig. 2.

Contour Maps and PDF's

The flame that was employed for the turbulent flame studies is shown in Fig. 3. The burner conditions under which this flame was produced were an inner fuel flow of

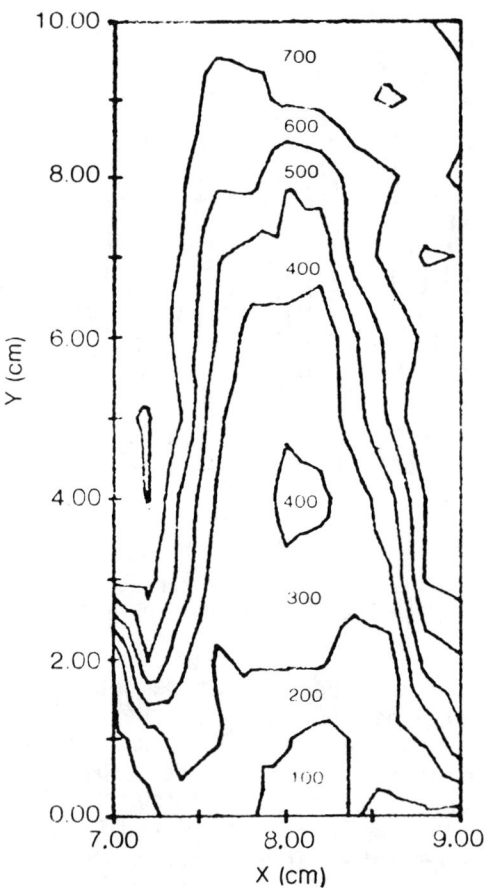

Fig. 6 Temperature-fluctuation contour of the $R_e \sim 6000$ turbulent diffusion flame. Position 0.0 is 2 cm above burner surface.

0.95 ℓ/min of propane and an outer air flow of 40 ℓ/min. The flame height was 14.5 cm, while the air-to-fuel mass flow ratio was 28 to 1. The flame was characterized primarily by turbulence, driven by the shear produced between the two fluid flows. Fast ciné photographs of the flame at 500 frames/s are shown in Fig. 4. At 2-ms intervals the effects of the shear of the flame can be observed.

A measurement grid consisting of an 11 × 11 matrix was superimposed on this turbulent flame. At each grid point 1500 single shots of data were taken and the data analyzed to obtain contour maps of the average values and their fluctuation and PDF's of the temperature and N_2 concentration.

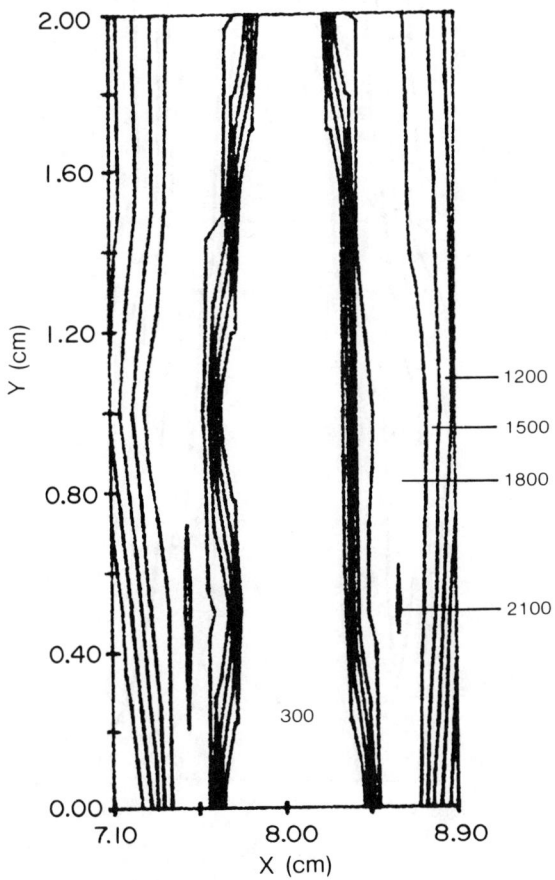

Fig. 7 Average-temperature contour of the flame-sheath region of the turbulent flame.

A contour plot of the average temperatures obtained in the turbulent flame is shown in Fig. 5. The isotherms of the contours cover a 300-K range and are individually labeled. The 0.0 position of the contour map is 2 cm above the burner surface. The propane fuel jet appears as a cold 300-K region in the bottom center of the flame. As one progresses up the centerline of the flame, the temperature rises to a maximum near Y = 6 cm and begins to fall off above Y = 8 cm. The maximum centerline temperature corresponds to the highly sooty central region displayed in Fig. 3. Near the base of the flame, the cold central fuel jet is surrounded by a blue flame sheath ∼ 2 mm in diam.

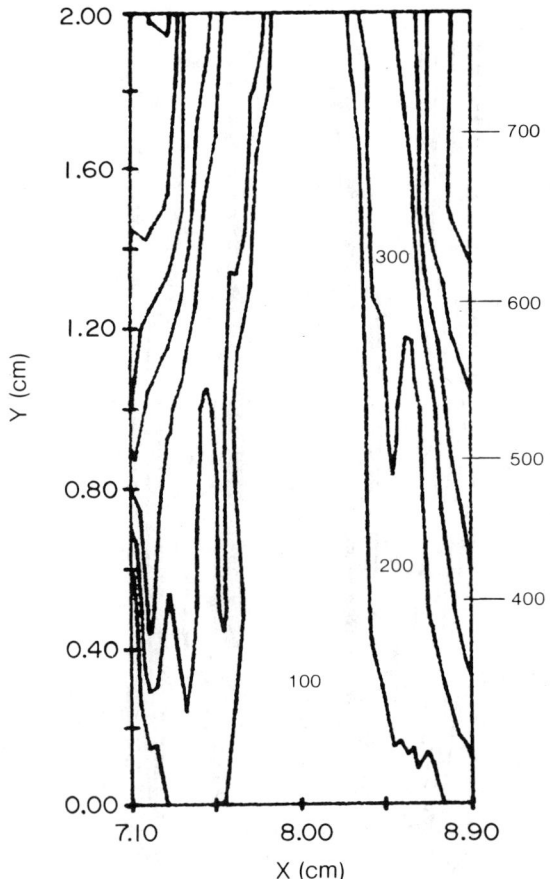

Fig. 8 RMS temperature contour of the flame-sheath region of the turbulent flame.

This sheath is extremely hot and will be discussed in more detail later. The overall picture of the flame given by the contour is one of a symmetric flame system in contrast to that indicated by the high-speed ciné photographs. While the average temperature contour is useful for identifying average flame characteristics, the fluctuation contour shown in Fig. 6 displays the true turbulent nature of the flame. In this contour the isotherms represent a 100-K range. The fluctuations are small at a point low in the flame but increase drastically as more and more air becomes entrained into the central fuel jet. The turbulence becomes quite large in the upper regions of the flame as the shear generated by the two different air flows dominates the flame. Thus, while the average temperature gives a picture of the time averaged flame, the fluctuation contour gives insight into the true turbulent nature of the flame system.

As mentioned earlier, surrounding the fuel jet near the burner surface is a blue flame sheath which is $\sim$ 2 mm in diam. In order to obtain a better understanding of this structure, a finer grid spacing was used for a profile. An 11 × 11 matrix was chosen to cover a 2 × 2 cm^2 region. The average temperature contour obtained in this region is shown in Fig. 7. The flame sheath, as indicated, is quite hot, reaching the near stoichiometric temperature of 2100 K, while the nearby fuel and air jets are cold. The flame sheath does not reach the stoichiometric temperature due to radiant heat loss of the small structure.[7] The fluctuation contour of this structure is shown in Fig. 8. The contour lines tend to map the air entrainment which is occurring at this location. While this study of the sheath indicates that the spatial resolution of the instrument is sufficient to obtain reasonable data, higher spatial resolution would be required for a more detailed study of this small structure.

Figures 9-14 display the individual PDF's obtained during profiling of the turbulent flame. Each PDF is the result of 1500 single shot measurements at a given location. The coordinates are the same as those used for the contour maps discussed above. Figure 9 displays the PDF's obtained at the Y = 0.0 cm position across the flame. As shown in the still photograph (Fig. 3) and in the average temperature contour map (Fig. 5), the flame sheath is located in this region. The first PDF at X = 7.3 cm corresponds to a location 2 mm on the air side of the sheath. Although no visible flame exists in this region, the temperature is higher than that of the ambient air due to heat losses of the flame sheath. At X = 7.4 cm, the outer edge of the sheath is being probed, resulting in a high average temperature. The center of the sheath was observed at X = 7.5 cm,

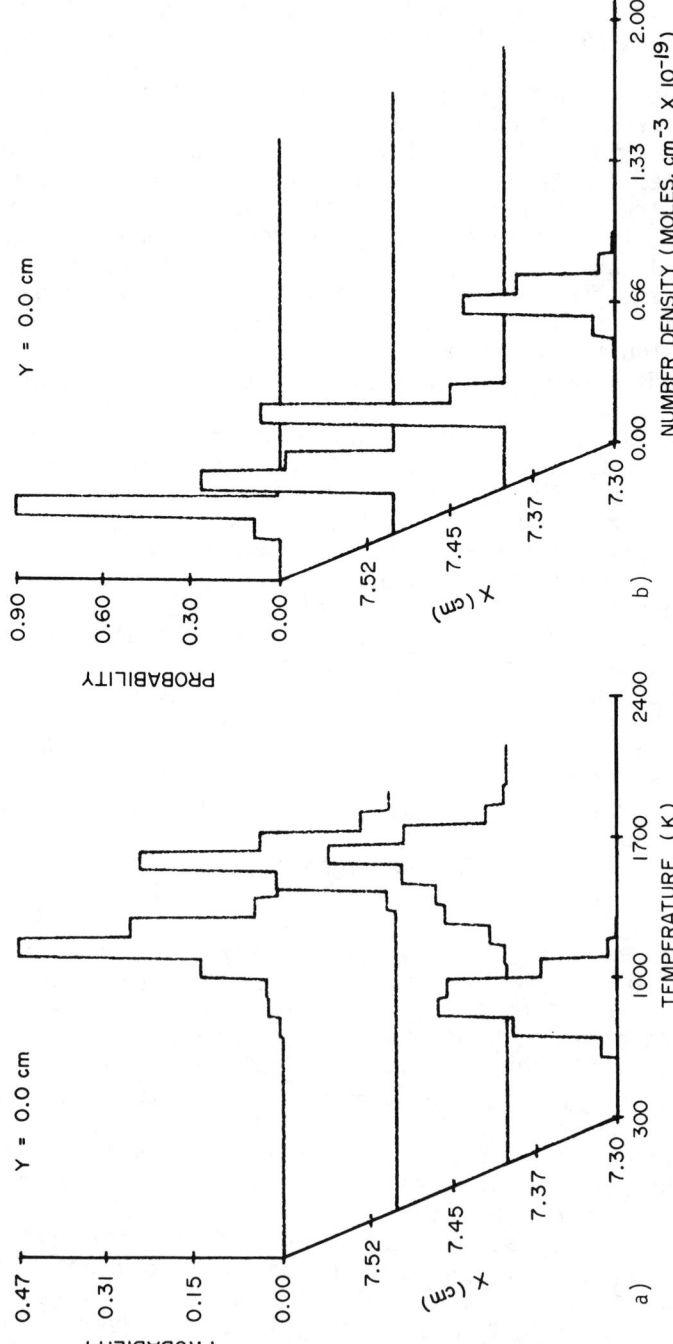

Fig. 9a) Temperature and b) N_2 concentration PDF's obtained at the Y = 0 cm position of the turbulent diffusion flame.

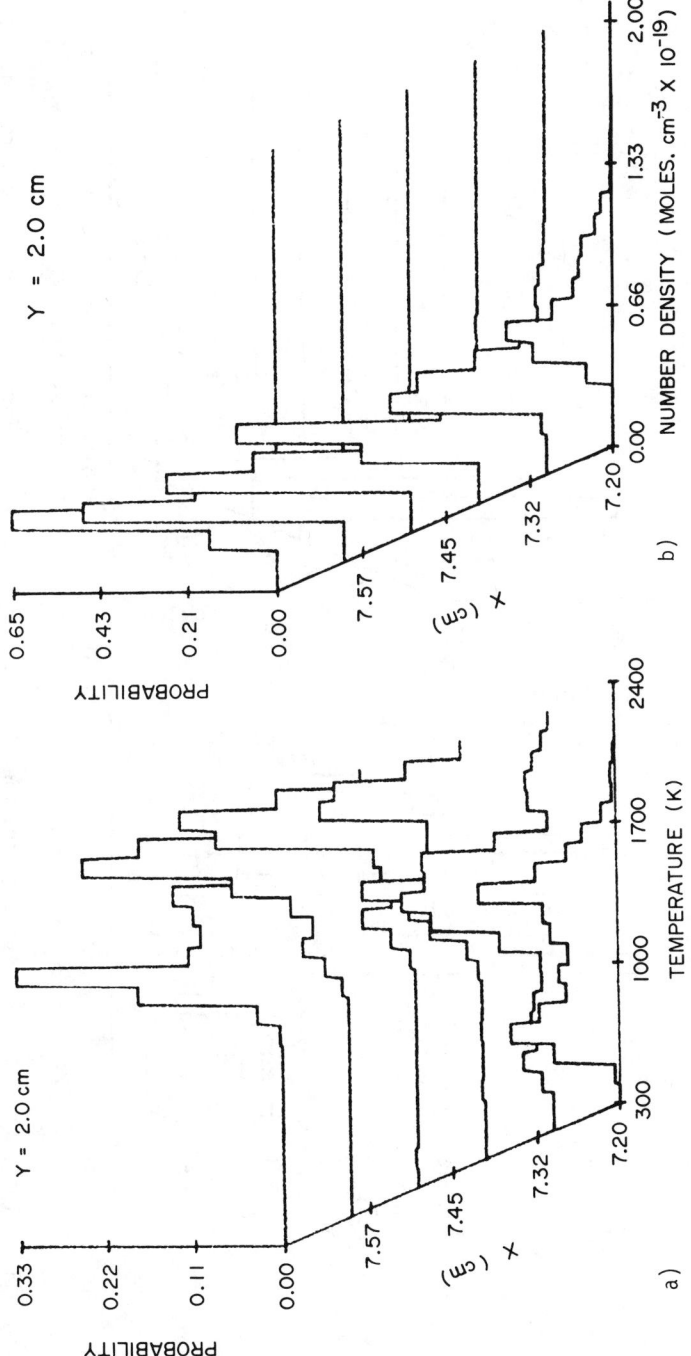

Fig. 10a) Temperature and b) N_2 concentration PDF's obtained at the Y = 2 cm position of the turbulent diffusion flame.

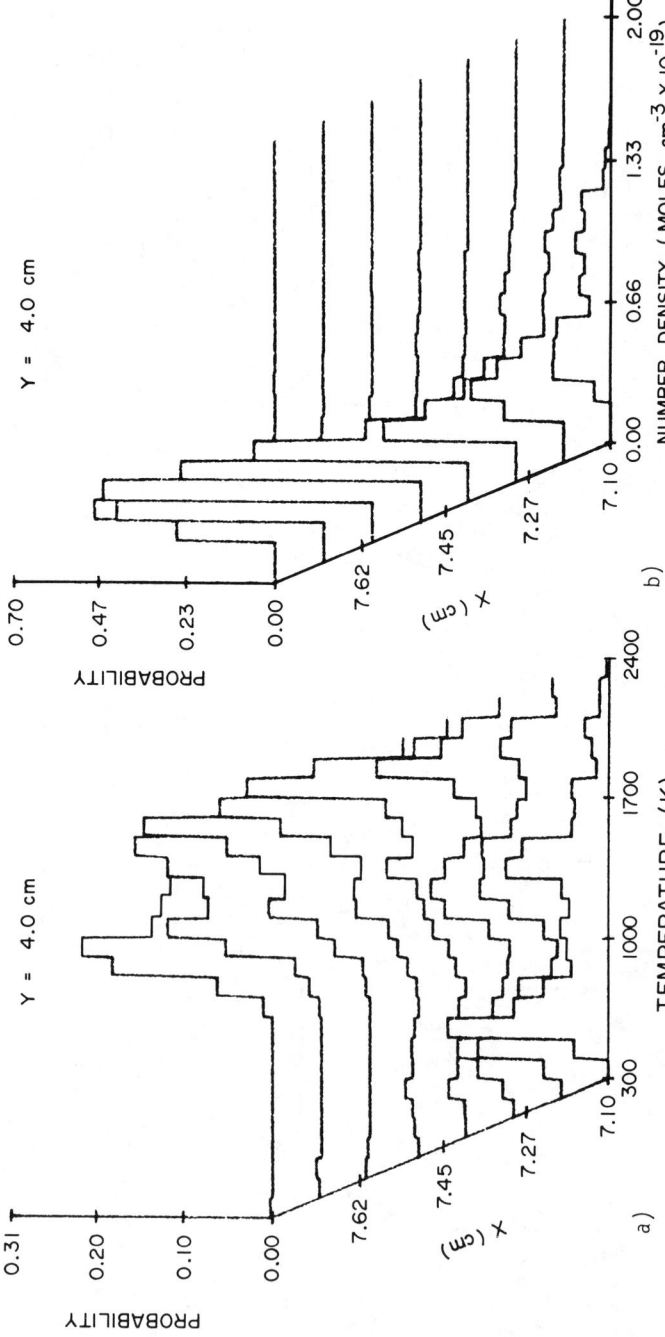

Fig. 11a) Temperature and b) N_2 concentration PDF's obtained at the Y = 4 cm position of the turbulent diffusion flame.

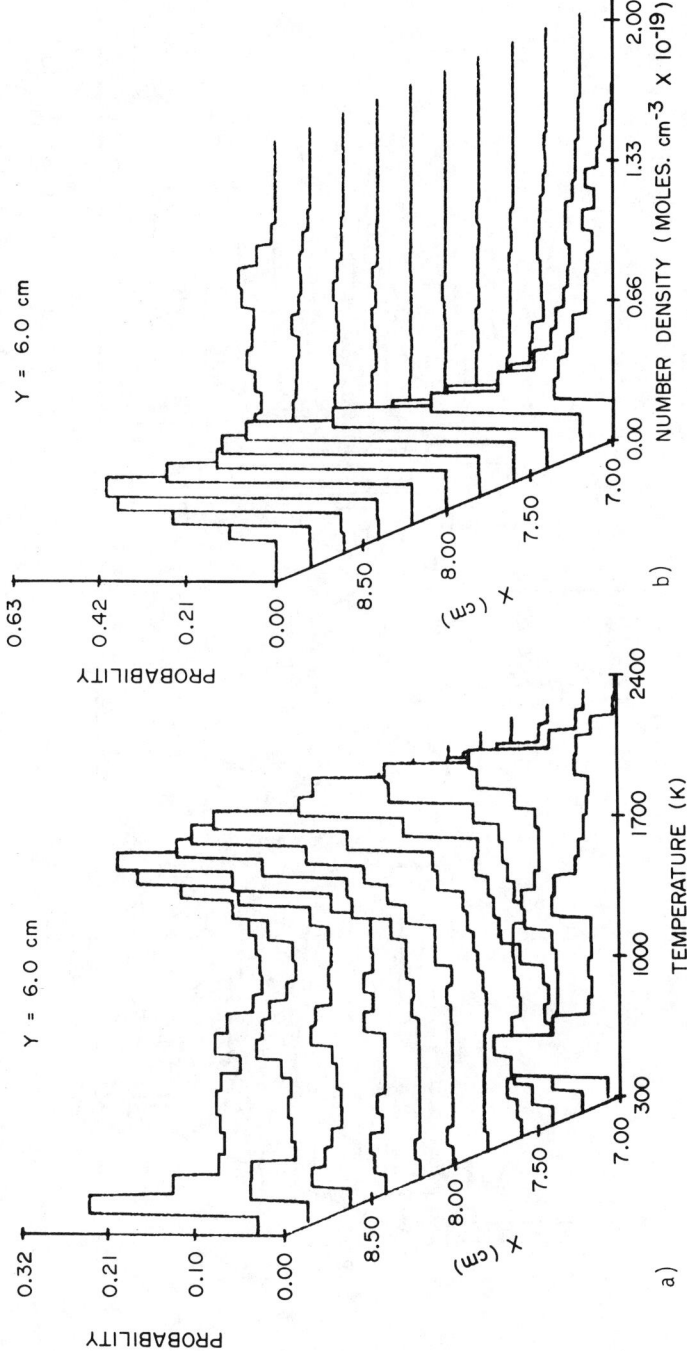

Fig. 12a) Temperature and b) N2 concentration PDF's obtained at the Y = 6 cm position of the turbulent diffusion flame.

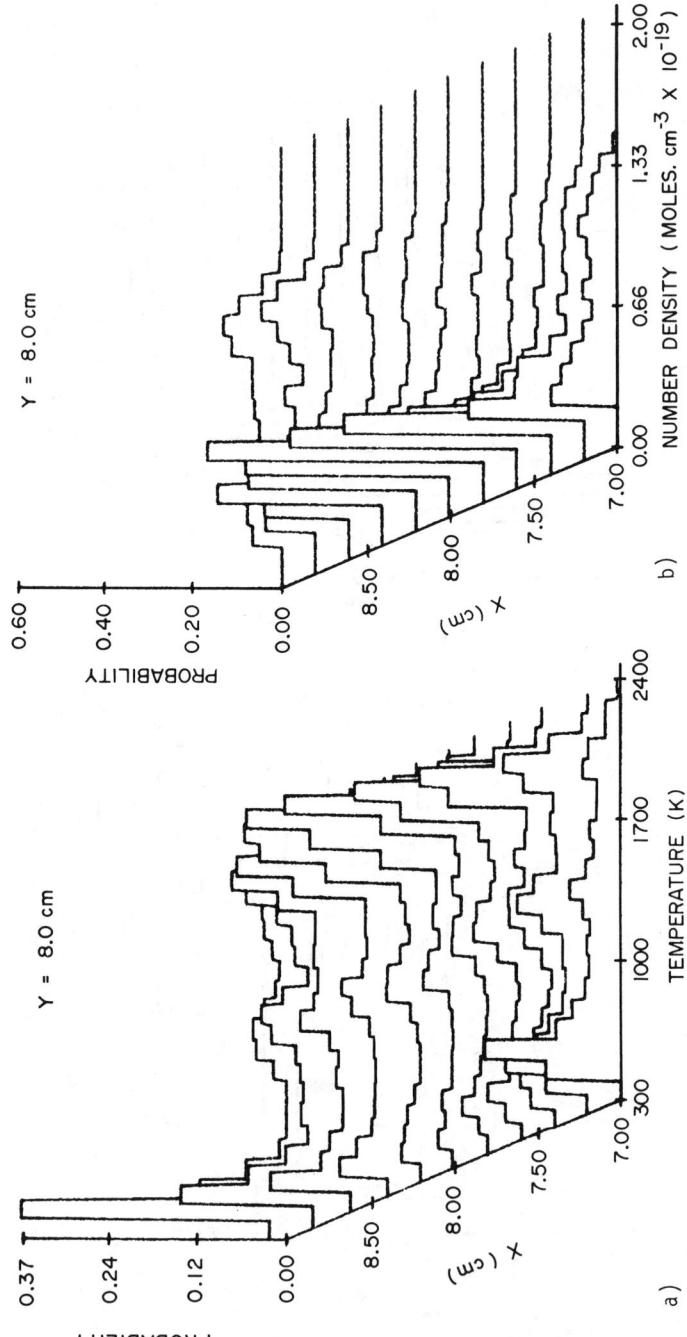

Fig. 13a) Temperature and b) N_2 concentration PDF's obtained at the Y = 8 cm position of the turbulent diffusion flame.

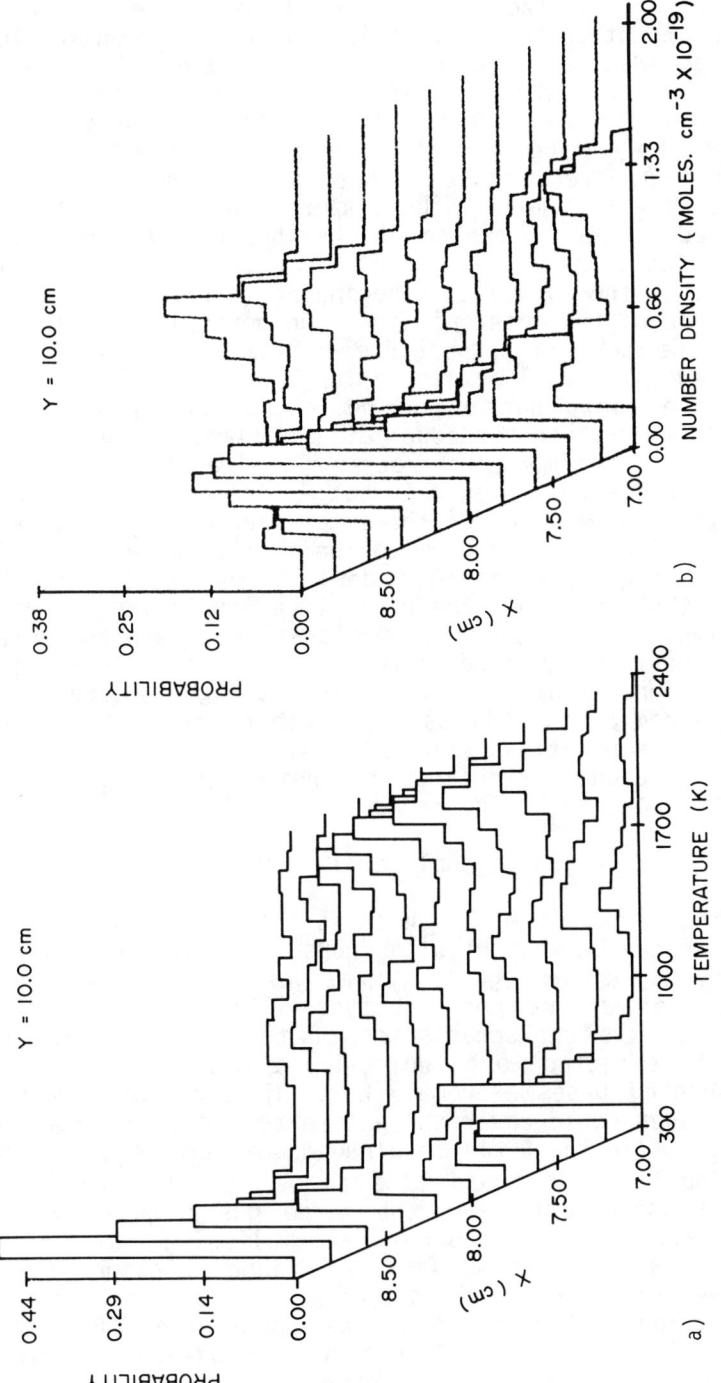

Fig. 14a) Temperature and b) N_2 concentration PDF's obtained at the Y = 10 cm position of the turbulent diffusion flame.

as indicated by the lower temperature at X = 7.6 cm. The N_2 concentration PDF's for this region are shown in Fig. 9b. Since an inverse relationship exists between the temperature and the N_2 concentration, the low N_2 number densities correspond to the hot temperature regions, which results in a narrow, low N_2 concentration mode for the PDF's in this region. As one proceeds higher up the sheath Y = 2-4 cm (Figs. 10-11), more high-temperature distributions are observed due to spreading of the sheath. The N_2 distributions display similar behavior. At the Y = 6 cm position (Fig. 12), the fuel jet is beginning to break up due to the dominance of the outer air jet. On centerline at this position, the maximum temperature outside the sheath area is observed. The fuel jet is no longer distinguishable, and a broadband sooty background dominates the flame emission. The PDF's of the outer edges of the flame at this location display a multimodal character, indicating intermittency due to the large air entrainment.[8] The N_2 PDF's are correspondingly broadened for the outer flame locations. As one proceeds from the Y = 6 cm to the Y = 10 cm position (Figs. 12-14), the temperature modes of the respective PDF's remain the same; only their relative population varies with position. At the upper flame locations, more and more air entrainment is observed as the cold outer air is mixed with the hot inner flame mixtures. The average temperature drops in this region as well as the width of the flame. At the Y = 8-10 cm positions (Figs. 13-14), the flame is dominated by the cold outer air flow, the respective temperature and N_2 PDF's becoming quite broad.

Conditional Sampling Results

The temperature PDF which is obtained at a given flame location can be thought of as consisting of three different components--unburnt gases, burnt gases, and burning gases. It is the various contributions of each of these which determine the characteristics of the PDF. Unburnt gases would be expected to be cold, burnt gases relatively hot, and burning gases extremely hot. Mixtures of unburnt and burnt gases could cover the entire temperature range, which is evident in the PDF's obtained downstream in the flame.

Ideally, one would like to be able to determine the various contributions of the three gas components to the total PDF. One method of accomplishing this for the burning gases is to monitor the flame emissions simultaneously with the CARS measurement. By monitoring the chemiluminescent emission of the flame, the CARS temperature data can be divided into two groups--one in which no visible flame emission is present and the second in which active combustion is occurring. Earlier work at this laboratory indicated that flame emission observed by a photomultiplier could be used to follow the dynamic behavior (flame turbules) of a turbulent propane flame[9] and conditionally sample velocity data in a large-scale combustor.[10] The emission system employed for the CARS study consisted of two orthogonally placed PMT's in which the image of the flame was projected (1 to 1) onto a 500-μ circular slit which defined the sample volume. Since the emission technique is a line-of-sight observation, two views were employed to confirm that the flame was actually present in the sample volume during the CARS measurement. The slits were aligned with the CARS sample volume by placing a card at the focus of the CARS beams, imaging the scattered light from the focus onto the slit, and aligning the slit with this image. This procedure was repeated for the second PMT to insure that both PMT's were monitoring the same position.

To determine which emission data were coincident with a laser shot, the Q-switch signal of the laser was monitored and used to set one bit of the 16-bit emission data word which was stored by the minicomputer. The most significant bit was chosen for this purpose since it indicates a negative value which could be easily distinguished. The two channels of emission data were transferred to the computer at a data rate of 20 kHz.

When combined with emission data, the temperature PDF can be divided into two conditionally sampled PDF's--one in the absence and one in the presence of visible flame emission. The cutoff point used to determine the absence of emission was 50 counts (out of 1000). The PDF plots obtained at three locations in the turbulent propane flame discussed above are shown in Figs. 15a-c. Position 1 corresponds to a centerline location 12 cm above the burner surface (position Y = 10 cm on the flame contour map). The temperature PDF for this location is shown in Fig. 15a. The average temperature was 1275 $\pm$ 605 K. The emission data taken simultaneously with the temperature data indicated that the flame was present only 29% of the time, which accounted for the low average temperature. The conditionally sampled PDF (shaded area of Fig. 15b) indicated that the visible flame is characterized by an average temperature of 1828 $\pm$ 170 K and a well-defined temperature distribution. The majority of the observed PDF is thus due to nonvisible gases (unburnt and burnt) having an average temperature of 1211 K.

At Position 2, located 10 cm above the burner surface (position Y = 8 cm on the flame contour map), the average temperature has increased to 1458 $\pm$ 455 K. The flame emission data indicated that the flame was present 43% of the

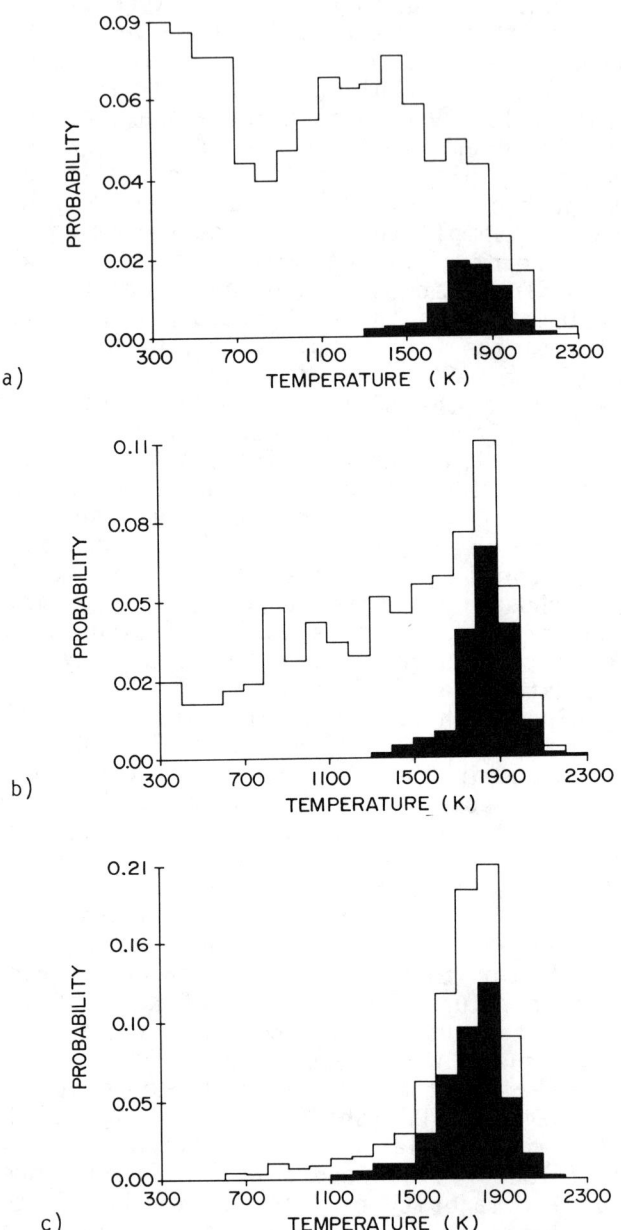

Fig. 15 Complete and conditionally sampled (black) temperature PDF's obtained on centerline a) 12 cm, b) 10 cm, and c) 8 cm above the burner surface.

time, which explains the higher temperature at this location. The temperature PDF for this location is shown in Fig. 15c. Notice that the general trends of the conditionally sampled data (flame present) are the same as those observed at Position 1, the major difference being that the flame is present more often at Position 2 (Fig. 15b). The average temperature (flame present) was 1801 ± 165 K.

At Position 3, 8 cm above the burner surface (position Y = 6 cm on the flame contour map), the average temperature has increased to 1704 ± 222 K. The emission data indicated that the flame was present 66% of the time. The region is thus dominated by the active combustion of the propane fuel. The conditionally sampled temperatures (flame present) display similar behavior to the other locations probed, with an average temperature of 1705 ± 185 K.

While the conditionally sampled data (flame present) displayed similar behavior regardless of the sampling position, the emission data (flame absent) exhibited widely varying behavior which was dependent on the mixing properties of the flame. The high temperatures in the absence of the flame are considered to be regions in which the flame turbules have just burned out but have not yet mixed with any cooler gases. The low-temperature regions consist of ambient air which is being entrained, and the intermediate temperature regions are the result of mixing of the burnt and unburnt gases. Presently 500-frame/s ciné photography is being incorporated which will be synchronized to the laser firing; this will aid in determining whether the interpretation of the emission data is correct.

Conclusions

A 10-Hz CARS instrument capable of simultaneous temperature and N_2 number density measurements has been employed for extensive profiling of a turbulent propane diffusion flame. Contour maps and PDF's constructed from the CARS data allow a more detailed study of the flame than that possible with conventional probes. Conditional sampling of the CARS temperature data by flame-emission monitoring indicates that the portion of the temperature PDF due to active combusting gases is consistent regardless of flame location, while the portions of the PDF due to non-combusting gases (either unburnt or burnt) are complex and vary with flame location.

Acknowledgment

This work was supported by US Air Force Contract F33615-80-C-2054.

References

[1] Goss, L. P., Trump, D. D., MacDonald, B. G., and Switzer, G. L., "10-Hz Coherent Anti-Stokes Raman Spectroscopy Apparatus for Turbulent Combustion Studies," Review of Scientific Instruments, Vol. 54, No. 5, May 1983, pp. 563-571.

[2] Goss, L. P., Switzer, G. L., Trump, D. D., and Schreiber, P. W., "Temperature and Species-Concentration Measurements in Turbulent Diffusion Flames by the CARS Technique," AIAA Paper 82-0240, AIAA 20th Aeroespace Sciences Meeting, Orlando, Florida.

[3] Kim, A, "Computer Programming in Physical Chemistry Laboratory: Least Squares Analysis," Journal of Chemical Education, Vol. 47, February 1970, pp. 120-122.

[4] Goss, L. P., Switzer, G. L., and Schreiber, P. W., "Flame Studies with the Coherent Anti-Stokes Raman Spectroscopy Technique," AIAA Paper 80-1543, AIAA 15th Thermophysics Conference, Snowmass, Colorado.

[5] Goss, L. P., and Switzer, G. L., "Application of Coherent Anti-Stokes Raman Scattering to Combution Media," Air Force Wright Aeronautical Laboratories, Wright Patterson Air Force Base, Ohio, AFWAL-TR-80-2122, Feb. 1981.

[6] Haraguchi H., and Winefordner, J. D., "Flame Diagnostics: Local Temperature Profiles and Atomic Fluorescence Intensity Profiles in Air-Acetylene Flames," Applied Spectroscopy, Vol. 31, November 1977, pp. 195-200.

[7] Gaydon, A. G., and Wolfhard, H. G., "The Structure of the Reaction Zone," Flames (3rd ed.), Chapman and Hall, Ltd., London, 1970.

[8] Lapp, M., Laser Probes for Combustion Chemistry, American Chemical Society Symposium Series, Vol. 134, edited by D. R. Crosley, American Chemical Society, Washington, D.C., 1980, Chap. 17.

[9] Roquemore, W. M., Britton, R. L., and Sandhu, S. S., "Investigation of the Dynamic Behavior of a Bluff Body Diffusion Flame Using Flame Emission," AIAA Paper 82-0178, AIAA 20th Aerospace Sciences Meeting, Orlando, Florida.

[10] Magill, P. D., Lightman, A. J., Orr, C. E., Bradley, R. P., and Roquemore, W. M., "Flowfield and Emission Studies in a Bluff Body Combustor," AIAA Paper 82-0883, AIAA/ASME 3rd Joint Thermophysics, Fluids, Plasma and Heat Transfer Conference, St. Louis, Missouri.

Comparison of CARS Combustion Temperatures with Standard Techniques

Richard R. Antcliff*
Systems Research Laboratories, Inc., Dayton, Ohio
and
Olin Jarrett Jr.†
NASA Langley Research Center, Hampton, Virginia

Abstract

This paper describes a comparative study of temperatures measured with Coherent Anti-Stokes Raman Spectroscopy (CARS) and those using standard measurement techniques. CARS temperatures from nitrogen vibrational-rotational spectra have been obtained in a premixed hydrogen-air laboratory flame supported on a flat flame burner over a temperature range from ~1000-2100K. The techniques used for comparison were sodium line reversal, thin wire thermocouples, and heat balance calculations. These studies show a good agreement over the entire temperature range studied.

Introduction

CARS thermometry is finding increased acceptance in the area of combustion diagnostics. The ability to make time- and spatially-resolved measurements makes the technique attractive in many practical environments including combustors,[1,2] internal combustion engines,[3,4] and gun-propellant flames.[5] To determine the useful range of the CARS temperature measurements, comparisons have been made with standard techniques. These techniques include thermocouple measurements,[1-3,6-12] sodium line reversal,[5,13]

Presented as Paper 83-1482 at the AIAA 18th Thermophysics Conference, Montreal, Canada, June 1-3, 1983. This paper is declared a work of the U.S. Government and therefore is in the public domain.
*Research Chemist, Research Applications Division.
†Aerospace Engineer, Hypersonic Propulsion Branch.

soot reversal,[14] and calculations based on combustion conditions.[15-17] These comparisons, however, have been limited to a small temperature range or a single point. In addition, only one technique is generally used for comparison. Notable exceptions are Refs. 1 and 12 in which corrected thermocouple measurements are compared with an extended range of CARS temperatures. The purpose of this study was to compare CARS thermometry with several standard techniques over the full range of temperatures encountered in a laboratory turbulent diffusion flame. Successful demonstration of CARS accuracy over such a wide range provides the confidence to use CARS as a diagnostic tool in the investigation of the complex turbulent combustion flowfields found in supersonic combustion ramjets (scramjets).

Background

Several excellent reviews include a treatment of the CARS theory;[18-20] therefore, only a brief summary following the development of Ref. 13 will be presented here. The vibrational-rotational levels in a gaseous molecule are affected dramatically by temperature. A measure of the population of these levels can therefore yield a precise temperature of the gas. Since nitrogen is a major component in air-fed combustion, and it has a vibrational-rotational level which can be easily probed with CARS, it was chosen as the thermometry species. To determine a CARS temperature, an experimental CARS spectrum was compared with a library of theoretical nitrogen spectra calculated at incremental temperatures. These theoretical spectra were calculated from the third order nonlinear susceptibility $[\chi^{(3)}]$.

The susceptibility can be expressed as

$$\chi^{(3)} = \chi_R + \chi_N \qquad (1)$$

where

$$\chi_R \propto \sum_{v,J} N \Delta(v,J)(d\sigma/d\Omega) / [\delta(v,J) - i \Gamma(v,J)] \qquad (2)$$

In this expression, N is the total number density, $\Delta(v,J)$ is the population difference between the states involved in the Raman transition, $\delta(v,J) = \omega_R - (\omega_1 - \omega_2)$, the frequency difference between the Raman transition (ω_R) and the CARS frequency ($\omega_1 - \omega_2$), $\Gamma(v,J)$ is the Raman half-width at half-max (HWHM) and $(d\sigma/d\Omega)$ is the Raman scattering cross section. χ_N is the nonresonant portion of the susceptibility. Using the modified values of Rado,[21,22] a temperature-dependent nonresonant function was calculated from adiabatic flame constituent concentrations. The total susceptibility was then used to calculate the CARS power as a function of frequency:

$$P_3(\omega_R) = \int P_1(\omega_1)d\omega_1 \int P_1(\delta) \, P_2(\omega_1-\Delta) \, |\chi^{(3)}(\Delta)|^2 \, d\Delta \quad (3)$$

where the CARS frequency is defined as $\Delta = \omega_1 - \omega_2$, and P_1 and P_2 are the powers in the pump (ω_1) and Stokes (ω_2) beams, respectively. This function was then convoluted over the configuration dependent slit function. The slit function used was a Gaussian with a full-width at half-max (FWHM) of 0.8 cm^{-1}. A $1/T^2$ dependence was assumed for the Raman half-width which was set to 0.05 cm^{-1} at 300K. These theoretical spectra were calculated every 20K and stored on tape. The experimental spectrum from each laser firing was fit to this library using a nonlinear Jacobian least-squares routine[23] adapted for CARS.

Experimental Techniques and Apparatus

A 6-cm-diam, sintered brass, water-cooled, porous plug burner was used to establish a uniform flame zone that could be readily varied over a wide temperature range. All temperature measurements were made approximately 1 cm above the burner surface. The burner was operated in thermal equilibrium in order that heat balance calculations could be used to determine the theoretical temperature.

CARS

The experimental arrangement is shown in Fig. 1. The primary laser was a Quanta Ray DCR-1A Nd:YAG. The 1.06-μm output was doubled with a KD*P Type II crystal to produce ~200 mJ of 532-nm output. A Pellin-Broca prism was used to separate the remaining 1.06-μm beam which was directed into a beam dump. Thirty percent of the green laser was split off to pump a broadband dye laser. The dye laser consisted of a broadband total reflector, a 1-cm-wide by 6-cm-long dye cell mounted at Brewster's angle through which Rhodamine 640 (7.8 × 10⁻⁵ molar in methanol) was circulated, and a 30% reflectance output mirror. This configuration produced about a 35% conversion efficiency with a FWHM of ~250 cm^{-1} centered at 606.5 nm. The broad band laser allowed generation of a complete CARS spectrum in a single 10-ns laser shot. This time resolution is desirable in rapidly fluctuating systems. A telescope was placed after the output mirror to control divergence and optically match the YAG beam at the sample volume. An optical flat was also inserted following the telescope to adjust the horizontal position of the dye beam for phase matching. The remainder of the green laser beam was split into two equal components to form the two ω_1 beams for BOXCARS[24] generation.

The BOXCARS arrangement allows the spatial resolution to be controlled by the interaction volume of the crossed laser beams. One of the ω_1 beams was made parallel with the dye laser (ω_2) beam using a dichroic mirror which had a maximum reflectance at 532 nm and maximum transmittance from 560-680 nm. A separation of 3 mm was produced between these two parallel beams. The other ω_1 beam was also made parallel with these beams producing a separation of about 2.54 cm between the two ω_1 beams. A 30-cm focal length, 7.6-cm-diam, plano convex, antireflective coated lens was used to focus the ω_1 and ω_2

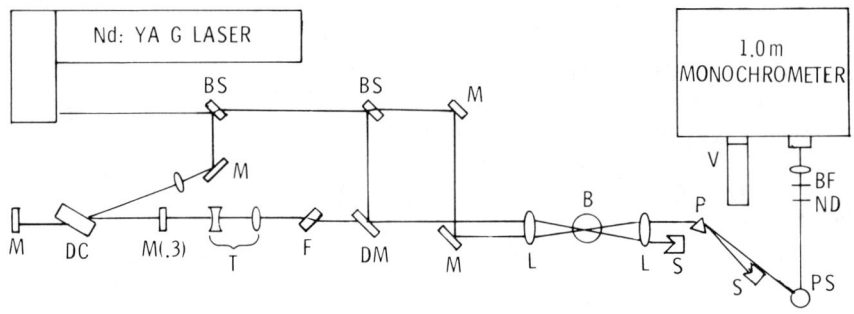

M = MIRROR
DC = DYE CUVETTE
M(.3) = 0.3 REFLECTANCE MIRROR
T = TELESCOPE
P = PRISM

F = FLAT
DM = DICHROIC MIRROR
L = LENS
B = H$_2$-AIR BURNER
BF = BLUE PASS FILTER

S = BEAM STOP
PS = PERISCOPE
BS = BEAM SPLITTER
ND = NEUTRAL DENSITY FILTER
V = VIDICON

Fig. 1 Schematic of the CARS measurement system.

beams into the sample region. The emerging beams were recollimated with a matched 30-cm focal length lens. Following the collimating lens one ω_1 beam and the ω_2 beam were directed into a beam dump. The other ω_1 beam and the generated CARS signal were predispersed by a 60 deg prism.

After the remaining ω_1 signal was dumped, the CARS signal was directed through a periscope, a series of neutral density filters, and a 473-nm band pass filter. The signal was then focused onto the entrance plane of a 1-m GCA McPherson 2051 monochromator equipped with a 1200 groove/mm grating. The multichannel detector was a silicon intensified vidicon (EG and G Model 1254). The vidicon controller was interfaced to a Hewlett Packard 9835A calculator for control of the laser firing and data acquisition. The spatial resolution was measured using a movable 0.25 mm jet of nitrogen within a 6 cm argon sheath. Measurable CARS signal to the 10% level was obtained within a 1-mm length region.

Thermocouple

A major problem encountered when using thermocouples as temperature probes has been the necessary corrections for heat losses by the probe. A novel approach for estimating the correction due to radiative losses has been used. Three thermocouples of different sizes were used simultaneously to measure the flame temperature. Using the theoretical basis of Moffat,[26] the radiation error correction is defined as

$$E_R = \sigma \epsilon A_R (T_J^4 - T_S^4) / h_c A_c \qquad (4)$$

where σ is the Stefan-Boltzmann constant, ϵ is the emissivity, T_J and T_S are the junction and surrounding temperatures, respectively. The effective convective area (A_c) and the effective radiative area (A_R) are normally equivalent. The coefficient of convective heat transfer (h_c) can be written, using the Nusselt-Reynolds number relationship as

$$h_c = K / D^{1/2} \qquad (5)$$

where K is a collection of constants and D is the wire diam. Substituting for h_c results in

$$E_R = D^{1/2} \sigma \varepsilon (T_J^4 - T_S^4) / K = K_2 D^{1/2} \qquad (6)$$

Using this relationship a plot of $D^{1/2}$ vs measured temperature can be extrapolated to zero diameter to obtain the radiation free temperature. To test the validity of the method, this technique was applied to some experimental thermocouple measurement using several wire diameters.[26] These data were found to be linear, and the extrapolation to zero wire diameter produced temperatures which agreed well with full heat loss corrected calculations.

The thermocouples used for the present study were fabricated of Pt-Pt/14% Rh. The wire sizes of the three thermocouples were 0.254, 0.076, and 0.025 mm. The wires were separated by ~1 cm. The radiation correction model assumes that the junction diameter is equal to the individual wire diameter. To satisfy this requirement, care was taken in fabricating the junctions to insure that the butt welds were nearly equal to the wire diameter. The wires were mounted perpendicular to the flow and parallel with the CARS beams.

This method only takes into account radiation losses and neglects probe induced conduction out of the flame. Although care was taken to correct the thermocouple data for radiation losses, the thermocouple measurements below 1400K were consistently lower than the CARS and heat balance temperatures. At higher temperatures, the corrected thermocouple data were in good agreement with the other techniques and are believed accurate to about ±100K.

Sodium Line Reversal

An excellent review of the sodium line reversal technique is found in Ref. 27. A brief summary of the technique will be given here. A doublet in sodium produces strong lines (the D lines) centered at 589.0 and 589.6 nm. When light from a continuous source is passed through a flame containing sodium vapor, the D lines will appear in absorption or emission depending on whether the brightness temperature of the source is higher or lower, respectively, than the flame temperature. When the temperature of the source and flame are matched (the reversal point) these lines disappear into the background.

A schematic of the sodium line reversal equipment used is shown in Fig. 2. The broadband source was a 6-volt, 18-A microscope illumination lamp. A 50-cm-diam,

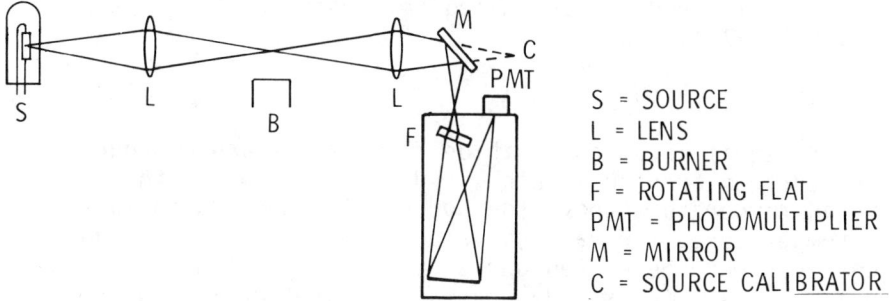

Fig. 2 Schematic of the sodium line reversal measurement system.

10-cm focal length lens was used to image the lamp about 1 cm above the burner surface and 20 cm from the lens. An identical lens was used to image light from the sample point (in the flame) into the monochromator. An iris with a 4-cm opening was located immediately preceding the second lens to limit ambient light. Calibration of the lamp image was accomplished by removing the steering mirror from the optical path to obtain a clear view of the lamp image. A Pyro Micro-Optical Pyrometer was used as a calibration standard. The mirror was mounted in a V block, allowing its removal without difficult realignment upon completion of the calibration.

A 1-m monochromator was used to observe the sodium and lamp signals. It is an f/10 instrument equipped with a concave holographic grating with a 980-mm focal length. A rotating optical flat behind the entrance slit repeatedly sweeps the image across the exit slit. The optical spectrum containing the two strong sodium D lines was displayed on an oscilloscope. The current to the lamp was adjusted to observe the reversal point.

The burner used in these studies was specially made to allow sodium line reversal measurements. A 1-mm tube installed in the center of the burner allowed aspiration of a small quantity of sodium hydroxide solution into the center of the flame. During CARS data collection, the solution of hydroxide was shut off as there was enough accumulation on the burner to produce a good sodium signal. CARS measurements were made 1 cm above the burner surface and 1 cm off-center so there would be no interference from the NaOH spray. Below 1700K a good sodium signal could not be produced even if the spray was present. Under normal conditions the temperature resolution was ~12K. Possible error sources include inaccuracy of the pyrometer measurements, and drift in the lamp calibration. The sodium determined temperatures should be accurate to ±75K.

Heat Balance Calculations

An equilibrium composition flame temperature program[28] for hydrogen burning in air was used to calculate the expected temperature above the porous plug burner. A matrix of temperatures for various values of fuel-air ratio and heat loss were generated and stored in the calculator. During a test, the steady-state flow rates of air and hydrogen were used to calculate the fuel-air ratio. The burner cooling water flow rate and temperature rise were measured and used to calculate the heat conducted into the porous plug burner and transferred to the imbedded cooling coil. These values of fuel-air ratio and heat loss were then used by inntexpolation to determine the temperature within the temperature matrix previous calculated.

Radiation from the gases was neglected. Hydrogen and air flows were measured with standard rotameters and the gases were premixed before entering the burner.

The heat balance temperature calculations, though simple in theory, depend upon many accurate measurements. Each of these measurements, including hydrogen flow and

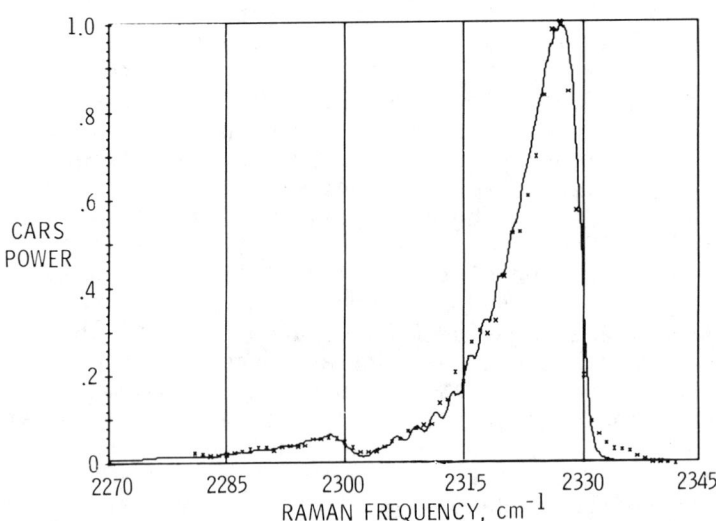

Fig. 3 Typical fit (solid line) of CARS experimental data points. Slit function is 0.8 cm^{-1}.

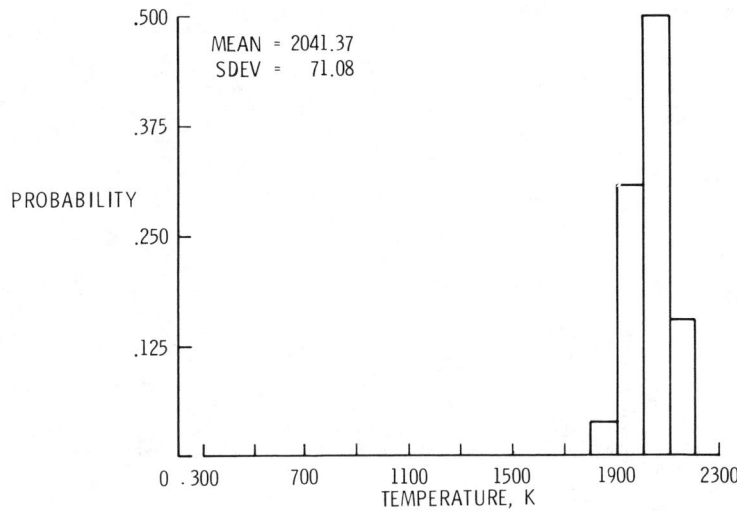

Fig. 4 Typical probability distribution histogram from 26 laser shots.

pressure, airflow and pressure, cooling water flow and the voltage generated by the temperature difference sensing thermisters have their own possible error sources. Analysis of the sensitivity of the calculated temperatures to the various measurement errors indicated that the temperature based on the heat balance calculations should be accurate to about ±100K.

Results and Discussion

To acquire the CARS data at 10 Hz, the calculator was used in the direct memory access mode. The memory size of the calculator limited the number of consecutive data scans to 30. Two scans were used for each laser shot because of the lag problems associated with the vidicon. Therefore, only 15 laser shots could be acquired consecutively. Data from each laser shot were fit individually as indicated previously. A typical fit of a single laser shot is shown in Fig. 3. The discrepancy on the high-frequency side of the 0 → 1 peak seems to be the result of some inaccuracy in the calculated spectra and showed up on all temperature fits. Normally two sets of 15 consecutive shots were acquired at each temperature. The temperatures obtained from the fitting of these data were used to generate probability distribution functions over this small sample. A typical histogram generated in this manner is shown in Fig. 4. The mean from each data set

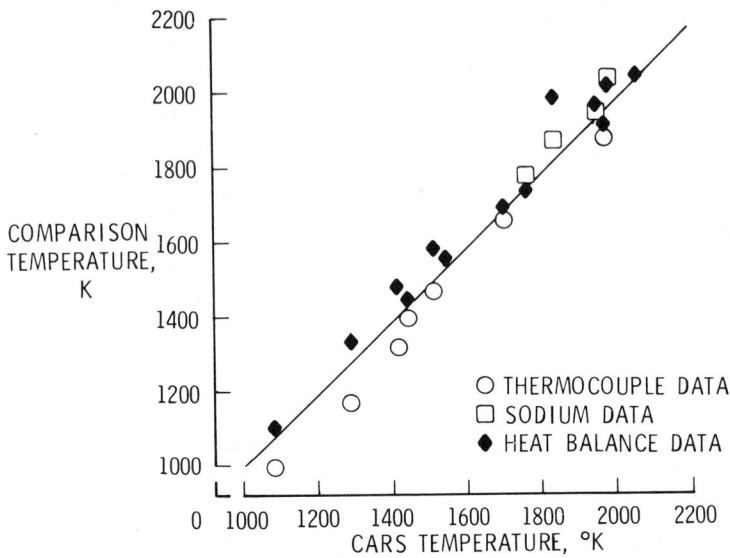

Fig. 5 Comparison of thermocouple, sodium line reversal and heat balance temperatures vs CARS derived temperatures.

was then used in comparison with the other techniques. The standard deviations of each temperature data set ranged from 60 to 90K. The standard deviation was largest for the lower temperature data, presumably because there is no hot band contribution at these temperatures and the fitting must rely only on the 0 → 1 bandwidth.

The comparison of the CARS temperature with those of the standard techniques is displayed in Fig. 5. An idealized one-to-one correlation is indicated by the solid line in the figure. In general, the agreement is quite good over the entire temperature range 1000-2000K. Below 1700K, thermocouple measurements were approximately 100 lower than those of CARS, presumably due to poor thermocouple heat loss corrections. Comparison of CARS with sodium line reversal above 1700K shows maximum differences of 50K. Finally, comparison of CARS with heat balance calculations indicated only small differences over the entire temperature range.

These results show that CARS thermometry agrees quite favorably with conventional thermometric techniques over a wide range of temperatures. The largest consistent deviations occur with the thermocouple data especially at low temperatures. The possible error sources in the CARS generated temperatures include Nd:YAG instability, temperature effects on the doubling crystal, flame variation, turbulence induced sample volume changes, fitting inaccuracies, detector lag problems, and dye laser fluctuations. Of these the major problem appeared to be the dye laser spectral instabilities which may stem from temperature fluctuations of the dye under local heating, mode beating, dye flow variations, and pump laser instabilities. Subsequent tests with an additional dye amplifier seem to help stabilize this dye laser system, and studies are presently underway to improve the dye laser stability.

Summary

A comparative study of CARS in the measurement of static temperature for a premixed laboratory flame was conducted. Sodium line reversal, Pt-Pt/14% Rh thermocouples, and heat balance methods were all used as bases of comparison. Over the entire temperature range 1000-2000K CARS gave results very close to the more standard techniques. This demonstrated the viability of the CARS technique at conditions typically found in the mixing-reacting flowfields of supersonic combustion ramjets. The potential for good accuracy, and the advantages of CARS in achieving both time- and spatially-resolved measurements, make it a particularly attractive diagnostic tool for scramjet applications.

Acknowledgments

The authors wish to express their appreciation to G. B. Northam, L. P. Goss, M. E. Hillard Jr., E. F. Germain, and W. P. Simmons.

References

[1] Switzer, G. L., Goss, L. P., Roquemore, W. M., Bradley, R. P., Schreiber, P. W., and Roh, W. B., "Application of CARS to Simulated Practical Combustion Systems," Journal of Energy, Vol. 4, Sep.-Oct. 1980, p. 209.

[2] Eckbreth, A. C., "CARS Thermometry in Practical Combustors," Combustion and Flame, Vol. 39(2), Oct. 1980, pp. 133-147.

[3] Stenhouse, I. A., Williams, P. R., Cade, J. B., Swords, M. P., "CARS in an Internal Combustion Engine," Applied Optics, Vol. 18, Nov. 1979, p. 3819.

[4] Klick, D., Marko, K. A., Rimai, L., "Temperature and Concentration Measurements by CARS Spectroscopy in a Firing Single-Cylinder Engine," SAE Paper 810227, International Congress and Exposition, Detroit, Michigan, February 23-27, 1981.

[5] Harris, L. E. and McIlwain, M. E., "CARS Spectroscopy of Gun Propellant Flames," Technical Rep. ARLCD-TR-81007, Sep. 1981.

[6] Moya, F., Druet, S. A. J., Taran, J-P. E., "Gas Spectroscopy and Temperature Measurement by Coherent Raman Anti-Stokes Scattering," Optics Communications, Vol. 13, Feb. 1975, p. 169.

[7] Rahn, L. A., Zych, L. J., Mattern, P. L., "Background-Free CARS Studies of Carbon Monoxide in a Flame," Optics Communications, Vol. 30, Aug. 1979, p. 249.

[8] Eckbreth, A. C., "CARS Investigations in Sooting and Turbulent Flames," Technical Report Project Squid UTRC-5-PU, Mar. 1979.

[9] Hall, R. J., "CARS Spectra of Combustion Gases," Combustion and Flame, Vol. 35(1), May 1979, p. 47.

[10] Switzer, G. L., Roquemore, W. M., Bradley, R. P., Schreiber, P. W., and Roh, W. B., "Evaluation of CARS in a Practical Combustion Environment," Technical Paper presented at 16th JANNAF Combustion Meeting, Monterey, California, Sep. 10-14, 1979.

[11] Murphy, D. V. and Chang, R. K., "Single-pulse Broadband Rotational Coherent Anti-Stokes Raman Scattering Thermometry of Cold N_2 Gas," Optics Letters, Vol. 6, May 1981, p. 233.

[12] Farrow, R. L., Mattern, P. L., and Rahn, L. A., "Comparison between CARS and Corrected Thermocouple Temperature Measurements in a Diffusion Flame," Applied Optics, Vol. 21, Sep. 1982, p. 3119.

[13] Goss, L. P., Switzer, G. L., Schreiber, P. W. "Flame Studies with the Coherent Anti-Stokes Raman Spectroscopy Technique," AIAA Paper 80-1543, AIAA 15th Thermophysics Conference, Snowmass, Colorado, Jul. 14-16, 1980.

[14] Eckbreth, A. C. and Hall, R. J., "CARS Thermometry in a Sooting Flame," Combustion and Flame, Vol. 36, Sep. 1979, p. 87.

[15] Taran, J. P. and Pealat, M., "Practical CARS Temperature Measurements," Technical Paper presented at 6th Symposium on Temperatures; Its Measurement and Control in Science and Industry, Washington, D.C., Mar. 14-18, 1982.

[16] McIlwain, M. E. and Harris, L. E., "Coherent Anti-Stokes Raman Spectroscopy of Gun Propellant Flames," Technical Paper presented at 17th JANNAF Combustion Meeting, Hampton, Virginia, September 22-26, 1980, Vol. 2, p. 379.

[17] Osin, M. N., Pushinin, P. P., Smirnov, V. V., Fabelinskii, V. I., and Tskhai, N. S., "Measurement of the Pressure and Temperature Distributions in a Supersonic Nitrogen Flow by Coherent Anti-Stokes, Raman Scattering," Sov. Tech. Phys. Lett., Vol. 6, Feb. 1980, p. 64.

[18] Nibler, J. W. and Knighten, G. V., "Coherent Anti-Stokes Raman Spectroscopy," Raman Spectroscopy of Gases and Liquids, Topics in Current Physics, edited by A. Weber, Vol. II Springer-Verlag, New York, 1979.

[19] Hudson, B., "CARS," ACS Symposium, Ser. 85, 1978, p. 171.

[20] Tolles, W. M., Nibler, J. W., MacDonald, J. R., and Harvey, A. B., "A Review of the Theory and Application of Coherent Anti-Stokes Raman Spectroscopy (CARS)," Applied Spectroscopy, Vol. 31(4), Jul.-Aug. 1977, p. 253.

[21] Rado, W. G., "The Nonlinear Third Order Dielectric Susceptibility Coefficients of Gases and Optical Third Harmonic Generation," Applied Physics Letters, Vol. 11, Aug. 1967, p. 123.

[22] Eckbreth, A. C. and Hall, R. J., "CARS Concentration Sensitivity with and Without Nonresonant Background Suppression," Combustion Science Technology, Vol. 25, No's. 5,6, 1981, p. 175.

[23] Kim, H., "Computer Programming in Physical Chemistry Laboratory: Least Squares Analysis," Journal of Chemical Education, Vol. 47, Feb. 1970, p. 120.

[24] Eckbreth, Alan C., "BOXCARS: Crossed-Beam Phase-Matched CARS Generation In Gases," Applied Physics Letters, Vol. 32, Apr. 1978, p. 421.

[25] Moffat, R. J., "Gas Temperature Measurement" Temperature - Its Measurement and Control in Science and Industry, edited by C. M. Herzfeld, Vol. 3, Reinhold Publishing Corp., New York, 1982, p. 553.

[26] Attya, A. M. and Whitelaw, J. H., "Velocity, Temperature, and Species Concentrations in Unconfined Kerosene Spray Flames," ASME Paper 81WA/HT-47, Winter Annual Meeting, Washington, D.C., November 15-20, 1981.

[27] Gaydon, A. G. and Wolfhead, H. G., Flames - Their Structure, Radiation, and Temperature, Chapman and Hall Ltd., London, p. 234., 1960.

[28] Mascitti, V. R., "A Simplified Equilibrium Hydrocarbon-Air Combustion Gas Model for Use in Air Breathing Engine Cycle Computer Programs," NASA TN D-4747, Sep. 1968.

Electronically Resonant CARS Detection of OH

James F. Verdieck,* Robert J. Hall,* and Alan C. Eckbreth*
United Technologies Research Center, East Hartford, Connecticut

Abstract

CARS (coherent anti-Stokes Raman spectroscopy) is a nonlinear spectroscopic technique capable of making remote, highly accurate measurements of temperature and concentration which are both temporally and spatially precise, in extremely difficult environments such as internal combustion engines and gas turbine exhausts. A major disadvantage of CARS methods is that application is limited to those species whose concentration is 1 % or greater. In order to extend CARS detectivity, electronically resonantly enhanced CARS can be employed, whereby one of the CARS laser frequencies is selected to be resonant with an electronic transition. This results in a large signal enhancement (possibly several orders of magnitude). This paper reports the results of applying resonant CARS to the detection of the hydroxyl radical in a methane/oxygen flame. Both the theory of electronically resonant CARS, predicted spectra, and the experimental procedure are described in detail. The variation of the resonant CARS spectra with different choices of electronic resonance is shown. A preliminary demonstration of saturation is also illustrated, as is the correlation of the resonant CARS spectrum with dependence of OH concentration on flame height.

Introduction

CARS (coherent anti-Stokes Raman spectroscopy) is a nonlinear optical process wherein three optical fields are combined in a material medium to generate a fourth optical field. As conventionally employed, two laser frequencies,

---

Presented as Paper 83-1477 at the AIAA Thermophysics Conference, Montreal, Canada, June 1-3, 1983. Released to AIAA to publish in all forms.
*Senior Research Scientist, Applied Laser Spectroscopy Laboratory.

ω_1 (the pump beam) and ω_2 (the Stokes beam), are mixed to produce the CARS frequency, ω_3, which appears as a coherent, collimated beam. This is illustrated in Fig. 1a. The CARS signal is large and easily detected (even in the prescence of particles or a highly luminous background) when the frequency difference between the two input frequencies corresponds to a Raman-active molecular vibration or rotation (or an electronic transition). Because CARS is a coherent, nonlinear process, laser beams are required to provide the proper phases of the optical fields in time and space, and the high intensity to enhance the nonlinear aspect of the process. Usually, high-peak power pulsed lasers are employed to generate CARS, particularly for combustion diagnostic applications. Photon energy is conserved in the CARS process (in contrast to the Raman effect), as illustrated in Fig. 1b. Note that $\omega_1 > \omega_2$. Fig. 1c exhibits two distinct methods of generating a CARS spectrum. One of these methods is to employ a broadband (100 cm^{-1} wide) Stokes laser, which results in the generation of the entire CARS spectrum of interest from each laser shot. This particular method is very useful for combustion diagnostics because single-shot (typically 10 nsec) thermometry can be performed. The second method uses two narrowband laser and tunes one of these, generating the CARS spectrum by scanning. It is this second, scanned technique which is utilized to generate resonantly enchanced CARS, by scanning ω_2 while holding ω_1 fixed on an electronic transition.

CARS applications are diverse and too numerous to catalog in detail here. It is an important spectroscopic technique well-suited for fundamental studies of molecules,

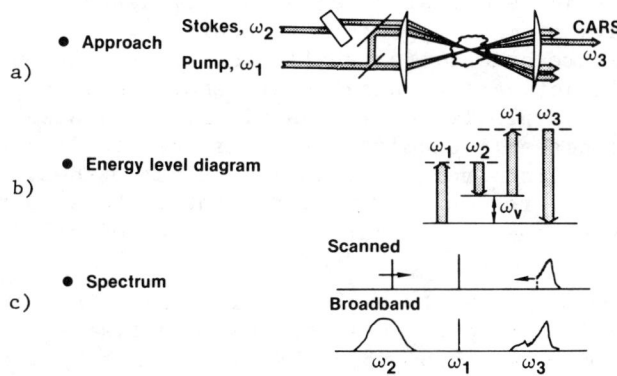

Fig. 1   Schematic diagram illustrating the basic concepts of CARS spectroscopy.

large and small. CARS has been repeatedly demonstrated to be a superior technique for remote diagnostics for combustion research, on both a laboratory scale and for large-scale practical devices. Examples include CARS temperature and concentration measurements in sooty flames,[1] plasmas,[2] internal combustion engines,[3] and gas turbine engines.[4] The theory and application of CARS have been extensively reviewed; these reviews may be consulted for detailed descriptions.[5-7]

A major disadvantage of CARS diagnostics methods is that, at atmospheric pressure, they are limited in application to species whose concentration is about 1 % or greater. A means of overcoming this limitation is to enhance the CARS signal through electronic resonance. Recall that the normal CARS process is vibrationally resonant; if in addition, one of the input CARS frequencies is resonant with an electronic transition, a large increase in the CARS signal occurs. This increase can be a factor of several hundred over the normal CARS effect. Electronically-resonant CARS (resonant CARS for short) has been observed for several large molecules in solution phase; however, to date, only a limited number of molecules have been observed to exhibit resonant CARS in the gaseous phase, namely I_2[8], NO_2[9], and most recently, C_2[10]. The reason for this limited application in gas phase is clear; there are very few simple, small molecules which absorb visible light(where dye lasers work efficiently). Indeed,the last cited case, C_2, is a radical which must be generated in a flame or a microwave discharge.[10]

The hydroxyl radical was chosen for study at this laboratory because of its ubiquitous presence and extreme importance in combustion chemistry. OH enters into the oxidation mechanisms of nearly all hydrogen-containing fuels, and plays an essential role in the combustion reaction pathways of most types of hydrocarbons. Hydroxyl radical reactions are usually very rapid (because OH is a radical, activation energies are small) and OH often enters into explosive chain reactions. OH is also important in atmospheric chemistry and has been implicated in acid rain formation from nitrogen and sulfur oxides. For these important reasons, it is important to be able to detect and measure the concentration of the OH radical in difficult environments.

Optical methods presently employed to measure OH concentrations are UV absorption spectroscopy and laser-induced fluorescence spectroscopy (LIFS). The absorption method is readily applied and useful where spatial resolution is not required. Even where applicable, gradient and

edge effects can degrade absorption measurement accuracy.
LIFS is a point measurement technique which enjoys good
success in carefully controlled laboratory devices operating cleanly at low pressure.[11] However, in practical
combustion environments, fluorescence methods can suffer
from interferences, such as fluorescence from other species
and from laser-induced particle incandescence. The major
problem with fluorescence techniques is collisional quenching, particularly for application in high pressure
devices. In contrast, CARS diagnostic measurements have
been demonstrated in several practical combustion systems
such as highly luminous sooting flames, internal combustion
engines, gas turbine combustors, solid propellant flames,
and alumina particle laden flows. The principal reason for
the success of CARS in these very difficult combustion
systems is the fact that the CARS signal emerges as a coherent beam, all of which can be captured. This provides a
unique advantage over incoherent techniques, such as fluorescence, particularly when the optical access is limited.
Because CARS methods have proven superior in these types of
environment, it is essential to extend the sensitivity for
CARS to radicals such as OH.

In the sections which follow, the progress on application of resonant CARS to the detection of OH is described.
The effort at UTRC has been and continues to be a combined
theoretical and experimental program. The first section is
a basic primer on OH spectroscopy, required for application
of resonant CARS. Following this section is a presentation
of the theory of resonant CARS with predicted resonant CARS
spectra for comparison with experimental spectra. Issues
of saturation and rotational state redistribution are not
considered because they are so difficult to model. A description of the experimental apparatus employed follows
the theoretical section. The results of the experiments to
date, are presented mainly through display of the resonant
CARS spectra obtained from different ω_1 electronic resonances, variation of incident laser power, and changing the
laser beam height relative to the burner surface. Conclusions reached from these investigations, and plans for
future studies are given in the final section.

Spectroscopy of the Hydroxyl Radical

It is essential to understand the electronic, vibrational, and rotational aspects of OH spectrosopy in order
to wisely select the resonant CARS pump laser frequency,
ω_1. Once ω_1 has been chosen, the Stokes frequency,
ω_2, is determined by the OH vibrational shift. The

spectral positions of the excitation frequencies, although a matter of some choice, depends strongly on the laser dyes available which yield high second-harmonic conversion efficiency in the appropriate region of the ultraviolet.

The spectroscopy of the OH radical is rather complicated for a diatomic molecule. This complexity arises mainly from the unpaired electron spin and the non-zero orbital angular momentum of the electronic ground state. Without going into fine detail, the result of spin-orbit coupling is a $^2\Pi$ electronic ground state which is split by ca. 126 cm^{-1}. This $^2\Pi$ state is inverted with the $\Omega = 3/2$ state lying below the 1/2 state. Following Dieke and Crosswhite,^{12} these states are designated F_1 and F_2 respectively, and are shown, along with the first excited state, $^2\Sigma$, in Fig. 2. The splitting between the upper state components f_1 and f_2 is exaggerated to illustrate the strongly vs weakly allowed transitions from the ground electronic state (generic label, X) to the first excited state(A). Also shown in this figure are the vibrational and rotational transitions from X to A. There are no vibrational selection rules for electronic transitions, but the Franck-Condon principle governs the strength of the vibronic transitions as the middle figure indicates. Note that the convention on labeling states is v' for the upper state and v" for the lower. Moreover, a particular transition is designated as v',v"; hence, the 1-0 vibronic transition implies a transition, in either absorption or emission, between v' = 1 of the A state and v" = 0 of the X state. For the case of a $^2\Pi$ state the rotational selection rule permits $\Delta J = 0, \pm 1$; therefore Q, P and R rotational branches are oberved as shown in Fig. 2. Note also that the quantum number J for this case is the total angular momentum; i.e., the sum of the orbital angular momen-

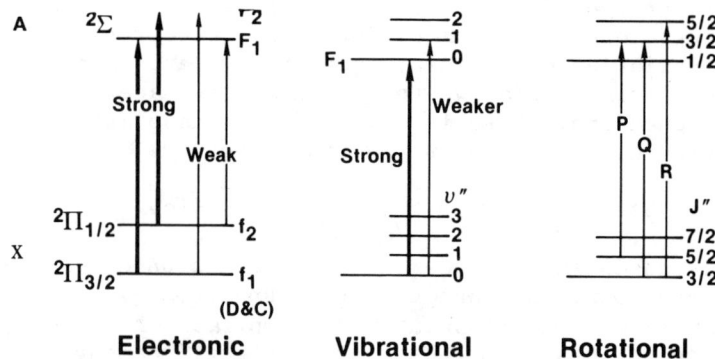

Fig. 2   Fundamental spectroscopic considerations for the OH radical.

tum, Ω, and the mechanical angular momentum of the molecule, N. For this reason, J" is half-integral and has an initial value of 3/2.

If one excludes the rotational fine structure and examines only the vibronic structure of the X to A transition of OH, a plot of vibronic bandheads results, as shown in Fig. 3. This spectrum is extremely important because, in conjunction with Fig. 4, the relative output of conventional laser dyes, it was used to determine the optimum vibronic band for the resonant CARS pump frequency, ω_1. The choice of the 1-0 band was made because it places ω_1 in the region of the very efficient laser dye, rhodamine 590, so that reasonable second harmonic power can be obtained. The Stokes frequency, ω_2, is required to lie in the range 3000-3500 cm^{-1} to the low frequency side (high wavelength) of ω_1. According to Fig. 4, the laser dye DCM covers the required range nicely, with good energy conversion. It must be noted that the 0-0 band, at 307 nm, much exploited for laser induced fluorescence, appears to be a poor choice for ω_1 because both of the laser frequencies would lie on the opposite edges of the DCM curve and the second harmonic power would be quite low.

The complete rotational structure of the 0-0 and 1-0 bands is shown in Fig. 5. This is the superposition of the P, Q, and R branches arising from both components of the $^2\Pi$ electronic ground state. In order to distinguish these components a subscript 1 or 2 is used for transitions from or to the F_1 or F_2 states, respectively. This notation is illustrated in the next figure, Fig. 6, which shows a portion of the 1-0 band greatly expanded. It is noted that there is considerable spacing--5 to 10 cm^{-1}--between lines such as $Q_1(14)$ and $P_1(9)$ in this region. A single transition is easily selected because the frequency bandwidth of the dye laser is less than 0.3 cm^{-1}.

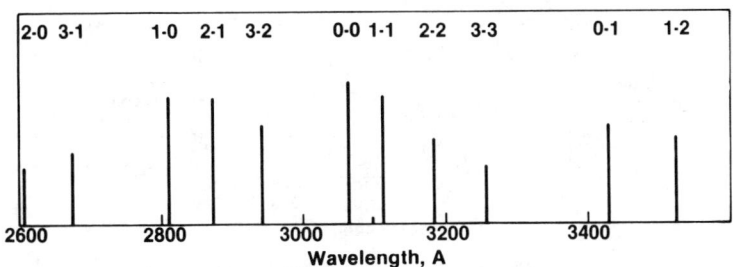

Fig. 3   Vibronic bandheads for the A to X transition of OH. Acetylene/oxygen flame at 3000 K. Adapted from Dieke and Crosswhite, Ref. 12.

In this region of the spectrum, 10 cm^{-1} corresponds to about 0.1 nm. The two transitions mentioned were employed for the first observation of resonance CARS in OH.

Resonant CARS Theory

The traditional approach to interpretation of resonance CARS spectra has been to employ the general, 48 term expression for the third-order electric susceptibility that has been derived by several authors.[8,13] This expression has been obtained either algebraically, using a perturbation expansion solution of the density matrix equation of motion, or diagrammatically, using "doubled" Feynman diagrams.

Both approaches yield equivalent results. The complete expression, showing all terms, has been presented by Bloembergen, Lotem, & Lynch.[13] Certain of the terms contain vibrationally resonant denominators which cause them to be large when the frequency difference between the Pump and Stokes sources encroaches upon Raman-active vibrational modes. If all other terms are lumped into the nonresonant background susceptibility, χ_{NR}, then the third-order electric susceptibilty may be expressed as a sum of this nonresonant term and a vibrationally resonant term which is a summmation of contributions from all Raman-active modes. That is (Druet, et al.),[8]

$$\chi^{(3)}(\omega_3) = \chi_{NR} + \sum_{a,b} \chi_R^{ba} \qquad (1)$$

where a and b represent the initial and final vibration-rotation quantum numbers of a particular Raman transition whose contribution to the resonant part of the suscepti-

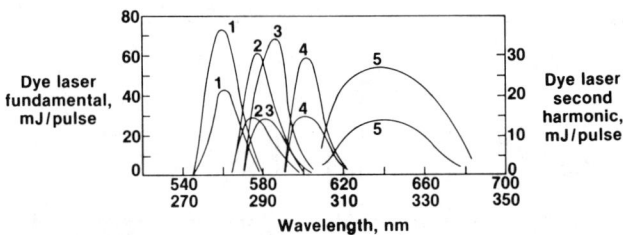

Fig. 4 Second harmonic conversion of some typical laser dyes using a 532 nm pump laser with 200 mJ/pulse. The dyes shown are: 1, Exciton R590; 2, Exciton R610; 3, Exciton Kiton Red; 4, Exciton 640; 5, Exciton DCM. Source, Quanta-Ray, Inc.

lity can be expressed as

$$\chi_{ab}^{(3)} = \frac{N}{\hbar^3} \frac{1}{(\omega_{ba}-\omega_1+\omega_2-i\Gamma_{ba})} \sum_{n'} \left(\frac{\mu_{an'}\mu_{n'b}}{\omega_{n'a}-\omega_3-i\Gamma_{n'a}} + \frac{\mu_{an'}\mu_{n'b}}{\omega_{n'b}+\omega_3+i\Gamma_{n'b}} \right)$$

$$\times \left[\sum_n (\rho_{aa}^{(0)}-\rho_{nn}^{(0)}) \left(\frac{\mu_{bn}\mu_{na}}{\omega_{na}+\omega_2-i\Gamma_{na}} + \frac{\mu_{bn}\mu_{na}}{\omega_{na}-\omega_1-i\Gamma_{na}} \right) \right.$$

$$\left. - \sum_n (\rho_{bb}^{(0)}-\rho_{nn}^{(0)}) \left(\frac{\mu_{bn}\mu_{na}}{\omega_{nb}-\omega_2+i\Gamma_{nb}} + \frac{\mu_{bn}\mu_{na}}{\omega_{nb}+\omega_1+i\Gamma_{nb}} \right) \right] \quad (2)$$

In Eq. 2, N represents the OH number density, ω denotes frequency; Γ denotes pressure-broadened linewidth; ρ, Boltzmann population; and μ, electric dipole matrix element. n and n' represent excited electronic states of the molecule, and it is apparent from examination of the denominators in Eq. 2 that electronic resonant enhancement will occur if either ω_1, ω_2, or ω_3 coincides with an allowed electronic transition. It is also apparent that selecting the pump frequency resonant with a particular a-n electronic transition "picks out" a particular initial ground vibration-rotation state a, making it possible to

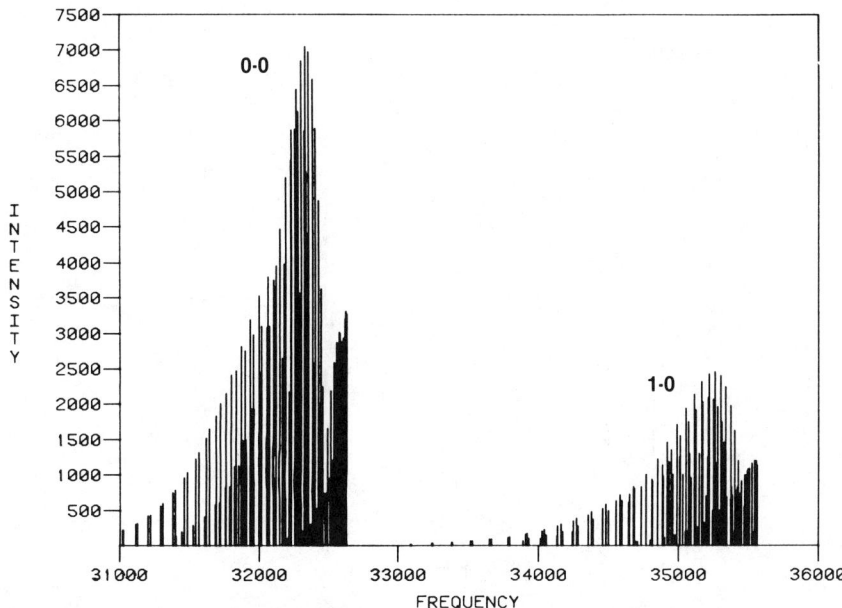

Fig. 5 Theoretical "stick" spectra of the 0-0 and 1-0 bands of the A-X transition in OH.

ignore the contributions of all other initial Raman states in the ground electronic vibrational manifold. Because the term which contains the denominator resonant in ω_2 is proportional to the population ρ_{bb} of the upper state in the Raman transition, it is reasonable to expect that its contribution will be small, particularly for a molecule like OH with a very large vibrational spacing. The rules for interpretation of resonant CARS spectra in terms of Eq. 2 have been discussed at length in the literature, with successful application to I_2,[8] N_2O,[9] and C_2.[10]

One notes that if the pump frequency is made to coincide with the frequency of a particular electronic transition a-n, and the Stokes frequency is tuned to sweep out the range of Raman frequencies, there will be double resonances which occur when $\omega_1 - \omega_2 = \omega_{ab}$, or $\omega_3 = \omega_{an'}$, and the possibility of triple resonances when $\omega_1 = \omega_{an}$, $\omega_1 - \omega_2 = \omega_{ab}$, and $\omega_3 = \omega_{an'}$. The criterion for the occurrence of a triple resonance is that the transitions a-n and a-n' be optically allowed and that $\omega_{ab} = \omega_{nn'}$. While the triple resonances do occur in molecules like I_2 and C_2 with closely spaced vibration-rotation states, they are expected to be extremely improbable in a molecule like OH with very large spacing between levels. If two distinct pump sources with slightly different frequencies are employed, then a triple

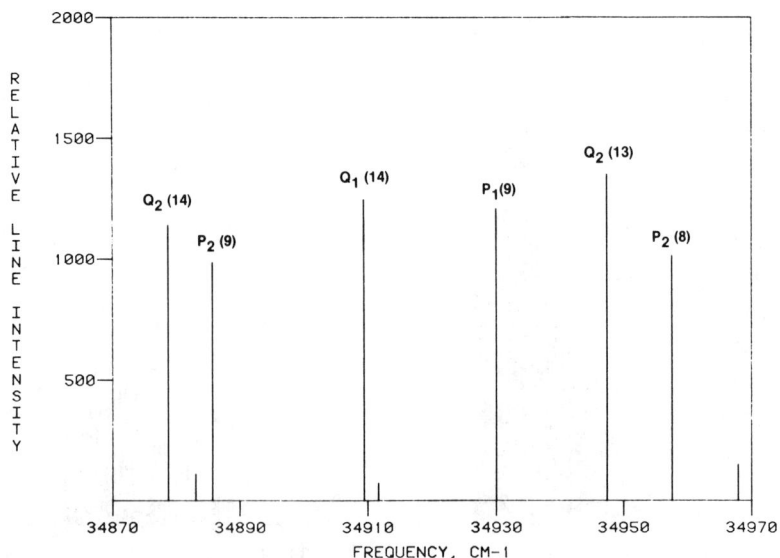

Fig. 6   High resolution "stick" spectrum of OH expanded about the $P_1(9)$ line of the 1-0 band.

resonance can be achieved in OH, but in general they should not be expected.

The electronic and vibrational spectroscopy of OH has been discussed previously. In the electronic absorption or emission spectrum, P, Q, and R branches are allowed, making possible 12 branches. However, the six satellite branches in which the F_1 or F_2 designation changes will be relatively weak, leaving six main branches. In OH vibrational spectroscopy O, P, Q, R, and S transitions are allowed. The weakness of the satellite branches in the electronic spectrum means that vibrational transitions in which the F_1 or F_2 designation changes can also be ignored. By examination of Eq. 2, and considering the selection rules for electronic and rotational transitions which must be obeyed for the resonant CARS process, it is possible to specify the types of resonances which will give rise to the spectral features in electronically resonantly enhanced OH CARS. As an example, if ω_1 is tuned to a Q-branch transition, then P, Q, and R Raman resonances will be observed, and the anti-Stokes term will be contributed by a Q-transition. This is illustrated in Fig.7. The general rules are illustrated in Table 1.

The fundamental interpretation of the resonant CARS spectrum of OH therefore involves straightforward counting of resonances, and it is anticipated that the spectrum will consist of contributions from "double resonances". Detailed calculations which will be discussed later show that double resonances of the a-n, a-n' type will make little contribution, and that the primary features are accounted for by a-n, a-b double resonances. Hence, one expects the resonant CARS spectrum of OH to consist of triplets if only the vibrational fundamental (v=0 to v=1) and no vibrational overtones ($\Delta v > 1$) are considered.

In order to perform quantitive evaluations of Eq. 2, it is necessary to have accurate spectroscopic information

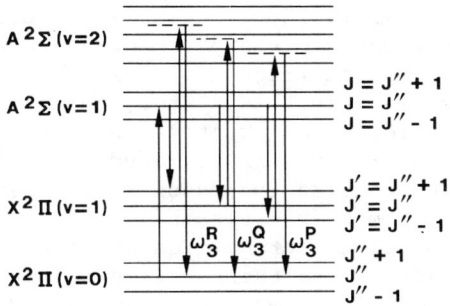

Fig. 7 Energy level diagram showing the origin of the triplet structure of the resonant CARS spectrum.

about vibration-rotation energy levels in the ground (X) and upper (A) electronic states, as well as values for the electronic transition dipole matrix elements. Fortunately, extensive investigations of this molecule have provided a great deal of information about these quantities. Specifically, tabulations of vibration-rotation energy levels have been published by Coxon,[14] and the absolute values of the electric dipole matrix elements can be deduced from the theoretical investigations of Chidsey and Crosley.[15] Values for the pressure-broadened linewidths also are needed, but reasonable estimates can be made here. Linewidths will be of great concern only if significant line-to-line variations are expected; at the extremely high gas temperatures of main interest in these investigations there are reasons for expecting this not to be the case. There is also a question of the proper phase to choose for the electric dipole matrix elements, because Franck-Condon factors are in general complex quantities. As Druet, et al.[8], have shown, the phase of the Franck-Condon factor is not of concern, if only one vibrational state in the upper A state contributes to the anti-Stokes summation in Eq. 2. Indeed, this is the case in OH.

Using the energy level tabulations of Coxon,[14] and the Einstein A-coefficients of Chidsey and Crosley,[15] the absorption spectrum of the 1-0 A-X system has been synthesized from the following expression for line intensity[16]

$$S_{v''J''}^{v'J'}(T) = \frac{1}{8\pi c \nu^2} \left(\frac{N}{p}\right) \frac{e^{-1.44E''/T}}{Q_{VR}} A_{v''J''}^{v'J'} (2J'+1)\left(1 - e^{-1.44\nu/T}\right) \quad (3)$$

where ν is the transition frequency and E'' is the energy in the ground state. Because the satellite bands have been ignored, six bands (two each of P, Q, and R) contribute to the synthetic spectrum shown previously (Fig. 5). The R-bands show the expected reversal at higher values of rotational quantum number. The 1-0 system has been chosen for this calculation because our experiments have been conducted with the pump source frequency chosen to coincide with those of various lines in this system, for the reasons discussed in the section on OH spectroscopy. A high resolution predicted spectrum was illustrated above (Fig. 6).

In order to synthesize the resonant CARS spectrum of OH, electric dipole matrix elements were computed by employing the rovibrational transition probabilities of Chidsey and Crosley,[15] and the rotational line strength formulas of Earls.[16] The complicated expression (Eqs. 1

and 2) for the third order resonant susceptibility undergoes a drastic simplification because one is concerned with only one a-n transition; the allowed Raman resonances are well separated and do not overlap; and only one or two (see (3)) anti-Stokes resonances a-n' will make contributions. If one assumes that only the lower vibrational state is significantly populated ($\rho_{bb} = \rho_{nn} = 0$), and if only one anti-Stokes resonance is important, then the resonant contribution to the third-order susceptibilty (Eq. 1) reduces to

$$\chi^{(3)}_{ab} \propto \frac{N\rho^{(0)}_{aa}}{\hbar^3 (\omega_{ba} - \omega_1 + \omega_2 - i\Gamma_{ba})} \frac{\mu_{an'}\mu_{bn'}}{(\omega_{n'a} - \omega_3 - i\Gamma_{n'a})}$$

$$\times \frac{\mu_{an}\mu_{bn}}{(\omega_{na} - \omega_1 - i\Gamma_{na})} \qquad (4)$$

A search was first undertaken for triple resonances; for each electronic transition in the 1-0 manifold, pump coincidence was assumed and the allowed Raman resonance

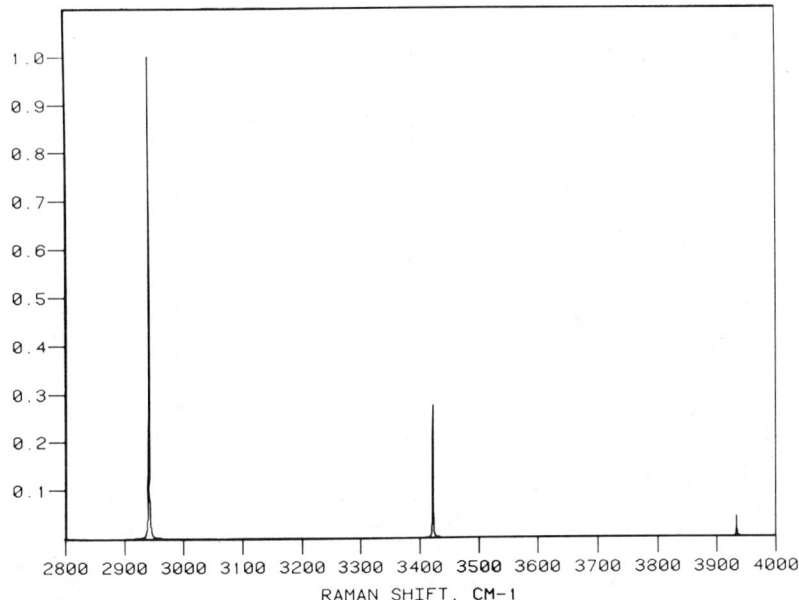

Fig. 8 Predicted resonant CARS spectrum for ω_1 tuned to $P_1(9)$.

frequencies calculated. The resulting anti-Stokes frequencies were then compared to the frequencies of the allowed transitions in the 2-0 manifold. This computer search confirmed the earlier expectation that no close triple resonances can be expected. The closest coincidence was about 4 cm^{-1} for the pump tuned to the $P_1(13)$ transition; no great enhancement is expected because the Boltzmann population factor for this transition is relatively small for the range of gas temperatures of interest.

Sample CARS calculations are presented in Figs. 8 and 9 for the pump tuned to various 1-0 A-X transitions. In

Table 1 Allowed resonant CARS transitions

Pump tuned to	Allowed Raman resonances as Stokes is tuned	Contributing anti-Stokes transition
Q	P,Q,R	Q
P	O,P,Q	P,R
R	Q,R,S	P,R

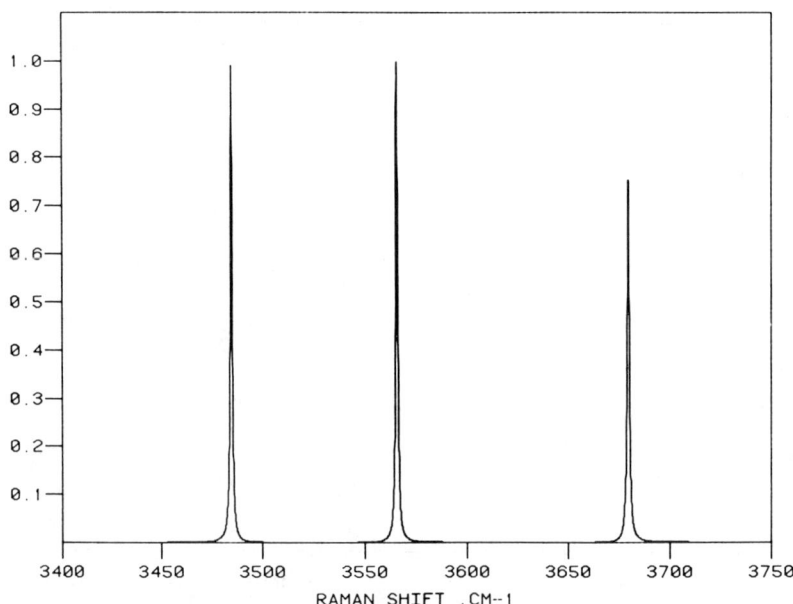

Fig. 9 Predicted resonant CARS spectrum for ω_1 tuned to $Q_1(2)$.

each case, a triplet is predicted; each sharp feature in the spectrum represents a double resonance of the type a-n, a-b. The relative strengths of the features are determined by the matrix elements which occur in Eq. 2, and also by the anti-Stokes denominators in that expression. Although no triple resonance is represented here, some of the Raman resonances are associated with anti-Stokes frequencies that lie much closer to anti-Stokes resonances (in the 2-0 A-X system) than others, and are accordingly enhanced. The relatively weak features in Figs. 8 and 9 correspond to Raman resonances whose anti-Stokes frequencies lie relatively far from the allowed anti-Stokes resonances. The large rotational constant ($B = 19$ cm^{-1}) of OH gives rise to the very large separations between outer lines that are evident in Figs. 8 and 9. This separation decreases for lower values of J probed by the pump laser, as evident by comparing the magnitudes of the outer line separations in Figs. 8 and 9.

The interpretation of the resonant CARS spectrum of OH in terms of third-order perturbation theory thus results in the prediction of a rather simple spectrum consisting of well-separated triplets. As has been emphasized by Druet and co-workers,[8] the perturbation expression (Eq. 1) for the third-order susceptibility will not be applicable if the pump or Stokes powers give rise to appreciable saturation. As will be seen, there is evidence of saturation in the experimental spectra obtained in these investigations, and this may preclude a simple interpretation of the experimental results. Under such conditions, corrective terms can, in principle, be applied (Druet, et al.[8]). But a better theoretical approach might be along the lines of the analysis of strong-field effects in CARS by Wilson-Gordon, Klimovsky-Baird, and Friedmann.[17] They perform a full solution of the density matrix equation of motion for a three level system, and obtain analytical solutions in var-

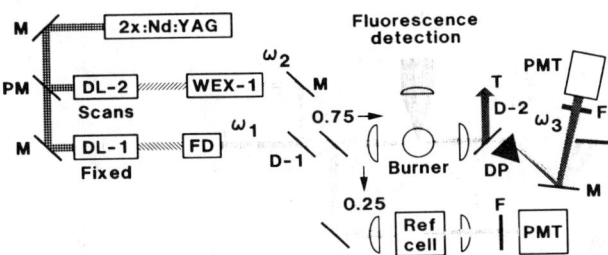

Fig. 10 Schematic diagram of the resonant CARS experiment. Symbols are described in the text.

ious limiting situations. Equally important is the question of collisional redistribution of population. As will be seen, the experimental results do confirm the basic triplet structure, but there is additional satellite structure around each basic component of the triplet that is suggestive of collisional transfer to adjacent rotational states. The question thus arises as to whether rotational redistribution within the upper A state and/or collisional electronic quenching to the ground X state could give rise to extra coherences. This problem may have connections with the "pressure-induced extra resonances" that have been predicted and observed by Prior, et al.[18] An interpretation of these extra resonances has been carried out by Grynberg,[19] who employs the formalism of the "dressed atom". These effects are areas of future investigation and are in the forefront of resonant CARS theoretical research.

Experimental

Several important factors must be carefully considered in the design of a resonant CARS experiment which attempts to detect a molecular free radical in the gas phase. The problems are compounded by the fact that the OH electronic transitions lie in the ultraviolet. The requirement of working in the UV introduces the additional complexities of frequency doubling two narrowband tunable dye lasers, working exclusively with fused silica optics, manipulating and detecting ultraviolet beams, and dealing with dichroic mirrors and filters that are only partially effective in re-

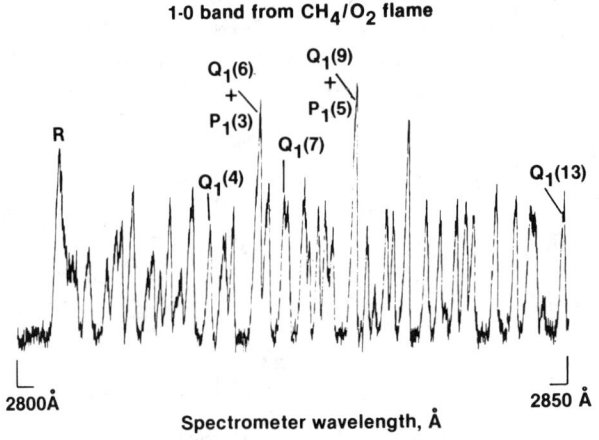

Fig. 11 OH emision spectrum employed for calibration purposes.

jection of unwanted wavelengths, while at the same time, severely attenuating the desired wavelength.

The configuration conventionally employed for resonant CARS makes use of two tunable, narrowband dye lasers whereby the ω_1 laser is tuned to a selected electronic resonance and held fixed, and then slowly scanning the Stokes laser, ω_2, which generates the resonant CARS spectrum, ω_3. The experimental configuration employed at UTRC is shown schematically in Fig. 10. The two tunable narrowband dye lasers, DL-1 and DL-2, are synchronously driven with the second harmonic (532 nm) of a Quanta-Ray pulsed neodymium YAG laser (DCR-1) operating at 10 Hz. The partial mirror (PM) allots about 40 percent of the 532 nm radiation to a commercial pulsed dye laser (Quanta-Ray PDL-1), denoted DL-2, and the remaining portion to a homemade dye laser (DL-1) of the Littmann grazing-grating configuration.[20] Each dye laser consists of side-pumped oscillator and preamplifier stages, followed by an end-pumped amplifier stage. Each dye laser produces laser radiation of frequency width less than 0.1 cm^{-1} and energy per pulse between 10 and 20 mJ, depending upon the laser dye chosen.

As stated previously, the outputs from both dye lasers must be frequency doubled. For the ω_1 laser, frequency doubling is a trivial task and requires only a manually adjustable second harmonic crystal, FD, in the figure. However, because ω_2 is scanned, an automatic angle-tuned device which maintains constant output power of the second harmonic frequency as the fundamental is tuned, must be employed. A WEX-1(Quanta-Ray) performs this task over the fundamental dye laser spectrum at scan rates in excess of 0.05nm/s. Second harmonic conversion efficiency appears to be about 10 % for both lasers; hence, ultraviolet energies are between 1 and 2 mJ per pulse, which corresponds to peak powers of 100 to 200 kW.

The two ultraviolet beams are combined with a dichroic mirror and focussed into the flame in the collinear CARS configuration. The beams emerging from the flame, now containing the CARS frequency also, are recollimated and directed to the detection system. ω_1 and ω_2 are partially removed with the dichroic D-2 and trapped in the trap, T. Because the dichroic is unsuccessful in completely removing ω_1 and ω_2 from ω_3, the three beams are spatially separated with the dispersing prism, DP. The unwanted beams are masked out and ω_3 is allowed to enter a filtered photomultiplier, PMT. The reference cell leg of the experiment is derived by picking off 25 % of the incident beams and focussing into a high pressure cell and detecting the nonresonant(non-vibrationally resonant) CARS

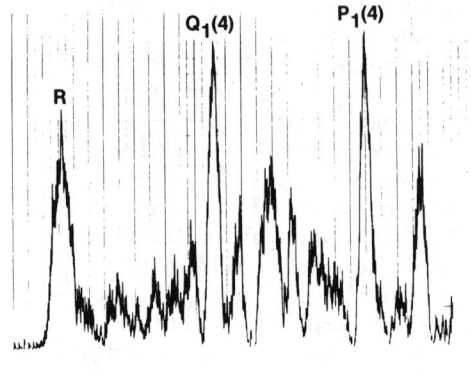

Fig. 12 Laser-induced fluorescence spectrum for ω_1 tuned to the $Q_1(4)$ line of OH.

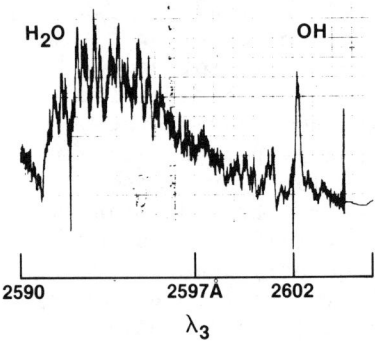

Fig. 13 First observation of resonant CARS in the OH radical. ω_1 tuned to resonance on $P_1(9)$.

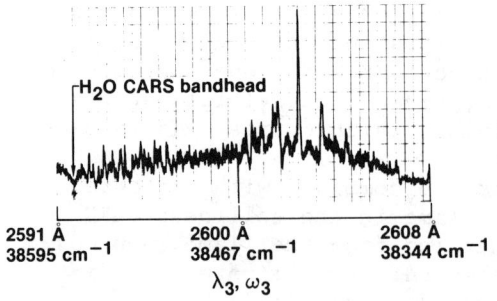

Fig. 14 Resonant CARS spectrum obtained for ω_1 tuned to $Q_1(14)$.

signal. Unfortunately the reference cell leg did not functions as well as anticipated, hence all the CARS spectra displayed below are uncorrected for laser power variation and dye laser profile.

A most essential feature of the experiment is the emission/fluorescence detection arm shown perpendicular to the CARS beams, above the burner, in Fig. 10. This portion of the experiment serves to calibrate a monochromator through the emission spectrum of OH from a methane/oxygen flame, obtained with the aid of a chopper and lock-in detector. As as example, a portion of the 1-0 emission from OH over a 50 A span is shown in Fig. 11. A few of the rotational lines are identitfied for illustration. More importantly, this subsystem serves to both indicate when resonance excitation is achieved, and to identify the particular transition which is resonant. This is done by recording the laser-induced fluorescence through use of a boxcar integrator following the monochromator-PMT. The LIF spectrum for ω_1 exciting the $Q_1(4)$ line of the 1-0 band of OH is shown in Fig. 12. This type of spectrum, much different from the emission spectrum, clearly and uniquely indentifies the resonance on $Q_1(4)$ because of the strong fluorescence from $P_1(4)$.

The burner used throughout these studies consisted of a staggered array of stainless steel capillary tubes(0.125 in o.d.) which were ground flush with an edge-cooled brass plate which held the tubes in place. The matrix of alternating fuel/oxidizer tubes was surrounded by an outer layer of tubes which provided a nitrogen sheath to stabilize the flame. During the course of this investigation, a methane-oxygen flame has been used rather than the more tempermental hydrogen/oxygen flame. This type of burner produces a flame which is very uniform in appearance. The burner is mounted on a translation stage which can be moved in the vertical direction so that the CARS laser beams can be varied in height relative to the burner surface.

Resonance CARS Results

The first demonstration of resonant CARS in OH was obtained with ω_1 resonant on the $P_1(9)$ line of the 1-0 vibronic band. The spectrum is shown in Fig. 13. This type of spectrum (a scanned CARS spectrum) is obtained by scanning the ω_2 dye laser, while recording the change in intensity of the ω_3 signal falling on the detector. The region of the spectrum to the left is the water CARS spectrum which is not an electronically resonant spectrum. However, because of the great abundance of water vapor in this flame, the "normal" CARS spectrum is easily observed. The sharp rise occurring around 2591 A is the bandhead of the water CARS spectrum. The resonant CARS of OH, occurring in the tail of the water CARS spectrum, has its strongest peak at 2602 A and represents a case where ω_1 has not been precisely tuned to the center of resonance. In contrast the resonant CARS spectrum of OH with ω_1 carefully tuned to the $Q_1(14)$ line (adjacent to the $P_1(9)$) is shown in Fig. 14. For this case, the water CARS spectrum is barely discernible, while the OH lines are

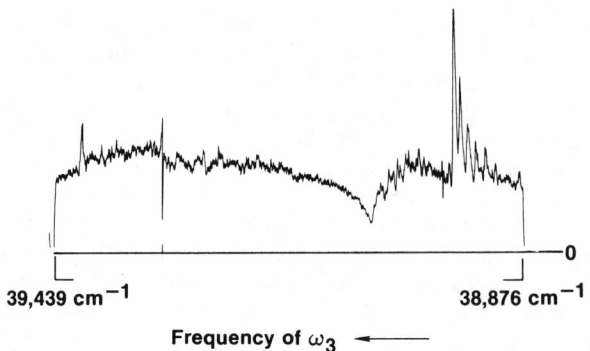

Fig. 15 Wide frequency scan of resonant CARS spectrum for resonance on $Q_1(4)$.

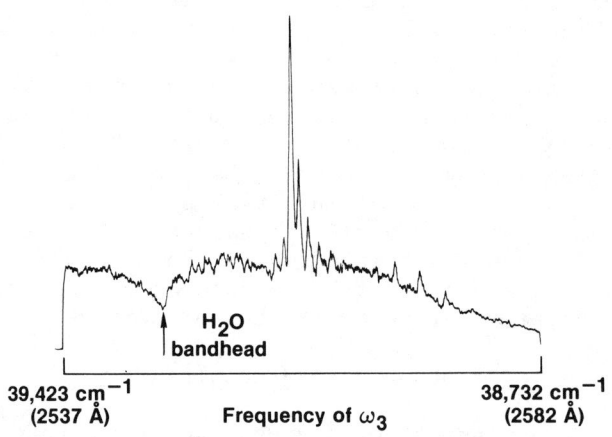

Fig. 16 Resonant CARS spectrum for ω_1 tuned to $Q_1(2)$.

much stronger, at least 5 times stronger. This is good evidence that resonant CARS is being observed, because the water concentration is at least four to five times that of OH in the CH_4O_2 flame, and because CARS signals scale as concentration squared, the normal CARS of OH would be less than 0.1 that of the H_2O CARS signal. In other words, it would be nearly totally lost in the water CARS band. The enhancement factor for the OH resonant CARS may even be larger than is apparent, because the true scaling of the CARS signal involves a consideration of linewidths and constructive/destructive interferences between lines.

The resonant CARS spectra obtained for the $P_1(9)$ and $Q_1(14)$ reonances are incomplete with respect to the predicted CARS spectra of Figs. 8 and 9. They are incomplete in the sense that the outermost lines are not observed in the experiment, because of the limited spectral tuning range of ω_2. The spacing between these components is $2B(2J + 1)$ where B is the rotational constant for OH, approximately 19 cm^{-1}. Hence, for high J values, the separation can be several hundred wavenumbers and far beyond the tuning range of a single laser dye. For these reasons, it was decided to move ω_1 toward resonances with lower J values, namely, $Q_1(4)$ and $Q_1(2)$. Examples of resonant CARS spectra for these two cases are shown in Figs. 15 and 16 respectively. As may be seen, the scans cover a reasonably wide frequency range, in excess of 300 cm^{-1}. For the case of the $Q_1(4)$ spectrum, considerable structure may be seen to the left of the water CARS bandhead. These lines have not yet been identified as positively originating from OH, however. For the $Q_1(2)$ resonance CARS, some structure is observed to the right of the strong central OH resonant CARS band, at approximately the proper spacing. This structure does not consist of a single line, but also exhibits some satellite lines, contrary to theoretical prediction.

A preliminary exploration of saturation was made by reducing the intensity of the ω_1 beam with fused silica flats and partially reflecting UV mirrors. The maximum attenuation so obtained was a factor of 5.5. In order to obtain a reasonable signal strength at this level of attenuation, the PMT supply voltage was increased. The results of this test are shown in Fig. 17. Note that there is a considerable change in the shape of the OH spectrum and in the number of lines. Note also that the water CARS spectrum is barely discernible in the attenuated ω_1 spectrum. The increased gain required to maintain the same signal level was approximately 10; because the CARS signal scales as the square of the ω_1 power, a reduction of about 30 should have been seen. This observation, along with the change in shape is evidence that saturation must be considered. A more thorough study is called for; one which also includes attenuation of the ω_2 laser beam to determine if both laser beams can contribute to saturation of the resonant CARS process.

Finally, an initial study of the sensitivity of the resonant CARS spectrum to changes in concentration was performed by changing the height of the incident CARS laser beams relative to the height of the burner surface. The results are shown in Fig. 18, where spectra for 0.5 cm increments are displayed. The scans shown cover just the spectral region of the water and OH CARS bands. Clearly, the OH concentration decreases as the CARS beams are moved

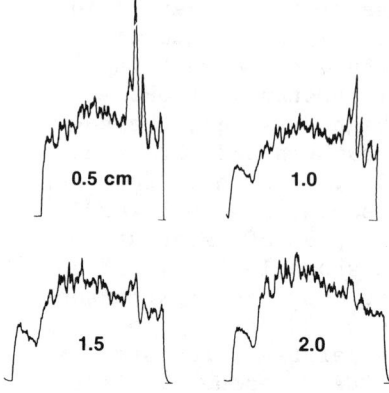

Fig. 18 Variation of the resonant CARS spectrum with change in laser beam height relative to burner surface.

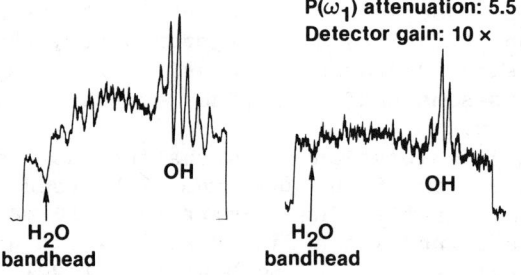

Fig. 17 Change in the resonant CARS spectrum with attenuation of the ω_1 intensity.

away from the burner surface; at the same time, the water concentration appears to be increasing slowly. At the largest separation, 2 cm, there is still some indication of OH, as a modulation on the background. Future studies will involve a detailed calibration of the concentrations of both OH and water throughout the flame, as well as the temperature distribution, so that resonant CARS as a quantitative diagnostic may be properly assessed. Saturation may prove to be a problem for quantitative measurements; however, even in the case of saturation, it may be possible to construct a working curve of concentration vs signal strength.

Conclusions

Electronically resonant CARS of the OH radical has been demonstrated in a very hot methane/oxygen flame. The resonant CARS spectrum was generated by use of two frequency-doubled, pulsed, narrowband dye lasers, one of which was tuned and then fixed on a selected OH X-A transition, while the other other was scanned over the appropriate range. The OH resonant CARS signal is strong compared to the water CARS spectrum, even though water is much more abundant than OH in the flame. Resonant CARS was observed for several different electronic resonances and was quite sensitive to precise tuning of the ω_1 frequency. The theory of resonant CARS in the OH radical has been treated and the predicted appearance of OH resonant CARS spectra presented. Saturation, for which there is some experimental evidence, has not been included in the theory. Agreement between theory and experiment is good, except for the experimental observation of satellite structure about the central line. This additional structure is believed to arise from collisional redistribution of population in rotational energy levels of the upper electronic state. The variation of the resonant CARS spectrum as a function of laser beam height in the flame was observed and offers promise that resonant CARS may be applicable to quantitative measurements. Future studies will involve refinement of the theory to include saturation. Further experimental work will assess the feasibility of making quantitative measurements in a flame, a thorough exploration of saturation effects, measurement of the resonance enhancement factor, and precise spectral line positions. Another interesing investigation which must be explored is that of pressure-induced extra resonances, which may occur in four-wave mixing processes such as CARS, a so-called PIER-4 process.[18] Additionally, triple resonances will be searched for, if a theoretical survey suggests that promising possibilities exist in OH.

Acknowledgments

The authors wish to acknowledge valuable discussions with Prof. Nicolaas Bloembergen of the Harvard University Physics Department and with Dr. D. R. Crosley of SRI International. Thanks are also due Edward Dzwonkowski and Normand Gantick for assistance with the experiments. The authors especially wish to thank the United States Air Force Office of Scientific Research for partial support of this research through Contract No. F49620-81-C-0063.

References

[1] Eckbreth, A.C., and Hall, R.J. "CARS Thermometry in a Sooting Flame," <u>Combustion and Flame</u>, Vol. 36, March, 1979, pp. 87-98.

[2] Pealat, M., Taran, J-P.E., Taillet, J., Bacal, M., and Brunetau, A.M., "Measurement of Vibrational Populations in Low-pressure Hydrogen Plasma by Coherent Anti-Stokes Raman Scattering," Journal of Applied Physics, Vol.53, April, 1981, pp. 2687-2691.

[3] Stenhouse, I.A., Williams, D.R., Cole, J.B., and Swords, M.D., "CARS in an Internal Combustion Engine," Applied Optics, Vol. 18, 15 November, 1979, pp. 3819-3825.

[4] Eckbreth, A.C., Dobbs, G.M., Stufflebeam, J.H., and Tellex, P.A., "CARS Temperature and Species Measurements in Augmented Engine Exhausts," AIAA Paper No. 83-1294, presented at the AIAAASME-SAE 19th Joint Propulsion Conference, Seattle, Washington, June 27-29, 1983.

[5] Druet, S., and Taran, J-P.E., "Coherent Anti-Stokes Raman Spectroscopy," in Chemical and Biological Applications of Lasers, C.B. Moore, Ed., Academic Press, New York, 1979, pp. 187-252.

[6] Tolles, W.M., Nibler, J.W., McDonald, J.R., and Harvey, A.B., "A Review of the Theory and Application of Coherent Anti-Stokes Raman Spectroscopy(CARS)," Applied Spectroscopy, Vol. 31, July-August 1977, pp. 253-272.

[7] Verdieck, J.F., Shirley, J.A., Hall, R.J., and Eckbreth, A.C., "CARS Thermometry in Reacting Systems," in Temperature, Its Measurement and and Control in Science and Industry, American Institute of Physics, New York, 1982, pp. 595-608.

[8] Druet, S., Attal, B, Gustafson, T.K., and Taran, J-P.E., "Electronic Resonance Enhancement of Coherent Anti-Stokes Raman Scattering," Physical Review A, Vol. 18., October 1978, pp 1529-1557.

[9] Guthals, D.M., Gross, K.P., and Nibler, J.W., "Resonant CARS Spectra of NO_2," Journal of Chemical Physics, Vol. 70, March, 1979, pp. 2393-2398.

[10] Attal, B., Debarre, D., Muller-Dethlefs, K., and Taran, J-P.E., "Resonance Enhanced Coherent Anti-Stokes Raman Scattering in C_2" Revues des Physique Appliquee', Vol. 18, p 39-50, January, 1984.

[11] Crosley, D.R., "Collisional Effects on Laser-Induced Fluorescence Flame Measurements," Optical Engineering ., Vol. 20, July-August 1981, pp. 511-521.

[12] Dieke, G.H., and Crosswhite, H.M.,"The Ultraviolet Bands of OH" Journal of Quantitative Spectroscopy and Radiative Transfer, Vol. 2, December 1962, pp. 97-199.

[13] Bloembergen, N., Lotem, H.L., and Lynch, R.T., "Lineshapes in Coherent Resonant Raman Scattering," Indian Journal of Pure and Applied Physics, Vol. 16, May 1978, pp. 151-158.

[14] Coxon, J.A., "Optimum Molecular Constants and Term Values for the X and for A States of OH," Canadian Journal of Phyics, Vol. 58, February 1980, pp. 933-949.

[15] Chidsey, I.L., and Crosley, D.R., "Calculated Rotational Transition Probabilities for the A-X System of OH," *Journal of Quantitative Spectroscopy and Radiative Transfer*, Vol. 23, March 1980, pp. 187-199.

[16] Earls, L.T., " Intensities in $^2\Pi$ - $^2\Sigma$ Transitions in Diatomic Molecules," *Physical Review*, Vol. 48, November 1935, pp. 423-424.

[17] Wilson-Gordon, A.D., Klimovsky-Baird, R., and Friedmann, H., "Saturation Effects in Coherent Anti-Stokes Raman Scattering," *Physical Review A.*, Vol. 25, April 1982, pp. 1580-1595.

[18] Prior, Y., Bogdan, A.R., Dagenais, M.W., and Bloembergen, N., "Pressure Induced Extra Resonances in Four-Wave Mixing," *Physical Review Letters*, Vol. 46, January 1981, pp. 111-114.

[19] Grynberg. G., "Resonances Between Unpopulated Levels in Non-Linear Optics," *Journal of Physics D.*, Vol. 14, May 1981, pp. 2089-2097.

[20] Littman, M.G., and Metcalf, H.J., " Spectrally Narrow Pulsed Dye Laser Without Beam Expander," *Applied Optics*, Vol. 17, July 1978, pp. 2224-2227.

Simultaneous CARS and Luminosity Measurements in a Bluff-Body Combustor

G. L. Switzer,* D. D. Trump,† and L. P. Goss‡
Systems Research Laboratories, Inc., Dayton, Ohio
W. M. Roquemore,§ R. P. Bradley, J. S. Stutrud,**
and C. M. Reeves††
Air Force Wright Aeronautical Laboratories
Wright-Patterson Air Force Base, Ohio

Abstract

High-speed ciné pictures of a diffusion flame stabilized by an axisymmetric, ducted bluff-body research combustor show that large-scale flame structures are formed downstream of the recirculation zone. These flame structures, referred to as flame turbules, are separated axially by nonluminous regions. Results of simultaneous temperature and flame-luminosity measurements in the region of these flame turbules are reported. The coherent anti-Stokes Raman spectroscopy (CARS) technique was used to make time and spatially resolved temperature measurements. Flame luminosity, along the line of sight passing through the CARS sampling volume, was recorded for each CARS measurement and used as a condition for sampling the CARS data. The characteristics of the flame turbules are interpreted in terms of the temperature probability distribution function (pdf). Temperature pdf's

---

Presented as Paper 83-1481 at the AIAA 18th Thermophysics Conference, Montreal, Canada, June 1-3, 1983. This paper is declared a work of the U.S. Government and therefore is in the public domain.

*Research Engineer, Research Applications Division.
†Associate Engineer, Research Applications Division.
‡Senior Chemist, Research Applications Division.
§Senior Research Scientist, Aero Propulsion Laboratory.
πSenior Research Engineer, Aero Propulsion Laboratory.
**Mechanical Engineer, Aero Propulsion Laboratory.
††Chemical Engineer, Aero Propulsion Laboratory.

obtained at different axial and radial locations downstream of the recirculating zone and for different air- and fuel-flow rates are also presented.

Introduction

The Air Force Wright Aeronautical Laboratories/Aero Propulsion Laboratory is sponsoring a research program directed toward experimental evaluation and theoretical development of combustion models.[1] As part of this program, a large-scale research combustor has been developed which has a simple geometry and clean inlet conditions for modeling, provides good measurement access to the combustion zones, and yet has some of the features of practical combustion devices. Both conventional probes and advanced laser techniques are being developed to study combustion processes in this combustor. Experiments conducted early in this program with a ruby-laser-based collinear optical system operating at 1 Hz demonstrated the ability of a CARS system to operate in the hostile environment of the research combustor.[2,3] These initial experiments also demonstrated the need to improve the spatial resolution of the CARS system and to increase the repetition rate of the measurements. This paper reports the results of initial checkout tests in the research combustor of a second-generation CARS system which has both an improved spatial resolution and a higher repetition rate. The time-averaged results of instantaneous CARS temperature measurements are compared with results obtained with a thermocouple in a parabolic flow region which was believed to be unperturbed by the presence of the thermocouple.

The flame in the research combustor is stabilized by the recirculation zone established by the bluff-body face of the combustor. High-speed ciné pictures show that large flame structures occur downstream of the recirculation zone.[4] These structures, denoted as flame turbules, are islands of flame surrounded by noncombusting regions and are believed to be the result of vortex shedding from the bluff-body.[5] Flame luminosity vs time records have been used for studying the dynamic characteristics of these turbules[4] and as a condition for sampling velocity data collected with a laser Doppler anemometer (LDA).[6] The conditionally sampled data show that the mean axial velocity is significantly higher in the flame turbules than in the nonluminous regions. Experiments conducted with a small laboratory Meeker burner have demonstrated a technique which employs flame luminosity for conditional sampling of CARS data.[7] The feasibility of using flame luminosity as a condition for sampling CARS temperature data taken in the research combustor downstream of the recirculation zone is examined in this paper. Also, the temperature probability distribution function (pdf) at different axial and radial locations downstream of the recirculation zone are presented along with pdf's obtained under different air- and fuel-flow conditions of the combustor.

Experimental Setup

Combustion-Tunnel Facility

A schematic diagram of the axisymmetric bluff-centerbody research combustor in the combustion tunnel is shown in Fig. 1. The centerbody is 79-cm long and has a diameter of 14 cm. Gaseous propane fuel is injected from the center of the bluff-body face through a 4.8-mm-diam nozzle. The annular air is conditioned using a honeycomb flow straightener and screens. The centerbody is mounted in a 25.4-cm-diam duct containing 30.5 × 7.6-cm viewing ports which provide optical and conventional probe access to combustion regions. Additional information on the combustion tunnel facility is given in Ref. 1.

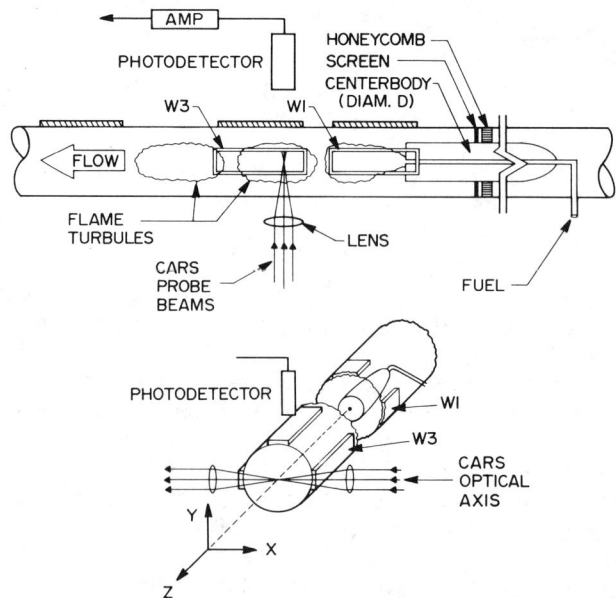

Fig. 1 Schematic diagram of combustor and optical configuration used to make simultaneous flame-luminosity and CARS-temperature measurements.

Thermocouple Description

Average flame temperatures were measured by a shielded thermocouple probe for which the radiation losses (at the indicated temperatures) were assumed to be negligible. The thermocouple probe was mounted vertically through a special slide arrangement located on the top of the duct. This permitted access to axial regions covered by window W3 in Fig. 1 without causing significant disturbance in the flow. The probe housing is water cooled and has an outside diameter of 6.35 mm. The platinum sensing head of the probe is made in the form of a T which is rotated on its side, with the base of the T pointing upstream. A Type-R thermocouple is contained in a 4.76-mm-diam shield that forms the base of the T. The shield is open on the upstream end, and bleed holes are provided downstream of the thermocouple junction to allow the gas to pass between the thermocouple and the shield. The T-shape design of the sensing head was adopted after experience showed that it tended to reduce the flow-interference effects of the probe. The Instrument Development Group at NASA Lewis Research Center designed and fabricated the probe using criteria derived from Ref. 8.

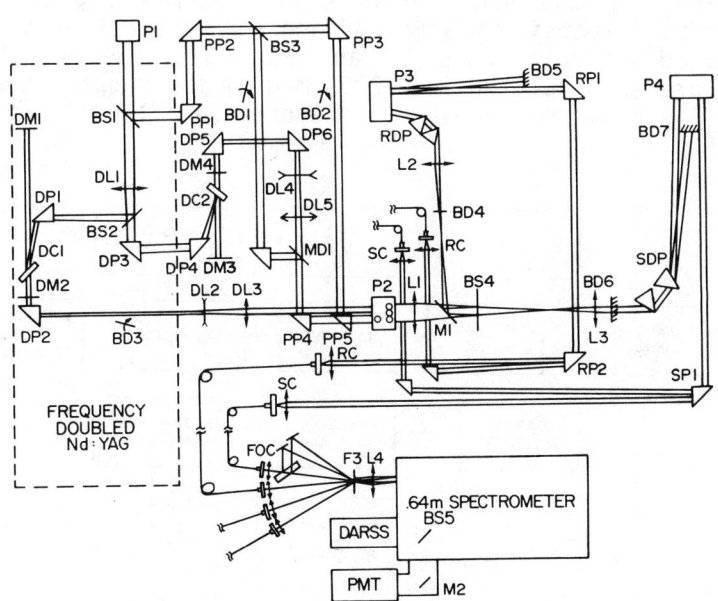

Fig. 2 CARS optical-system schematic.

CARS System Description

Optical Configuration. The CARS system is based upon a frequency-doubled neodymium-YAG laser operating at a 10-Hz repetition rate and in a folded BOXCARS[9] optical geometry. This combination allows high temporal [10-ns full width half-maximum (FWHM)] and spatial (< 1 mm^3) measurement resolution within the combusting medium. A detailed description of the basic optical configuration and design features of this system can be found elsewhere.[10] Figure 2 is the optical schematic of the CARS system as configured for these measurements to obtain simultaneous temperature and N_2 and O_2 concentration data. Two dye lasers, consisting of flowing dye cells DC1 and DC2 and their associated dichroic mirrors (DM1-4), produce the broadband (150-cm^{-1} FWHM) Stokes radiation for N_2 and O_2, respectively. These two beams are combined with two equal-intensity beams of the 532-nm pump radiation, and the four beams are elevated via periscope P2. The CARS beams are then directed to form a folded BOXCARS geometry in which the pump beams converge at a full angle of $\simeq 6$ deg and the N_2 and O_2 Stokes beams converge with full angles of 1.5 deg and 0.5 deg, respectively, from the plane containing the pump beams. Anti-Stokes signals for both gas species are produced in the sample volume and also in a resonant CARS reference cell containing a mixture of N_2 and O_2. These reference signals are used for laser-power normalization. Sample and reference anti-Stokes signals are isolated by dispersing prisms SDP and RDP, respectively, and de-elevated by periscopes P4 and P3. The two pairs of signals are then focused into 200-μm-diam fiber-optic transmission lines by the sample- and reference-collection optics SC and RC.

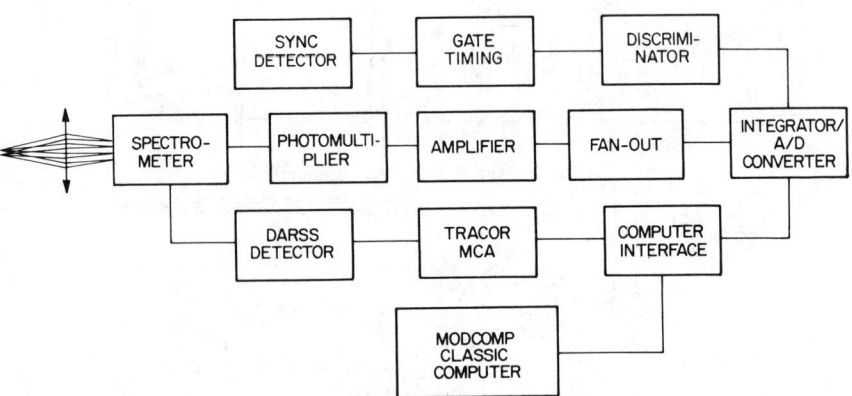

Fig. 3 Signal-detection and data-collection electronics configuration.

The four CARS signals are transmitted through separate fiber-optic delay lines of varying lengths to a remotely situated spectrometer where each is detected and analyzed for spectral content by a diode array rapid scan spectrometer (DARSS) Reticon detector and for integrated intensity by a photomultiplier. In addition to the four beams transmitted by the fiber optics, two beams are generated as a result of splits of 1 and 10% of the N_2 sample beam before it enters the spectrometer. These two beams plus the remaining 89% of the N_2 sample CARS signal form the basis of a scheme to expand the dynamic range of the DARSS to the level necessary to accommodate the large signal fluctuation encounted in a dynamic environment.[11] The combined effects of the fiber-optic core diameter, the 1200 grooves/mm spectrometer grating, and the 50-μm interchannel spacing of the DARSS construction contribute to a DARSS dispersion of 2.6 cm^{-1}/channel and an overall system resolution of 7.5 cm^{-1} HWHM.

Data-Collection Electronics

Once the sample and reference CARS optical signals have been converted by the detectors, the data are collected and stored for subsequent analysis by the data-collection electronics outlined in Fig. 3. The spectral information obtained by the DARSS is digitized by the Tracor-Northern 1710 Multichannel Analyzer System. A single photomultiplier and gated analog-to-digital conversion provide integrated intensity information for each of the four CARS signals generated.

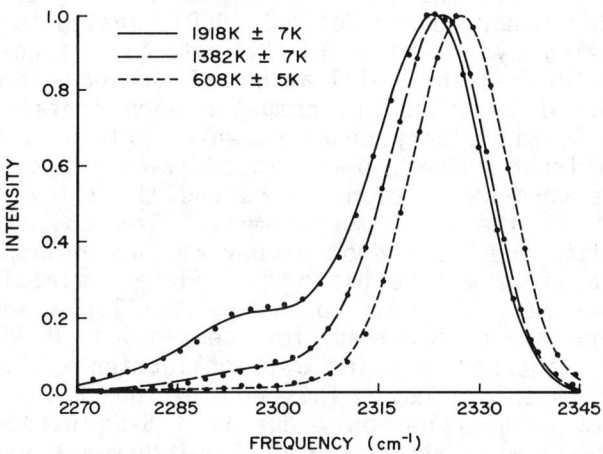

Fig. 4   Best-fit temperature results for three single-shot N_2 spectra. Experimental data (·).

Data Reduction

The N_2 spectral information stored during acquisition is reduced with the aid of a computer program to arrive at a "best-fit" temperature. The reduction program incorporates a nonlinear least-squares fitting[12] routine which iteratively adjusts the frequency and temperature estimate of the experimental spectrum by comparing it to precalculated spectra from a library of N_2 spectra produced at various temperatures. Three such best-fit single-shot temperatures, representing three distinct ranges of temperatures encountered during the reported measurements, are illustrated in Fig. 4. Once the temperature determination has been made, the integrated intensities of the N_2 and O_2 CARS signals can be used to determine the concentration of these species present during the measurement pulse. However, for the purpose of this study, concentration measurements were not made.

Luminosity-Data-Collection System

The luminous emissions from the combusting gases were monitored through the use of an EG&G UV-100B photodiode. As indicated in Fig. 1, the diode, a collecting lens, and a field stop were positioned normal to the CARS optical axis and approximately 10 cm above the sampled volume. This configuration collected light emitted from within a 1.1-deg half-angle cone of view, resulting in a cross sectional diameter of 9 mm at the sample point.

Signals from the photodetector were amplified and selectively filtered by bandpass filters ranging in value from 100 Hz to 100 kHz. The amplified signal was digitized by a 0-10 V, 12-bit binary-coded-decimal (BCD), analog-to-digital converter being cycled at a 10-kHz rate by a free-running pulse generator. Each digitization of luminous intensity was accompanied by a 16-bit computer word containing the output of a 12-bit binary counter, which maintains a count of the CARS laser pulses, plus two additional bits—one to identify the word as a "count" word and the other to mark the "event" of the CARS measurement. The collection of luminosity data and laser shot number was accomplished continuously to yield a time history of flame luminosity (see Fig. 8). However, in order to reduce the large amount of disk storage space required for continuous 10-kHz data acquisition, windowed-emission data collection was employed during most of this study. This mode of operation enabled emission-data acquisition only during a 5-ms window which opened symmetrically about the CARS measurement event and resulted in a factor of 200 reduction in necessary disk

Table 1 Air-flow conditions

Air-flow (kg/s)	Inlet temperature (K)	Annulus velocity (m/s)	Duct velocity (m/s)	Air Re. No.[a] ($\times 10^{-5}$)
0.50	294	11.6	8.2	0.76
1.00	294	23.3	16.4	1.52
1.50	294	34.9	24.6	2.27
2.00	294	46.6	32.8	3.02

[a]Calculated using duct velocity and centerbody diameter.

storage space, thus allowing large sample sizes without sacrificing data pertinent to conditional sampling. The correlation between a given CARS measurement and the luminous emission during that measurement was accomplished through the laser-shot counter/event marker. That is, although the CARS and luminosity data-acquisition cycles proceed asynchronously, they are initiated from a common manual command. From the commencement of acquisition, synchronization of a CARS measurement to the coincident emission is accomplished through the laser-shot counter contained in each data stream.

One of the difficulties in interpreting the luminosity data was the determination of the signal level which represents the base line or zero flame emission. Depending upon such factors as amplifier gain, filter bandwidth, and the amount of scattered light, the photodiode baseline signal varied from 0 to $\approx$ 200 counts of the 999-count range of the three-digit BCD emission word. For the purpose of the conditional sampling of data presented in this paper, the dividing line between the presence and absence of flame was defined as the percentage of the maximum value of recorded emission above which the density of emission values decreased uniformly throughout the spectrum of observed temperatures.

Test Conditions

Experiments were conducted with the combustor operating under the various air- and fuel-flow conditions given in Tables 1 and 2, respectively. The annulus-air and fuel-exit velocities for each experiment are given in the Results and Discussion Section. The combustor was operated at an ambient pressure of 0.98 bars. Measurements to determine the optimum sample size for the combustion environments examined indicated that although the average of the temperatures measured did not change significantly after 500 samples, their pdf's became stable only after approximately 1500 samples. Thus, the CARS data presented are the result of more than 1500 individual measurements.

Results and Discussion

Comparison of CARS and Thermocouple Measurements

To establish the reliability of the CARS temperature measurements, comparisons were made between the CARS and thermocouple derived radial and axial temperature profiles. Figure 5 presents a comparison (at an axial position of

Table 2 Fuel-flow conditions

Fuel-flow (kg/hr)	Inlet temperature (K)	Fuel velocity (m/s)	Fuel tube Re. No. ($\times 10^{-4}$)
6.0	400	69.6	4.20
8.0	391	91.0	5.71
10.0	382	110.8	7.29
13.0	378	141.4	9.50
16.0	366	168.1	11.9

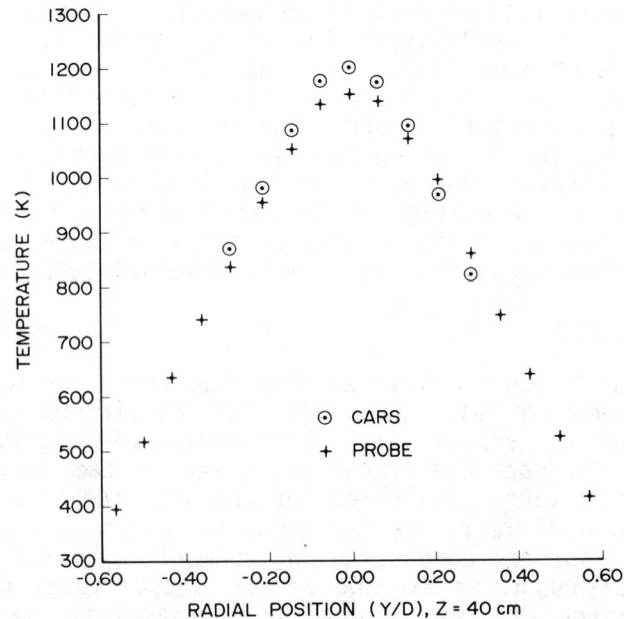

Fig. 5 Comparisons of CARS and thermocouple radial temperature profiles at axial location of 40 cm (2.86D) for an annulus-air velocity of 23.3 m/s and a fuel-exit velocity of 69.6 m/s.

Z/D = 2.86) of the average CARS temperature profile obtained along an X-axis radial scan and the average thermocouple-indicated temperature profile obtained on a Y-axis radial scan. This plot indicates not only that the temperatures obtained by the two methods agree, well within an anticipated ≃ 10% precision of the CARS measurements in the indicated temperature range, but also that the combustion process is proceeding on a time-averaged basis symmetrically about the tunnel axis. Flame symmetry has also been verified through comparison of thermocouple X- and Y-axis profiles which agree to within 50 K ($\pm$ 2 standard deviations) at the centerline temperature. The centerline average-temperature profiles compared in Fig. 6 also show quite good agreement between the two methods.

Ciné Pictures

The radial and axial profiles of average temperature given in Figs. 5 and 6 are well-behaved functions and appear to be similar to those measured in many nonrecirculating combusting flows. Unfortunately, such profiles do not provide clear insights into the underlying, more basic dynamic processes that give rise to the time-averaged results. A view of the dynamic process responsible for the time-averaged profiles is shown in the high-speed ciné pictures in Fig. 7. The color ciné pictures were made at a rate of 500 frames/s;

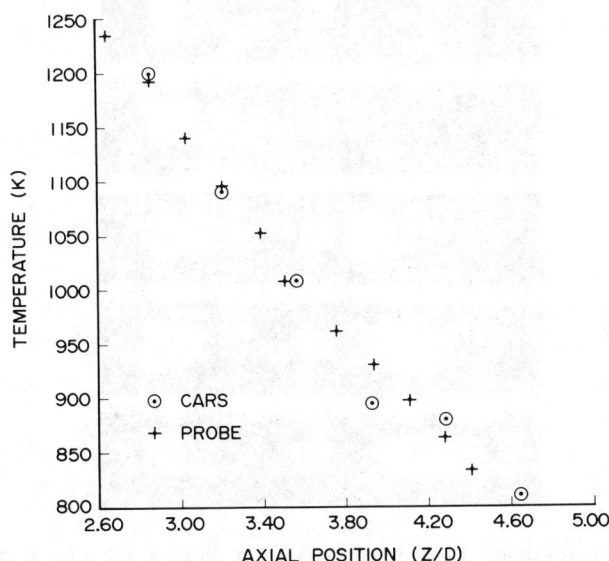

Fig. 6   Comparisons of CARS and thermocouple centerline axial temperature profiles for annulus-air velocity of 23.3 m/s and fuel-exit velocity of 69.6 m/s.

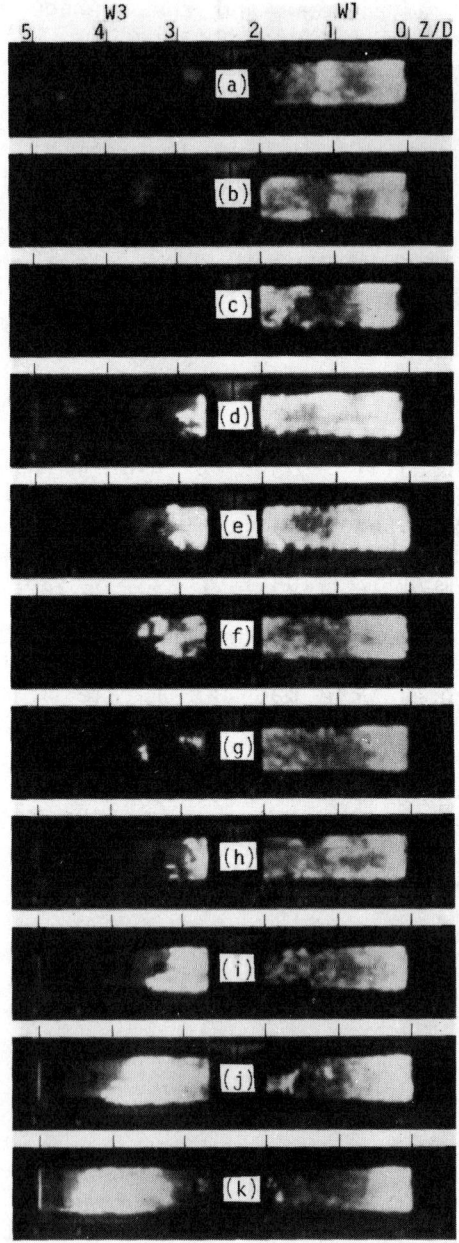

Fig. 7 High-speed (500 frames/s) ciné photographs of flame for annulus-air velocity of 23.3 m/s and fuel-exit velocity of 69.6 m/s. Alternate frames are shown.

however, only every other frame is presented in the figure to illustrate the pertinent flame structures. The recirculation zone established behind the bluff-body was confined to the region viewed through window W1 (see Fig. 1). The time-averaged temperature profiles in Figs. 5 and 6 were made downstream of the recirculation zone in the region viewed through window W3. This region was characterized by large flame structures (flame turbules) as shown between locations 3D and 5D in Fig. 7k and small flame structures as at 3D in Fig. 7f. The large structures can maintain their identity over many centerbody diameters downstream. The bright part of the flame was due to blackbody radiation of soot particles which show up on the color photographs as yellow. Sometimes, blue flames, which were free of soot particles, were observed in window W3. Such small blue-flame structures exist at the axial locations 3D and 4D in the original color photographs for Figs. 7a-7b, but they are barely noticeable on the black-and-white frames. Careful study of the high-speed ciné pictures shows that the flame turbules formed downstream of the recirculation zone were often the direct result of toroidal vortices being shed from the bluff-body.[5] At times, the flame turbules were actually shed vortices that had picked up burning fuel near the end of the recirculation zone as they moved downstream. Figure 7 shows that the average temperatures obtained in this flame are the result of a combination of various size flame structures and the interconnecting noncombusting regions which alternately pass through the measurement location.

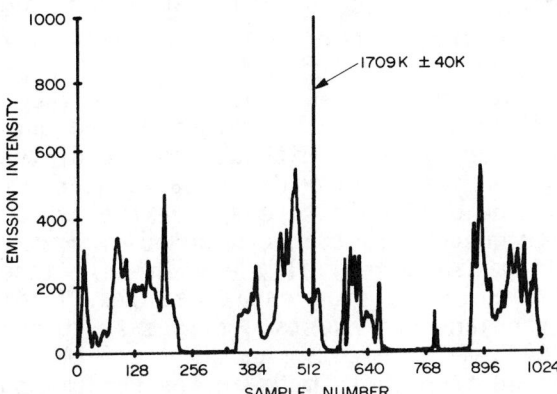

Fig. 8 Continuously sampled (10 kHz) flame-luminosity data at axial location of 40 cm (2.86D) for annulus-air velocity of 23.3 m/s and fuel-exit velocity of 69.6 m/s. Labeled marker indicates CARS measurement event.

Flame Luminosity

Figure 8 is an intensity vs time trace of flame luminosity recorded at an axial location of 40 cm (2.86D) by the emission photodiode. The temporal relation of the CARS measurement event and flame luminosity is indicated by the marker within this 102.4-ms emission data sample. A flame turbule can be identified by the electrical pulse that is produced as the turbule is convected passed the field-of-view of the photodetector. The nonluminous regions between the flame turbules are clearly evident by the constant voltage regions between the pulses. It should be noted that although the spectral sensitivity of the photodiode extends from 350 to 1000 nm, it is possible that comparatively low intensity emission (e.g., blue flame) could be indistinguishable from the background-emission noise level. The different time widths noted in Fig. 8 are due to the different sizes or velocities of the turbules. The high-frequency fluctuations riding on some of the pulses may be due to the small flame structures shown in Figs. 7e-7g. These small structures are close together and are moving with the same velocity as the larger flame turbule, thus possibly producing high-frequency luminosity fluctuations.

PDF Variation with Axial Location

A Lagrangian description of the dynamic processes shown in Fig. 7 seems appropriate in that the downstream motion of the flame and nonflame regions can be tracked over rather large distances. However, thermocouple, flame-luminosity, and CARS devices provide an Eulerian description of the processes since they record events as they are convected past a stationary measurement point. The interpretation of the Eulerian-type measurements in terms of the Lagrangian-type processes is not always straightforward and can often be confusing. For example, without the aid of the ciné pictures in Fig. 7, which provide a Lagrangian view of the flow, it would be difficult to envision the probable cause of the high-frequency fluctuations noted in Fig. 8. Conditional sampling can potentially aid in this interpretation by identifying Eulerian-type data that result from certain specific Lagrangian-type events. It is easy to envision from Fig. 8 that CARS measurements made in the flame turbules can be distinguished from those made in the nonluminous regions. Such a distinction can aid in obtaining a Lagrangian description of the dynamic processes responsible for the time-averaged temperature profiles given in Figs. 5 and 6.

An experiment was conducted to: (1) determine how the temperature pdf's changed with axial location and (2) to

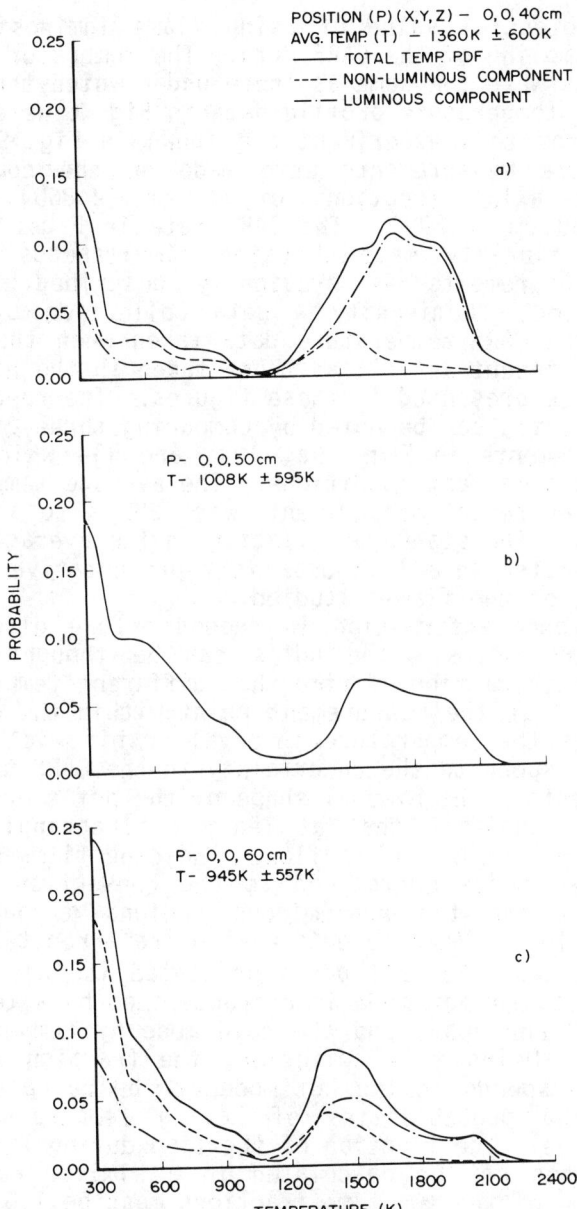

Fig. 9 Conditionally sampled axial temperature pdf's for annulus-air velocity of 23.3 m/s and fuel-exit velocity of 69.6 m/s at axial locations of (a) 40 cm (2.86D), (b) 50 cm (3.57D), and (c) 60 cm (4.29D).

examine the potential value of using flame luminosity for conditional sampling of the CARS data. The combustor operating conditions were the same as those under which the time-averaged axial temperature profile data in Fig. 6 were taken. The results from this experiment are shown in Fig. 9. The CARS temperature measurements were made on the combustor centerline at axial locations of 40 cm (2.86D), 50 cm (3.57D), and 60 cm (4.29D). The CARS data in Figs. 9a and 9c were conditionally sampled using simultaneous flame-luminosity measurements as previously described in the section entitled, "Luminosity - Data Collection System." The pdf's of the CARS temperature data taken when the flame turbules were present as well as those taken in the nonluminous regions are presented in these figures. The reproducibility of the pdf's can be noted by comparing three different sets of measurements in Figs. 9a, 10a, and 11a which were made under the same test conditions. The average temperature from these three repeat measurements was 1289 K to a precision of $\pm$ 67 K. The standard deviation in the average temperatures indicated in all figures is representative of the dynamic nature of the flames studied.

Some background information is needed before discussing the temperature pdf's. The pdf's can be thought of as representing the fraction of time that different temperature intervals exist at the measurement point within the combusting flow, with the temperature interval or bin width being chosen to correspond to the uncertainty in the CARS temperature measurements. The bimodal shape of the pdf's presented in this paper suggests that at least two Lagrangian-type processes are occurring in the flow. The ciné films suggest that these two modes correspond to the convection of the flame turbules and the nonluminous regions through the measurement volume. These events will be referred to as the hot and cold modes. For all pdf's presented in this paper, it appears that the hot mode is characterized by a temperature of 1100 K and above and the cold mode by temperatures below 1100 K. Using this criterion, the fraction of time which the flow spends in the hot mode can be calculated by summing up the probabilities of having temperatures of 1100 K and above. The fraction of the time during which the cold mode exists can be calculated in a similar way. Of course, the sum of the two time fractions must be 1.0.

The axial profile in Fig. 6 shows that the average centerline temperature decreases with increasing axial location. The pdf's in Fig. 9 give some insight into the processes responsible for this effect. Inspection of Fig. 9 shows that the cold mode always peaks at 300 K, whereas the hot-mode peak--or, more appropriately, the hot-mode average temperature--shifts to somewhat lower values farther downstream. Also, the fraction of time that the hot mode exists decreases with increasing axial location. Thus, the axial decay in average temperature is the combined effect of the decrease in the hot-mode temperature and in the fraction of time during which these temperatures exist. This implies that cold air is being entrained and mixed with products within the flame turbules as they move downstream, thus lowering the average temperature in the turbule. Also, some of the flame turbules burn out or the reactions are quenched as they move downstream, thus reducing the fraction of time during which the hotter temperatures exist. It should also be noted that the average temperature occurs in the trough between the hot and cold modes, indicating that the flow-field is actually at this temperature a small (less than 3%) fraction of the time.

The conditionally sampled data in Figs. 9a and 9c provide some additional insight into the dynamic processes responsible for the time-averaged temperatures. Although it is generally true that when the flame is present, the temperatures are in the hot mode and when the flame is absent, the temperatures are in the cold mode, this is not always the case. The conditionally sampled data in Fig. 9 show that at times, no flame is present but the temperatures are in the hot mode, and at other times flame is present but the temperatures are in the cold mode. This apparent anomaly is undoubtedly due, in part, to the fact that the flame luminosity can be detected at any location along the line-of-sight of the photodetector, whereas the CARS measurement is made at a point contained within the field-of-view of the detector. With this in mind, some explanations of the conditionally sampled data can be offered--but unfortunately cannot be verified.

The conditionally sampled data suggest that the hot mode is composed of at least two types of events, one corresponding to the presence of flame and the other to the absence of flame. The hot mode which is associated with the nonluminous regions peaks at about 1500 K in both Figs. 9a and 9c. This mode is believed to be caused by the hot product gases from a flame turbule that have consumed all or almost all of the fuel just prior to passing through the measurement location. The width of the hot mode associated with the nonluminous events is believed to be determined by the time available for mixing before the product gases pass through the measurement location. Relatively long mixing times could account for the positive skewness of the cold, nonluminous mode. The width of the hot mode, associated with measurements taken when the flame was present, suggests that the large flame turbules may consist of smaller packets of burning fuel separated by nonreacting regions of product gases and entrained air. The degree of mixing between the entrained air and the product gases within a flame turbule could account for the broad distribution of the hot mode and the distribution in the cold mode associated with the flame measurements.

PDF Variation with Radial Location

Figure 10 shows pdf's obtained at X locations of 0, -2, and -4 cm at an axial location of 40 cm. The air- and fuel-flow-rate conditions are the same as those in which the time-averaged radial profile data in Fig. 5 were collected. These pdf's have a very similar bimodal shape to those obtained at various axial stations. In fact, much of the previous discussion with reference to Fig. 9 seems to apply to Fig. 10. The average temperature is located between the hot and cold modes, and the flowfield is at this temperature only a small fraction of the time. The decrease in average temperature with increasing radius appears to be primarily due to a decrease in the fraction of time during which the hot mode is present. Some shifting of the hot-mode distribution to slightly lower temperatures takes place as the radius is increased, but this appears to be a secondary effect. These pdf's suggest that the entrainment and mixing decrease significantly as the flow moves toward the centerline. The distribution of the cold-mode temperatures reflects the various degrees of mixing of products and entrained air that have taken place prior to their passing through the measurement volume.

PDF Variations with Air- and Fuel-Flow Rates

Figure 11 shows how the temperature pdf's change with fuel-flow rate at a fixed axial location and a fixed air-flow rate. The shaded regions represent the temperature distributions in the absence of flame. The bimodal structure is apparent for each of the three fuel-flow rates. The decrease in the standard deviations with increasing fuel-flow rate is a result of the increase in the hot-mode contribution of the pdf's, whereby the hot mode becomes more like a single mode. The nonluminous contribution to the hot mode increases in magnitude as well as shifts to higher temperatures with increasing fuel-flow rate. The increase in fuel-flow rate does not appreciably change the fraction of time during which unmixed inlet air is present. However, the degree of mixing and heat diffusion is significantly reduced as the fuel-flow rate is increased, as indicated by the increased probabilities of the cold-mode temperatures other than the 300-K bin.

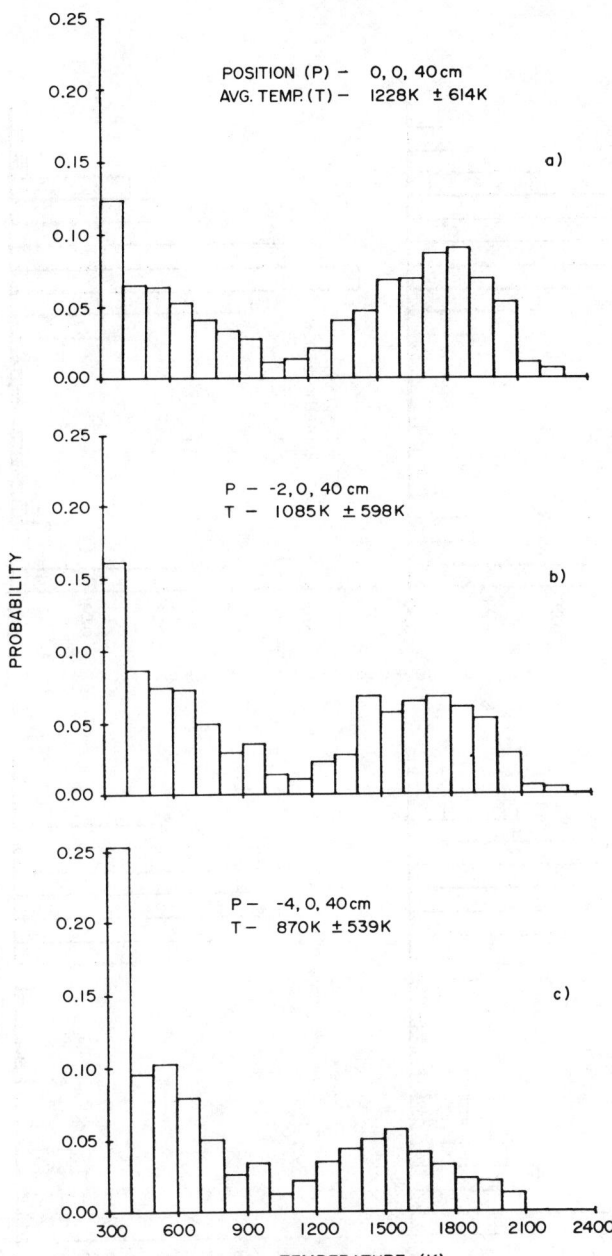

Fig. 10 Temperature pdf's for annulus-air velocity of 23.3 m/s and fuel-exit velocity of 69.6 m/s for axial location of 40 cm (2.86D) and radial (X) locations of (a) 0.0 cm, (b) -2.0 cm (0.143D), (c) -4.0 cm (0.286D).

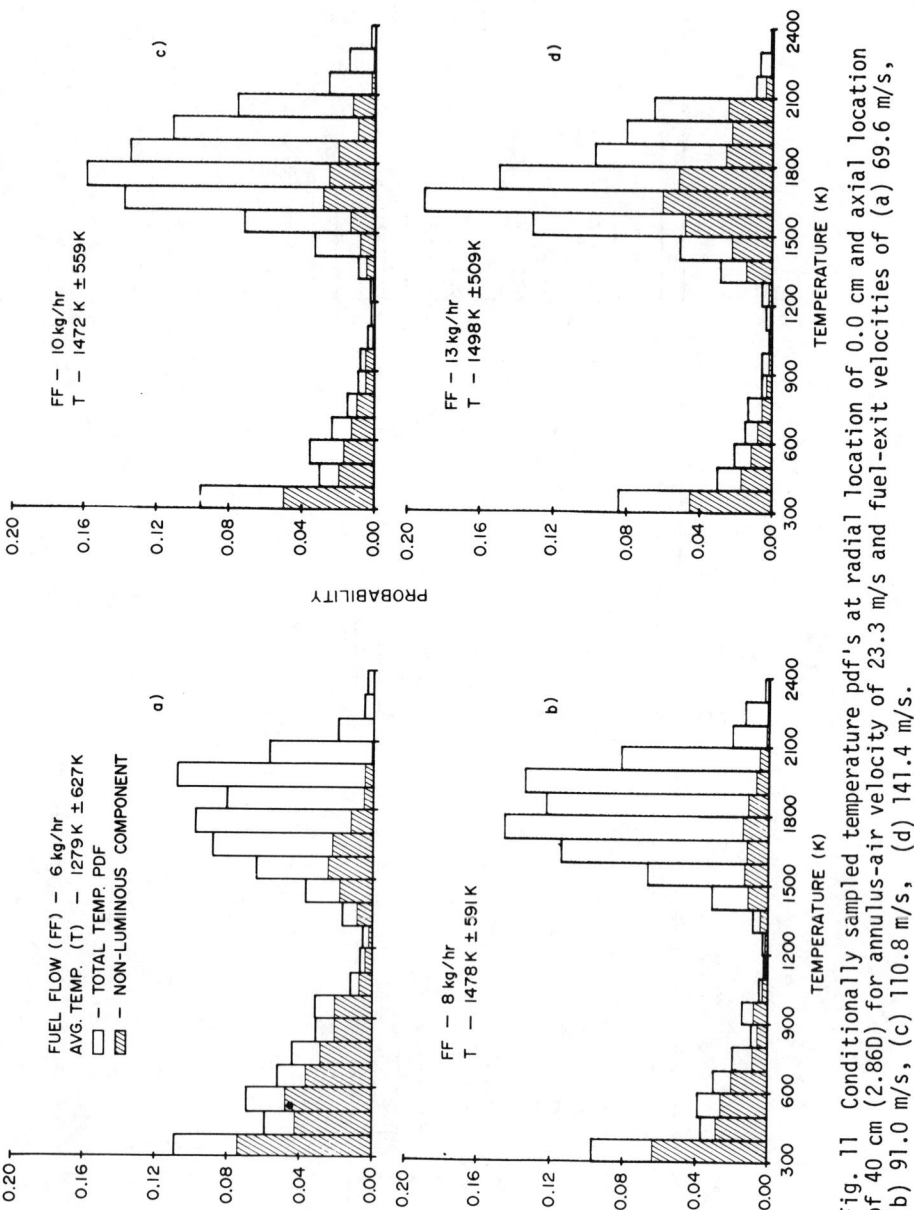

Fig. 11 Conditionally sampled temperature pdf's at radial location of 0.0 cm and axial location of 40 cm (2.86D) for annulus-air velocity of 23.3 m/s and fuel-exit velocities of (a) 69.6 m/s, (b) 91.0 m/s, (c) 110.8 m/s, (d) 141.4 m/s.

SIMULTANEOUS CARS AND LUMINOSITY MEASUREMENTS

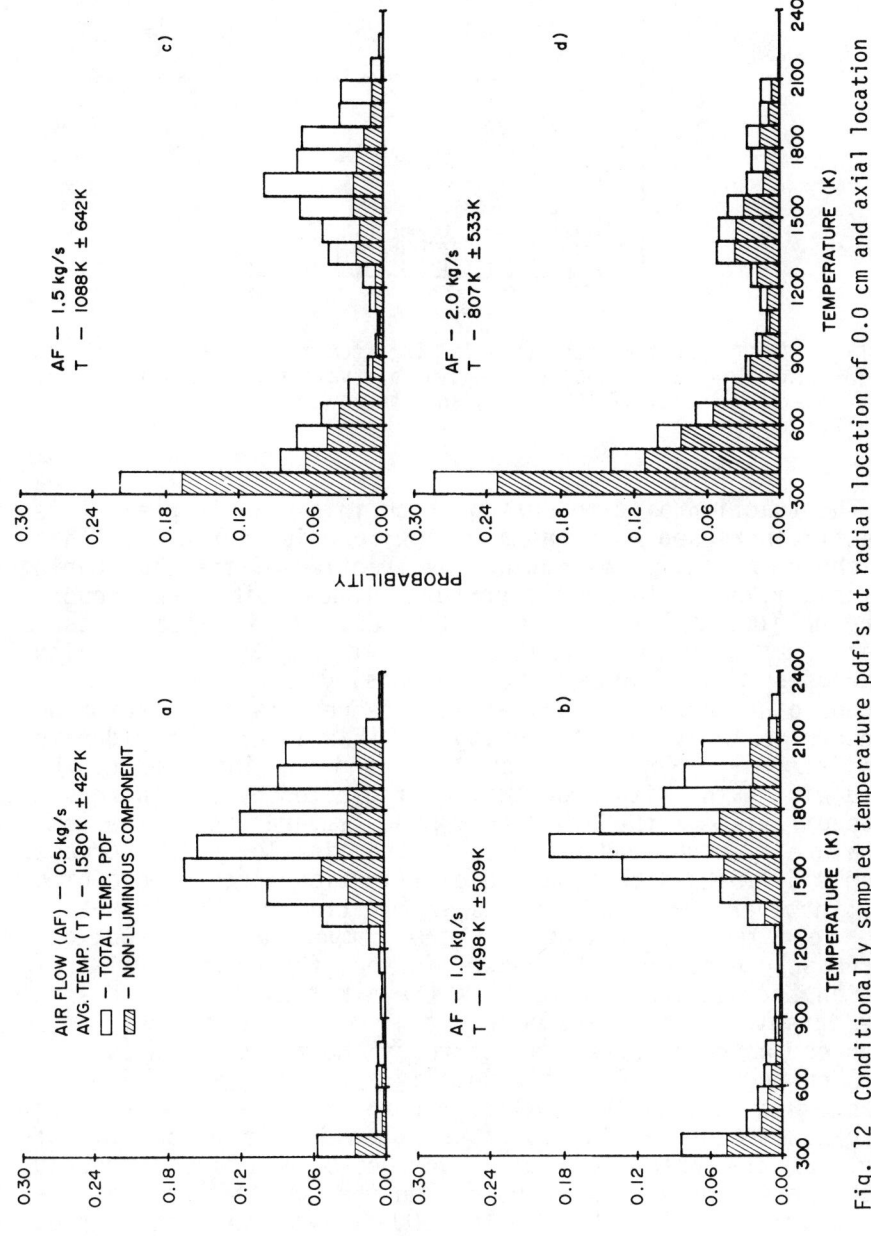

Fig. 12 Conditionally sampled temperature pdf's at radial location of 0.0 cm and axial location of 40 cm (2.86D) for fuel-exit velocity of 141.4 m/s and annulus-air velocity of (a) 11.6 m/s, (b) 23.3 m/s, (c) 34.9 m/s, and (d) 46.6 m/s.

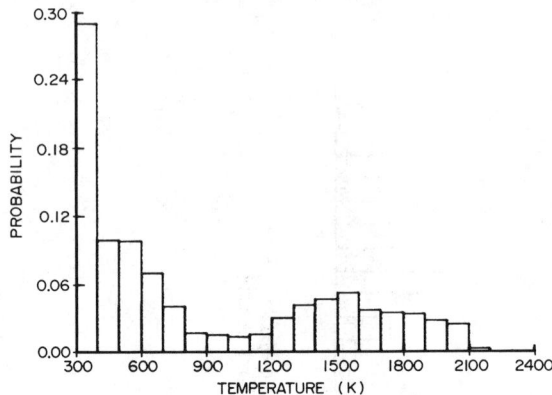

Fig. 13 Temperature pdf at radial location of 0.0 cm and axial location of 40 cm (2.86D) for annulus-air velocity of 46.6 m/s, fuel-exit velocity of 168.1 m/s, and average temperature of 876 K ± 587 K.

The fraction of time during which inlet air is present is greatly increased when the air-flow rate is increased. This is shown in Fig. 12 by noting the fraction of the time during which combustor inlet temperatures (300-K bin) are present for the four different air-flow rates. It is also evident that the nonluminous contribution to the 300-K peak also increases significantly with increasing air-flow rate. A factor of 4 increase in air-flow rate results in a factor of 2 decrease in average temperature. This is primarily due to the increase in entrained cold air and the increase in the cold air mixing with products, as indicated by the increased probabilities of the other cold-mode temperatures. The temperature fluctuations increase as the air-flow rate increases from 0.5-1.5 kg/s because the distribution becomes more bimodal; however, as the air-flow rate continues to increase, the cold mode dominates and the temperature fluctuations decrease. A surprising feature is that the hot-mode distribution becomes more uniform as the air-flow rate increases. The cause is not understood, but it is probably related to changes in the internal structure of the flame turbules.

A comparison of Fig. 12d and Fig. 13 shows how the pdf's change with increasing fuel-flow rate at an air-flow rate of 2 kg/s. Trends similar to those noted in the discussion of Fig. 11 are observed. The average temperature increased with increasing fuel-flow rate, and the temperature fluctuations decreased slightly. The 300-K peak does not change, whereas the probabilities for the other cold-mode temperature decrease with increasing fuel-flow rate. This indicates that the degree of mixing is reduced.

Summary and Conclusions

The feasibility of using a BOXCARS-configured CARS system for making instantaneous (10-ns) temperature measurements at a point (1 mm^3) at a rate of 10 Hz in a large-scale research combustor has been demonstrated. CARS measurements of axial and radial temperature profiles in a highly sooting flame compared favorably with profiles measured with a Type-R shielded thermocouple probe. The CARS system was also used to investigate the dynamic processes in the bluff-body, turbulent diffusion flame which are responsible for the time-averaged temperature profiles. High-speed ciné pictures suggest that the average temperature results from the convection of large flame structures and nonluminous regions past the measurement point. Flame-luminosity measurements made simultaneously with the CARS temperature measurements were used to distinguish the CARS data taken within the flame structures from those taken in the nonluminous regions. The temperature pdf's determined at different axial and radial locations for different air- and fuel-flow rates were bimodal. The cold mode peaked at the combustor inlet temperature (300 K), and the hot mode was distributed between 1100 and 2200 K. The conditionally sampled data showed that both the hot and cold modes could be resolved into two additional modes, one consisting of temperatures measured in the presence of flame and the other consisting of temperatures measured in the absence of flame. However, it is generally true that the hot mode is dominated by those temperatures measured in the presence of flame and that the cold mode is made up predominantly of temperatures measured when the flame is not present. The results of this study show that the CARS technique is a valuable tool for studying dynamic combustion processes and that conditional sampling of the CARS data using flame luminosity enhances its value.

Acknowledgments

The authors wish to thank Ron Britton for setting up and participating in the luminosity measurements and for preparing several figures used in this paper. The authors would also like to thank Melvin Russell for operating the combustion tunnel. This work was performed as part of Air Force Contract F33615-80-C-2054 and Aero Propulsion Laboratory In-House Work Unit 30480590.

References

1. Roquemore, W. M., Bradley, R. P., Stutrud, J. S., Reeves, C. M., and Krishnamurthy, L., "Preliminary Evaluation of a

Combustor for Use in Modeling and Diagnostic Development," ASME Publication 80-GT-93, March 1980.

2. Switzer, G. L., Roquemore, W. M., Bradley, R. P., Schreiber, P. W., and Roh, W. B., "CARS Measurements in a Bluff-Body Stabilized Diffusion Flame," Applied Optics, Vol. 18, July 1979, pp. 2343-2345.

3. Switzer, G. L., Goss, L. P., Roquemore, W. M., Bradley, R. P., Schreiber, P. W., and Roh, W. B., "Application of CARS to Simulated Practical Combustion Systems," Journal of Energy, Vol. 4, Sept.-Oct. 1980, pp. 209-215.

4. Roquemore, W. M., Britton, R. L., and Sandhu, S. A., "Investigation of The Dynamic Behavior of a Bluff-Body Diffusion Flame Using Flame Emission," AIAA Paper 82-0178, AIAA 20th Aerospace Sciences Meeting, Jan. 11-14, 1982, Orlando, FL.

5. Roquemore, W. M., Bradley, R. P., Stutrud, J. S., Reeves, C. M., Britton, R. L., Sandhu, S. A., and Archer, R. S., "Influence of the Vortex Shedding Process on a Bluff-Body Diffusion Flame," AIAA Paper 83-0335, AIAA 21st Aerospace Sciences Meeting, Jan. 10-13, 1983, Reno, NV.

6. Magill, P. D., Lightman, A. J., Orr, C. E., Bradley, R. P., and Roquemore, W. M., "Simultaneous Velocity and Emission Measurements in a Bluff-Body Combustor," AIAA Paper 82-0883, AIAA/ASME 3rd Joint Thermophysics, Fluids, Plasma and Heat Transfer Conference, June 7-11, 1982, St. Louis, MO.

7. Goss, L. P., Trump, D. D., Switzer, G. L., and MacDonald, B. G., "CARS Thermometry and N_2 Number Density Measurements in a Turbulent Diffusion Flame," AIAA Paper 83-1480, AIAA 18th Thermophysics Conference, June 1-3, 1983, Montreal, Canada.

8. Glaw, G. E., Holnda, R., and Krause, F. N., "Recovery and Radiation Corrections and Time Constants of Several Sizes of Shielded and Unshielded Thermocouple Probes for Measuring Gas Temperatures," NACA-TP-1099, Jan. 1978.

9. Eckbreth, A. C., "BOXCARS: Crossed-Beam-Phase-Matched CARS Generation in Gases," Applied Physics Letters, Vol. 32, April 1978, pp. 421-423.

10. Switzer, G. L. and Goss, L. P., "A Hardened CARS System for Temperature and Species Concentration Measurements in Practical Combustion Environments," Temperature Its Measurement and Control in Science and Industry, Vol. 5, edited by James F. Schooley, 1982, pp. 583-587.

11. Goss, L. P., Switzer, G. L., and Trump, D. D., "Temperature and Species Concentration Measurements in Turbulent Diffusion Flames by the CARS Technique," AIAA Paper 82-0240, AIAA 20th Aerospace Science Meeting, Jan. 11-14, 1982, Orlando, FL.

12. Kim, A., "Computer Programming in Physical Chemistry Laboratory: Least Squares Analysis," Journal of Chemical Education, Vol. 47, May 1970, pp. 120-122.

Chapter II. Laser-Induced Fluorescence

Nonintrusive Pressure Measurements with Laser-Induced Iodine Fluorescence

James C. McDaniel*
University of Virginia, Charlottesville, Virginia

Abstract

Laser-induced fluorescence from iodine molecules seeded into a flowfield is shown to provide the capability for the nonintrusive measurement of pressure with both spatial and temporal resolution. It is shown theoretically that, in the limit of high pressure, fluorescence from iodine, excited at linecenter by a tunable, narrow-bandwidth laser, is inversely proportional to pressure and insensitive to other flowfield variables. The high-pressure limit is shown to apply for pressures greater than about 150 Torr in an iodine-seeded flowfield. Data collected in a static cell and in an underexpanded-jet flowfield verify the theoretical prediction. A photographic negative is shown to record two-dimensional pressure distributions directly when exposed by fluorescence from a sheet of laser radiation. Iodine transitions which are optimum for pressure measurement and accessible with a tunable dye laser are identified. The effect of fluid velocity on the linecenter fluorescence signal is evaluated. Preliminary measurements in a non-steady flow demonstrate that the technique can be used to resolve pressure fluctuations on the order of 1% at frequencies up to a few kHz with good spatial resolution.

Introduction

Pressure is a variable of fundamental importance in all fluid flowfields. The surface pressure distributions on solid bodies produce the integrated forces of primary

---

Presented as Paper 83-1468 at the AIAA 18th Thermophysics Conference, Montreal, Canada, June 1-3, 1983. Copyright © American Institute of Aeronautics and Astronautics, Inc., 1983. All rights reserved.
*Assistant Professor, Department of Mechanical and Aerospace Engineering.

importance in aerodynamics. The generation of sound pressure waves and their transmission through turbulent shear layers are fundamental unsolved problems in aeroacoustics. Knowledge of pressure fields in turbulent boundary layers is essential to understanding mechanisms for drag reduction and the structure of the turbulent boundary layer itself.[1]

Due to the importance of pressure in fluid mechanics, a technique for in situ pressure measurement without perturbation to the flowfield is greatly needed. Distributions of time-averaged static pressure on surfaces can be measured with small pressure taps, connected to manometers, without significant disturbance to the flow. Time-resolved measurements of pressure at the wall of a turbulent boundary layer have been reported using piezoelectric crystals mounted flush with the wall[2] and flexible pressure transducers built into the wall and ready optically.[3] Measurements of time-averaged pressure away from the wall have been made with pitot-static and total-head tubes for many years.[4] Probes have also been used for time-resolved pressure measurement away from walls in turbulent flows using disc[5] and small tube[6] probes. If the flow speeds and turbulence intensities are low enough, the use of such standard intrusive probes may be satisfactory, as long as their size is small compared to the relevant scales of the flowfield. However, for high-speed and highly turbulent flows, the complicated effects of probe geometry, Reynolds number dependence, and lateral turbulent velocity components make the quantitative use of intrusive probes very difficult.[6] Quantitative pressure measurements in turbulent flowfields require detectors smaller than the smallest turbulent eddies (often well under 1 mm) and with frequency response usually to a few kHz.[6] The detector must also be sensitive enough to detect fluctuations whose magnitude is often less than 1% of the mean.

The ideal pressure detector would be based on an optical technique. An optical probe does not present a mechanical obstruction to the flow, can be made very small for good spatial resolution, can have very good frequency response, and is free of probe geometry and Reynolds number complications and effects of lateral velocity components in highly turbulent flows. Since spatial resolution is usually a requirement for quantitative measurements, any technique based on absorption or refractive index variations integrated along an optical path is not acceptable. A sensitive light scattering technique would provide both spatial and temporal resolution without perturbation to the flowfield. Laser-induced fluorescence is an extremely sensitive nonintrusive light scattering technique which has been demonstrated for absolute measurements with number densities as low as 100 atoms/cm^3 (Ref. 7). Laser-induced fluorescence is orders of magnitude stronger (at pressures of usual interest in fluid mechanics) than other light scattering processes, such as Rayleigh or spontaneous Raman scattering, and is not limited by problems associated with particle seeding when employing Mie scattering (marker shot noise and particle lag, for example).

In order to use fluorescence for flowfield measurements, a laser is tuned into resonance with an absorption line of a gas and the resulting fluorescence is collected from one or more points in the flowfield. Electron beams have also been used to induce gases to fluoresce but are useful only in very low-density flowfields.[8] Until tunable uv lasers are available to selectively excite the molecular components of air, a suitable seed gas must be added to the flow. Atomic sodium has been used for quantitative measurements in high-speed flows.[9] Pressure was inferred from the collisional broadening of a numerically-fitted sodium lineshape in this work. However, sodium must be injected at very high temperatures, thereby perturbing the flow, and is severely limited in seeding concentration by laser absorption and radiative trapping effects. Quantitative density visualization has been reported in a transonic compressor rotor using 2,3 butanedione as the seeding gas.[10] Iodine is an attractive molecule for seeding a flowfield since it absorbs and fluoresces strongly in the visible part of the spectrum and has high enough vapor pressure to be seeded at room temperature without perturbation to the flowfield. The use of laser-induced iodine fluorescence for visualization of the mixing of supersonic streams[11,12] and for the quantitative measurement of density[13] and velocity[14] in compressible flows has been reported. The difficulty involved in using laser-induced fluorescence for flowfield measurements (once a suitable seed gas has been chosen for the particular application) is that it has been difficult to quantitatively relate the fluorescence signal to individual flowfield parameters.[13] This difficulty arises due to collisional processes that broaden the absorption line and compete with the radiative process for decay of the excited state. A quantitative model has been developed for laser-induced fluorescence in iodine that includes the important effects for measurements in a compressible flowfield.[15] The theoretical result of this model and a limiting form which enables a simple interpretation of the fluorescent signal in terms of pressure is presented in the next section. The section entitled "Application to Pressure Measurement" discusses the practical application of the laser-induced fluorescence technique to pressure measurement and presents numerical calculations and preliminary experimental results.

Theory of Laser-Induced Fluorescence

Rate Equation Model

Absorption of a photon from a laser, tuned into resonance with an allowed transition of a molecule (or atom), induces a transition of the molecule from a ground state to an excited state. The excited-state molecule is far from thermal equilibrium and, therefore, must decay to a ground state. There are several competing decay mechanisms: spontaneous emission, or fluorescence, and collisionally-induced processes including quenching (nonradiative decay), population redistribution in the excited state, and predissociation (transfer to an unbound state). These mechanisms compete with fluorescence for decay of the excited state and, combined with the broadening mechanisms that affect the initial absorption of the photon, introduce a pressure and temperature dependence that complicates the interpretation of laser-induced fluorescence. However, these processes also present the possibility of pressure or temperature measurement providing that they can be correctly modeled.

A quantitative model, based on a classical rate-equation approach, has been developed to describe laser-induced fluorescence in iodine.[15] This model was derived for the overlapping P13, R15 rotational lines in the (43-0) vibrational band of the B-X electronic transition of iodine. These transitions coincide with the 514.5-nm line of the argon laser. However, the model is valid for any iodine transition, or any other molecular transition, that can be modeled as a two-level system (that is, no additional energy levels are strongly-coupled to the two levels resonant with the laser). The steady-state solution to the rate-equation model, with excitation by a laser having a linewidth which is small compared to the molecular linewidth and an intensity much less than the saturation intensity of the transition, is given as[15]

$$S_F = C \frac{A_{21}}{A_{21} + Q} \frac{V(D,B)}{\Delta\nu_d} f_1 N_{I_2} \qquad (1)$$

where C is a constant that depends on molecular and optical parameters but is independent of parameters of the flow-field. The first factor after the constant in Eq. (1) is the fluorescence efficiency factor. This factor is the ratio of the radiative decay rate A_{21} to the total decay rate $A_{21} + Q$, where Q is the collisional quenching rate. The second factor is the absorption lineshape where

$$V(D,B) = \frac{B}{\pi} \int_{-\infty}^{+\infty} \frac{\exp(-y^2)dy}{B^2 + (D-y)^2} \qquad (2)$$

is the Voigt integral which includes the effects of homogeneous and inhomogeneous broadening and Δv_d is the Doppler linewidth. The Voigt integral contains a detuning parameter

$$D = 2\sqrt{\ln 2} \; \{[v_\ell - v_o(1 - \frac{u}{c})]/\Delta v_d\} \qquad (3)$$

where v_ℓ is the laser frequency and v_o is the center frequency of the molecular transition, Doppler shifted by an amount proportional to the ratio of the fluid velocity component in the direction of the laser beam u and the speed of light c. The Voigt integral also contains a broadening parameter,

$$B = \sqrt{\ln 2} (\Delta v_c / \Delta v_d) \qquad (4)$$

where Δv_c is the collisional linewidth. The next factor in Eq. (1) is the fraction f_1 of the total ground electronic state population that is in the level excited by the laser. The final factor is the total iodine number density N_{I_2}.

The pressure and temperature dependence of the fluorescent signal is quite complicated. The collisional linewidth and the quenching rate both vary with the collision frequency and are, therefore, given by

$$\Delta v_c = C_b (p/\sqrt{T}) \qquad (5)$$

and

$$Q = C_q (p/\sqrt{T}) \qquad (6)$$

where C_b and C_q are the collisional broadening and quenching cross sections. The Doppler linewidth is a function of temperature,

$$\Delta\nu_d = \sqrt{\frac{8(\ln 2)kT}{mc^2}} \, \nu_o \qquad (7)$$

where k is the Boltzmann constant and m is the molecular mass. The ground electronic state is assumed to remain in Boltzmann equilibrium throughout the process of laser-induced fluorescence. The population fraction is, therefore, written as the product of the vibrational and rotational fractions as

$$f_1 = \frac{1}{Z_v} \exp(-E_v/kT) \frac{B(2J+1)}{kT} \exp\left[-\frac{BJ(J+1)}{kT}\right] \qquad (8)$$

where Z_v is the vibrational partition function defined by

$$Z_v = \sum_v \exp(-E_v/kT) \qquad (9)$$

E_v is the vibrational energy, B is the molecular rotation constant, and v and J are the vibrational and rotational quantum numbers of the energy level. Therefore, due to the quenching factor, the absorption lineshape, and the Boltzmann population distribution, the fluorescence signal from iodine is a complicated function of pressure and temperature and is not simply proportional to the iodine number density N_{I_2} in a flowfield.

<u>High-Pressure Limit</u>

The general relationship between the iodine fluorescent signal and flowfield parameters in Eq. (1) is quite complicated. In order to relate the fluorescence to a single flowfield parameter and remove dependence on other unknown parameters, limiting cases of Eq. (1) can be utilized. For the measurement of density, detuning of the frequency of a narrow-bandwidth laser away from the molecular linecenter was used to achieve the limit of large detuning.[13] In this limit the temperature and pressure dependence of the resulting off-resonant fluorescent signal was removed. In order to see how laser-induced fluorescence can be used to quantitatively measure pressure, the high-pressure limit of Eq. (1) will be examined. In the limit of high pressure, the quenching rate becomes large compared to the radiative rate and the fluorescence efficiency factor becomes A_{21}/Q. The Voigt function becomes

Lorentzian at high pressure,

$$\lim_{B \gg 1} V(D,B) = \frac{1}{\sqrt{\pi}} \frac{B}{D^2 + B^2} \tag{10}$$

If the laser is tuned to the center of the absorption line, the value of the lineshape at the molecular linecenter at high pressure is

$$\lim_{B \gg 1} V(0,B) = 1/\sqrt{\pi} B \tag{11}$$

Therefore, for linecenter excitation in the high-pressure limit Eq. (1) reduces to

$$S_F = \frac{C \, A_{21}}{\sqrt{\pi \ln 2}} \frac{1}{Q} \frac{\Delta\nu_d}{\Delta\nu_c} \frac{1}{\Delta\nu_d} f_1 N_{I_2} \tag{12}$$

Inserting the pressure and temperature dependencies given in Eqs. (5) and (6) above gives

$$S_F = C'(T/p^2) \, f_1 N_{I_2} \tag{13}$$

where C' is a new constant given by

$$C' = \frac{C \, A_{21}}{\sqrt{\pi \ln 2}} \frac{1}{C_q C_b}$$

For the case of constant temperature and constant iodine concentration, Eq. (13) predicts that the linecenter iodine fluorescent signal will vary as $1/p^2$ at high pressure.

For flowfield measurements iodine is thoroughly premixed into the flowfield reservoir with a seeding fraction f_s given by the ratio of the reservoir pressure to the iodine vapor pressure (determined by the reservoir temperature). The iodine density in the flowfield is then given by the product of the seeding fraction (assumed uniform throughout the flowfield) and the total flowfield number density N,

$$N_{I_2} = f_s N \tag{14}$$

The fluorescent signal from a point in the iodine-seeded flowfield is, therefore, given by

$$S_F = \frac{f_s C'}{k} f_1 \frac{1}{p} \tag{15}$$

where the ideal gas law has been used to replace the density. The linecenter fluorescent signal at high pressures is related to the flowfield pressure and is independent of other flowfield parameters, except for the temperature dependence of the fractional population. (This temperature dependence can be removed, as will be shown in the next section.) The linecenter fluorescent signal in a uniformly-seeded, high-pressure flowfield is predicted to be inversely proportional to pressure.

Application to Pressure Measurement

Discussion of the Approach

In the previous section it was shown that in the limit of high pressure the fluorescence excited at linecenter by a narrow-bandwidth laser can be simply related to pressure, with the temperature and density dependence removed. This method for pressure measurement does not require that the entire lineshape be resolved or numerically fitted but only that the laser frequency be fixed at the center of an absorption line. Since maximum fluorescence is collected for linecenter excitation, the technique can produce a measurement with better signal-to-noise ratio than the off-resonance technique used for density measurement.

The relationship between the fluorescent signal and pressure in Eq. (15) is a direct one, albeit nonlinear. However, it is a simple matter in practice to achieve the linear relationship by recording the inverse of the measured fluorescent signal, with any background contribution to the signal subtracted before the inversion. Such a set of inverted values then directly gives the relative pressure distribution. Conversion to an absolute distribution requires knowledge of the pressure at one point in the distribution, for example, by measurement with a transducer at the wall. The inversion can also be achieved at multiple points simultaneously by illuminating the flowfield with a thin sheet of laser light and recording the resulting fluorescent distribution on photographic film. The intensity transmitted by such a developed negative τ is related to the intensity present during exposure I by an inverse power law[16]

$$\tau = \kappa I^{-\gamma} \quad (16)$$

where γ is the slope of the linear region of the Hurter-Driffield curve for the photographic negative and κ depends on the film and the exposure time. By correct choice of the film and control of the exposure and development processes, γ can be made equal to unity. In this case the intensity transmitted by the developed film negative is inversely proportional to the fluorescence intensity during exposure, or directly proportional to the relative pressure distribution. Conversion to absolute pressure is made by measuring the transmitted intensity at a location where the pressure is known. The negative is, therefore, a quantitative picture of the two-dimensional pressure field and can be either digitized for direct quantitative analysis or analyzed spectrally using optical or digital information processing techniques.[16] The inverse proportionality between the fluorescent signal and pressure is also inconsequential for the measurement of fluctuations of pressure about a mean value since it only introduces a 180-deg. phase shift between the measured fluorescent signal and the pressure fluctuation. This phase shift can be easily accounted for when correlating the pressure with another fluctuating flowfield parameter, such as velocity.

Before actually applying the laser-induced fluorescence technique to the quantitative measurement of pressure, several questions must be addressed. The first regards the remaining temperature dependence in Eq. (15) due to the fractional population factor f_1. The second is, what pressure is needed in order that the high-pressure limit be valid? A third question that must be considered is the effect of the fluid velocity on the linecenter fluorescence. These three questions will be addressed next.

Choice of Optimum Iodine Absorption Line

The temperature dependence remaining in Eq. (15) is due solely to that of the fractional population of the absorbing energy level for the transition excited by the laser. This temperature dependence can be effectively removed for a wide range of temperature by proper choice of the absorption line. The fractional population of an energy level is a function of its vibrational and rotational quantum numbers and the temperature, as given by Eqs. (8) and (9). A plot of the fractional population for four iodine energy levels in the ground electronic state is given in Fig. 1 as a function of Mach number for a steady, isentropic flow from a room temperature reservoir. The iodine transition which can be excited by the argon laser at 514.5 nm originates from the ground vibrational level, $v''=0$, and a low rotational level, $J''=15$. It is seen in Fig. 1 that the fractional population in this level is a strong function of temperature, increasing by a factor of about 5 as the temperature decreases from 298-106 K with

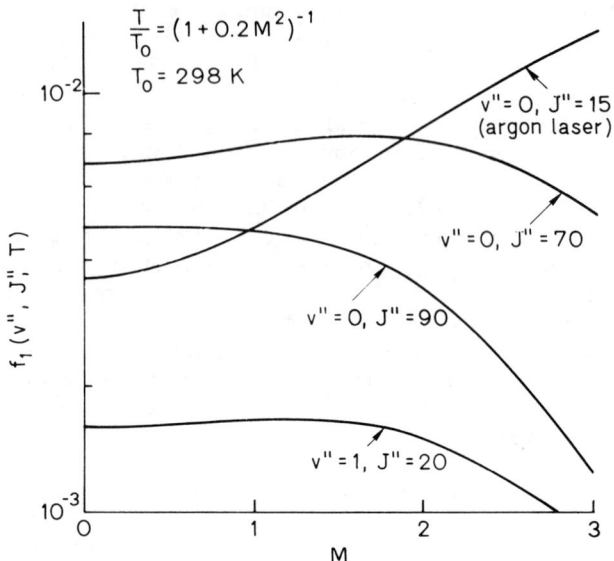

Fig. 1 Mach number variation of the fractional population of four iodine energy levels in the ground electronic state.

increasing Mach number from 0-3. By choosing a transition originating from an energy level with a higher rotational quantum number, this temperature dependence can be effectively removed. The increase in vibrational population fraction with Mach number is then canceled by a decrease in rotational population fraction. For $J''=70$, f_1 varies by ±6% over a Mach number range of 0-2.5. For $J''=90$, the variation from Mach numbers of 0-1 is within ±1%. A transition from the first excited vibrational level may also be utilized for measurements over a Mach number range of 0-1.75 if a low rotational level is chosen, as shown for $v''=1$ and $J''=20$ in Fig. 1. For this energy level the decrease in vibrational population fraction with increasing Mach number is canceled by an increase in rotational population fraction. In either case the idea is to choose a transition for which the product of vibrational and rotational population fractions of the absorbing energy level is essentially independent of temperature over the range of temperature variation in the flowfield.

The three temperature-insensitive iodine energy levels shown in Fig. 1 can be easily accessed by a tunable dye laser. For wavelengths near 550 nm, the peak output for R110 dye, many transitions originating from ground vibrational energy levels with rotational quantum numbers in the

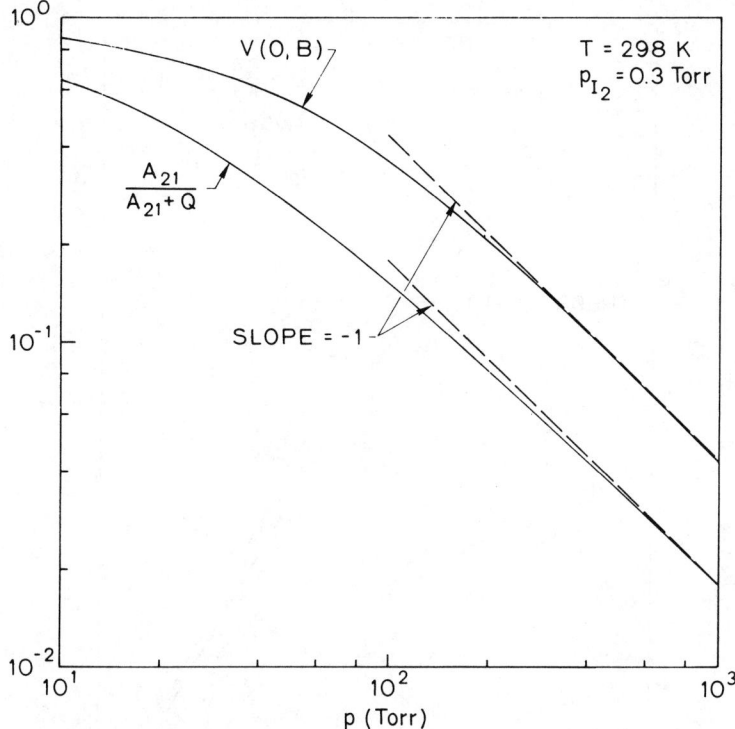

Fig. 2 Pressure variation of Voigt function, at linecenter, and Stern-Volmer factor for iodine at room temperature.

range of 70-90 have been identified.[17] The transitions at this wavelength are also among the strongest of the iodine absorption lines due to the maximum values of the Franck-Condon factors for the vibrational bands found in this region.[18] Therefore, the temperature dependence remaining in Eq. (15) can be completely removed, for all practical purposes, by proper choice of the iodine absorption line and will not be considered in the further analysis.

Validity of the High-Pressure Limit

Another important question about the practical applicability of Eq. (15) is the pressure needed for the high-pressure limit to be valid. The two necessary and sufficient conditions are that the quenching rate be much larger than the spontaneous emission rate $Q \gg A_{21}$ and that the lineshape at linecenter be predominantly Lorentzian. These two conditions are examined in Fig. 2 by plotting the pressure variation at room temperature of the Voigt func-

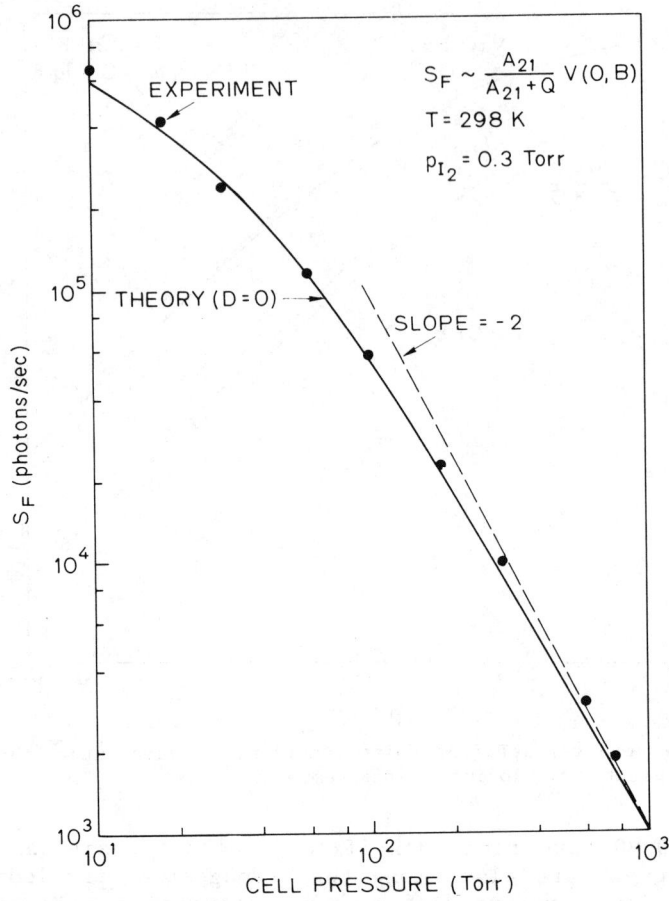

Fig. 3 Pressure variation of linecenter iodine fluorescent signal at room temperature with constant iodine density.

tion at linecenter and the Stern-Volmer factor. The quenching and collisional broadening constants used in this calculation are from Ref. 15. In the high-pressure limit, both factors vary at $1/p$ for fixed temperature; that is, the curves asymptotically approach the dotted lines drawn with a slope of -1 in this figure. The deviation between the linecenter value of the Voigt function and the $1/p$ asymptote is 17% at 100 Torr, 6% at 200 Torr, and 1.8% at 400 Torr. The difference for the Stern-Volmer factor decreases from 14% at 100 Torr to 3.8% at 400 Torr. Roughly speaking, both the linecenter Voigt function and the Stern-Volmer factor exhibit the high-pressure limit (to within 10%) for pressures greater than 150 Torr at room temperature.

Since the fluorescent signal is proportional to the product of the Voigt function and the Stern-Volmer factor, it should vary at $1/p^2$ in the high-pressure limit for the case of constant temperature and fixed iodine density. The pressure variation of the linecenter iodine fluorescent signal is plotted in Fig. 3 for room temperature and a fixed iodine vapor pressure of 0.3 Torr. Data from a static cell experiment[15] with fixed iodine density and variable nitrogen pressure are shown in this figure. The theoretical curve is a fit of Eq. (1) to the data. It is seen that the theory agrees well with the experiment for excitation at linecenter over this range of pressure. The fluorescence does asymptotically approach the $1/p^2$ high-pressure limit with a deviation that is essentially the sum of the separate deviations of the Voigt function and the Stern-Volmer factor shown in Fig. 2. The static cell experimental data exhibit the high-pressure limit (to within 10%) by approximately 300 Torr.

In an iodine-seeded flowfield with constant seeding fraction the iodine number density is proportional to the total flowfield density [Eq. (14)] and, therefore, to pressure through the ideal gas law. This additional pressure dependence cancels that of one of the two previous inverse pressure dependencies to produce the $1/p$ variation in Eq. (15). The pressure variation of the product of the Stern-Volmer factor, the Voigt lineshape function at linecenter and the iodine density is shown in Fig. 4 for two fixed temperatures. These curves represent the variation of the fluorescence signal with pressure, given in Eq. (1), for the case of constant temperature and with the population fraction not included. For room temperature, the product asymptotically approaches the $1/p$ line (dotted) with a deviation given by the sum of the two factors in Fig. 2 or the data in Fig. 3. However, for measurements in a high Mach number flowfield the temperature decreases considerably. (For example, for flow at Mach 3 from a room temperature reservoir the temperature is about 100 K.) The pressure variation for a fixed temperature of 100 K is also shown in Fig. 4. It is seen that the high-pressure limit extends to lower pressure when the temperature is reduced. This occurs because a temperature decrease reduces the Doppler broadening contribution to the Voigt lineshape and increases the quenching rate at a fixed pressure. The linecenter fluorescent signal reaches the high-pressure limit (to within 10%) by about 100 Torr for a temperature of 100 K. It is also seen in Fig. 4 that the fluorescence is temperature independent, as predicted by Eq. (15), in the same limit.

Figures 2-4 are plots for the case of constant temperature which show that the high-pressure limit is valid (to within 10%) for pressures above 300 Torr at 298 K or 100 Torr at 100 K and independent of temperature in this asymptotic limit. In a compressible flowfield the pressure, temperature, and density all vary with spatial location in a steady flow and also with time is an unsteady flow. One way to exhibit the combined effect of pressure, temperature, and density on the linecenter fluorescent signal is to plot the fluorescence as a function of Mach number for a steady, isentropic flow in which the flowfield parameters are all uniquely related to the Mach number.[19] Figure 5 is such a plot of the fluorescent signal normalized by its value at the reservoir conditions for flow from a room temperature reservoir at three reservoir stagnation pressures. The dotted curve is the inverse pressure distribution, normalized by the reservoir pressure. It is seen that the curves all agree well at low Mach numbers where the pressure is high. However, the curves fall below the inverse of the pressure distribution as the Mach number increases and pressure decreases. Increasing the stagnation pressure increases the pressure

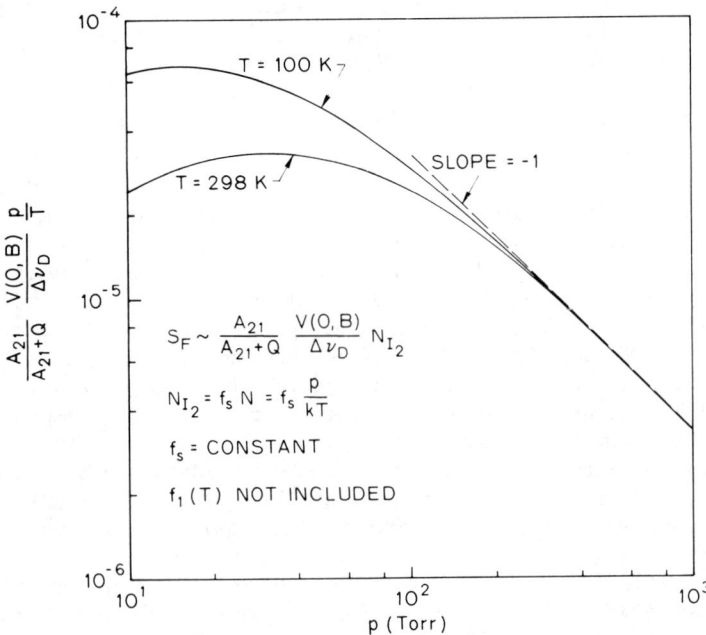

Fig. 4 Pressure variation of linecenter iodine fluorescent signal in a uniformly-seeded, variable-density flowfield.

correspondingly at any value of the Mach number and, thereby, decreases the deviation at that Mach number.

The percentage difference between the inverse pressure distribution and the normalized fluorescent signal is shown in Fig. 6 as a function of Mach number for the three values of stagnation pressure. It is seen that increasing the reservoir pressure decreases the difference dramatically at high Mach numbers. The difference at Mach 2 is 11% for a reservoir pressure of 1.5 atm and decreases to 2% for 5 atm and 0.5% for 15 atm. The 10% difference point is at Mach

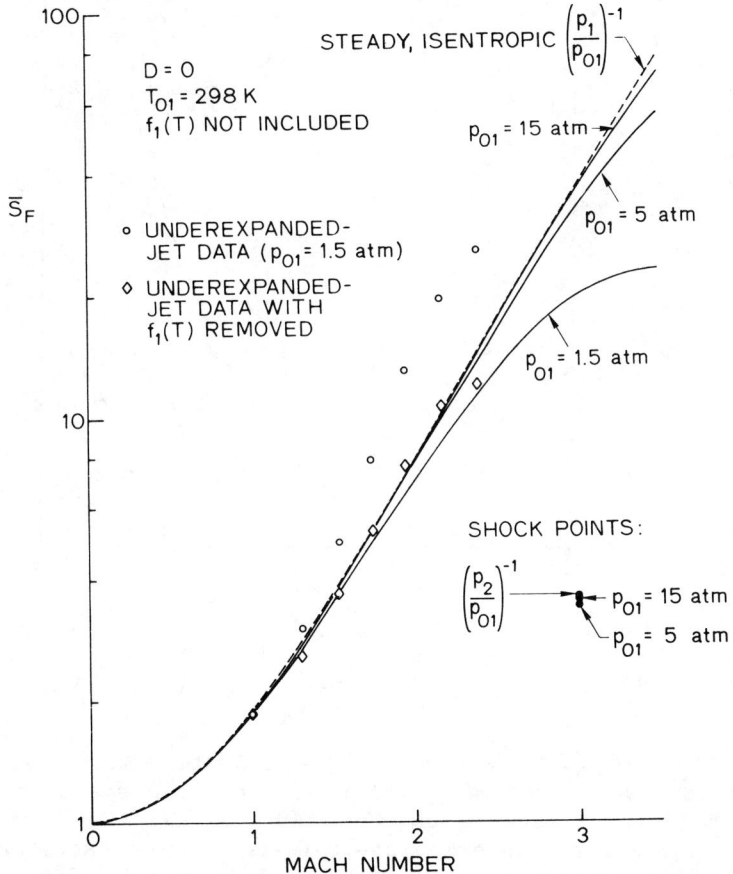

Fig. 5 Mach number variation of normalized iodine fluorescent signal in a steady, isentropic flowfield for three reservoir pressures. Inverse of normalized pressure distribution shown by dotted curve. Circles are data from an underexpanded-jet flowfield. Diamonds are same data with population fraction variation removed. Change in fluorescence and pressure across a normal shock at Mach 3 also shown.

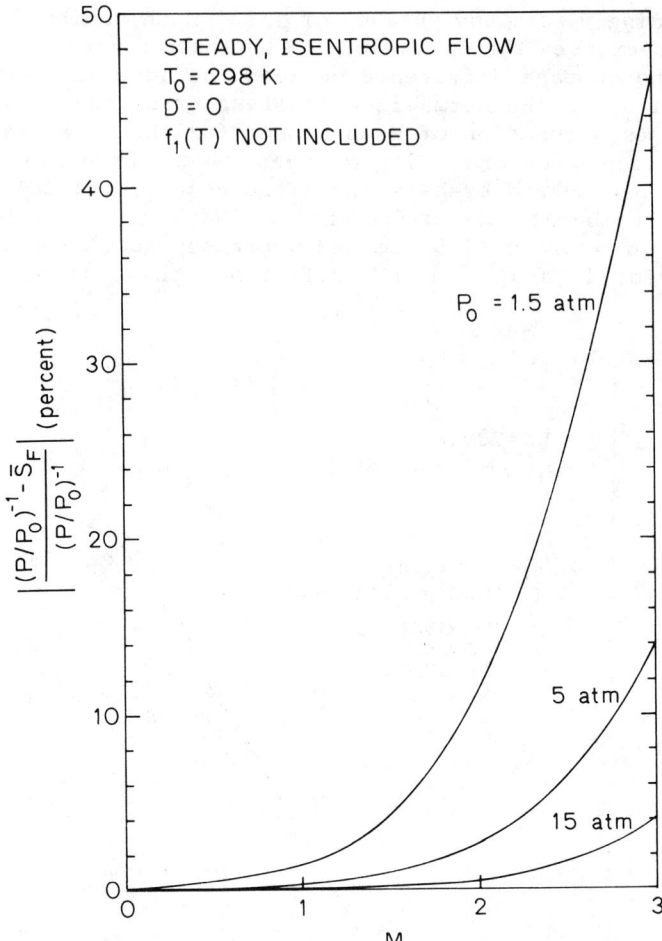

Fig. 6 Percentage difference between inverse pressure and normalized fluorescence as a function of Mach number for three reservoir pressures.

1.9 for a stagnation pressure of 1.5 atm, Mach 2.8 for 5 atm, and well over Mach 3 for 15 atm. The static pressure for approximately 10% deviation in this Mach number range is 150 Torr, intermediate between the two cases shown in Fig. 4 since the temperature lies between 298-100 K. In any practical application, the reservoir pressure should be chosen to be the minimum consistent with the desired accuracy in the pressure measurement. Minimizing the pressure both maximizes the fluorescent signal level and minimizes the mass flow rate requirement of the flow facility.

To emphasize that no isentropic assumption has been made in leading to Eq. (15), the variation of the fluorescent signal and the pressure across a normal shock discontinuity at Mach 3 is shown in Fig. 5. It is seen that the agreement is very good since the shock increases the pressure and, therefore, shifts further in the direction of the high-pressure limit. Measurements of pressure in viscous flows or flows with heat addition and conduction pose no complications for the laser-induced fluorescence technique.

Figure 5 also includes some preliminary data that demonstrates that the theory and the technique are applicable to measurement in a practical compressible flowfield. The data was collected pointwise along the centerline of an underexpanded jet with a reservoir pressure of 1.5 atm.[15] An argon laser was tuned to the linecenter of the iodine transition at 514.5 nm. The laser beam intersected the jet normal to its centerline in order to remove the effect of the Doppler shift (to be discussed in the next section). The data points (circles) are shown for a Mach number range of 1, at the sonic exit plane of the jet, to 2.4. It is seen that the data deviates from the inverse pressure distribution with increasing Mach number along the jet

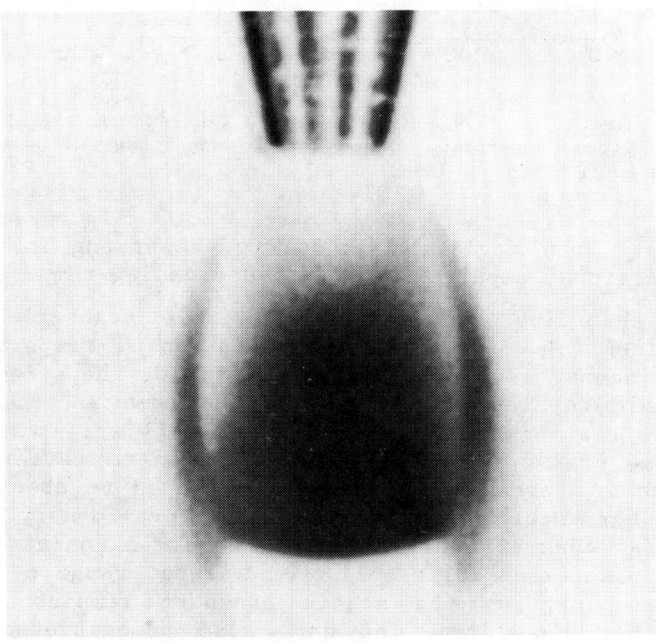

Fig. 7   Direct print of photographic negative exposed by fluorescence from iodine-seeded, underexpanded-jet flowfield.

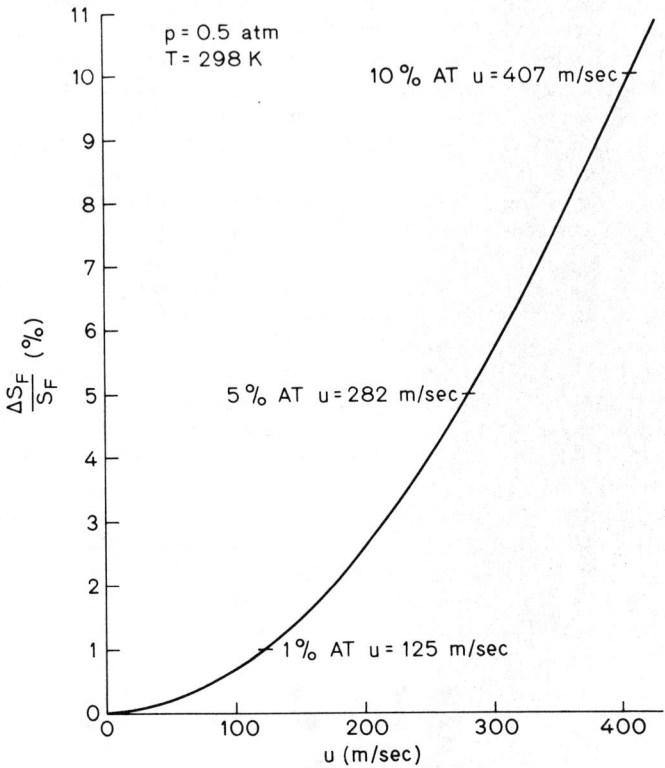

Fig. 8 Percentage change in linecenter fluorescent signal from Doppler-shifted absorption line versus magnitude of velocity component in direction of laser beam.

centerline. This deviation is due to the strong increase in the fractional population f_1 with decreasing temperature for the absorbing energy level of this transition at 514.5 nm (see Fig. 1). Since the variation of temperature with Mach number is known in this flowfield, the effect of the variable population fraction can be removed from the data. The resulting points (diamonds) fall very well on the inverse pressure distribution and begin to deviate at about Mach 2 where the pressure has dropped to about 150 Torr, in agreement with theory. This preliminary data demonstrates clearly that the technique for a nonintrusive pressure measurement works well over a large range of Mach numbers as long as the static pressure remains above approximately 150 Torr. The data also demonstrates the necessity for choosing the iodine transition carefully in order to remove the effect of the fractional population on the resulting fluorescent signal.

Figure 7 is a direct print of a photographic negative exposed by the fluorescent distribution from the underexpanded-jet flowfield.[15] The flow is from left to right, originating in a 1.5-atm reservoir and expanding into a background pressure of 53 Torr through a 1.5-mm-diam. sonic nozzle. A supersonic flowfield is produced with surrounding barrel and normal shock waves. The film and development used to produce this negative yielded a value of γ in Eq. (16) slightly less than one. Thus, this negative approximately produced the inversion of the intensity distribution required to provide a record of the pressure in this two-dimensional cross section of the flowfield. The pressure decrease with distance as the flow accelerates and the large pressure increase as the flow suddenly decelerates across the normal shock wave at Mach 4.4 are readily apparent. The negative in Fig. 7 is not quantitatively proportional to pressure in the region from about one nozzle diameter downstream of the nozzle exit to the normal shock wave since the pressure falls below 150 Torr in this region. Also the effects of the temperature-dependent population fraction of this transition at 514.5 nm and the Doppler shift for points away from the jet centerline (to be discussed in the next section) cause this photograph to be not exactly proportional to the pressure distribution. However, the photograph illustrates well that a two-dimensional pressure field can be recorded on a single photographic negative. This capability can be invaluable for short duration flow facilities and a tremendous time-saving technique for continuous flow facilities. A quantitative representation of the overall pressure field can be extremely useful for the interpretation of data collected at isolated points. Such a negative can be digitized for the quantitative recording of the two-dimensional pressure distribution and also for digital image processing, if desired. The negative can also be used as the object in an optical information processing system for optical manipulation of its frequency spectrum. Such techniques are often useful, for example, in enhancing desired flowfield structures.

Effect of Fluid Velocity

When measuring pressure by tuning a laser to the center of an iodine transition, the effect of the fluid velocity on the fluorescence must be considered. If the fluid has a velocity component in the direction of the laser beam the absorption line will be Doppler-shifted in frequency [see Eq. (3)]. The iodine transitions are quite broad due to their hyperfine splitting.[20] The broad linewidth minimizes the effect of the Doppler shift. For example, a 300-m/s velocity component in the laser beam direction is required to shift the iodine absorption line by half its width.[14] The percentage change in the linecenter fluorescence signal resulting from a flowfield velocity component in the direction of the laser beam is shown in Fig. 8 for a pressure of 0.5 atm and room temperature. For these conditions it is seen that the fluorescence signal is decreased by 1% for a 125-m/s velocity component and 10% if the component is increased to 407 m/s. For measurements in high velocity flows where the Doppler shifts become considerable, the shift produced by the average velocity can be removed by tuning the laser to locate the center of the Doppler-shifted line and fixed at that frequency for a pressure measurement. Small velocity fluctuations about the average value will not affect the linecenter fluorescent signal. This figure also shows that laser frequency drifts, usually in the range of 50 MHz for an unstabilized dye laser over a period of several minutes, do not influence the measurement of pressure. A 50 MHz frequency drift corresponds to approximately a 50-m/s velocity component in Fig. 8 and a resulting change in the fluorescent signal which is well under 1%.

Time-Resolved Pressure Measurement

The preceding discussion has focused primarily on the measurement of time-averaged pressure with spatial resolution in steady flowfields. However, the same technique can be used for measurement of time-resolved pressure fluctuations in nonsteady or turbulent flowfields. For the resolution of pressure fluctuations the question of the fluorescent signal level available and the resulting signal-to-noise ratio of the measurement arises. The fluorescent signal from a strong iodine transition at room temperature and 0.5-atm pressure and with optimum collection from a 1-mm^3 volume is given in Ref. 15 to be 10^6 photons/s. If the fluorescent signal is sampled at a 1-kHz rate, on the order of 10^3 photons are collected per measurement interval. The signal-to-noise ratio, as limited by photon statistics, is equal to the square root of the signal, or 32 for these conditions. Thus, fluctuations in pressure of 3% could be resolved at frequencies near 1 kHz with this calculated fluorescent signal level.

A preliminary experiment has been conducted to evaluate the feasibility of using laser-induced iodine fluorescence for the measurement of pressure fluctuations.[21] The sound pressure field in an acoustic cell, excited at its fundamental resonant frequency by a loudspeaker, was used to establish a pressure fluctuation of known amplitude and frequency. A single-mode argon laser was tuned to the center of the iodine transition at 514.5 nm using a temperature-controlled intracavity etalon. For a cell pressure of 300 Torr and a 1% sound pressure fluctuation established at 570 Hz, the collected fluorescent signal displayed a waveform that was identical to that of a microphone measurement at the same location in the cell, but shifted in phase relative to the microphone output by 180 deg. The signal-to-noise ratio for the fluorescent signal, collected on a single sweep of the oscilloscope, was better than 25:1 at every point along the resulting waveform. This result is <u>extremely</u> encouraging and establishes that pressure fluctuations on the order of 1% at frequencies up to a few kilohertz should be measurable with good spatial resolution using laser-induced fluorescence from iodine.

The argon laser was used in the preliminary experiment to study the feasibility of time-resolved pressure measurement. However, the transition accessible with the argon laser at 514.5 nm is not ideal for the quantitative measurement of pressure fluctuations since simultaneous temperature fluctuations will influence the measurement through the population fraction. In order to examine the magnitude of this effect, the fluctuation in fluorescent signal is related to that of the pressure and the population fraction by forming the logarithmic derivative of Eq. (15), resulting in

$$\frac{\Delta S_F}{S_F} = -\frac{\Delta p}{p} + \frac{\Delta f_1}{f_1} \qquad (17)$$

The fluorescent signal fluctuates 180 deg. out of phase with the pressure, due to the 1/p variation and also depends on the magnitude of the change in the population fraction due to the simultaneous temperature fluctuation. Evaluating the variation of the population fraction with temperature about a mean temperature of 298 K for the iodine transition at 514.5 nm (see Fig. 1) and relating the magnitude of the temperature fluctuation to that of pressure with an isentropic assumption, the fluorescent fluctuation is related to the pressure fluctuation by

$$\frac{\Delta S_f}{S_F} = -1.44 \frac{\Delta p}{p} \qquad (18)$$

Therefore, when using the argon laser to induce iodine fluorescence there is a 44% contribution to the signal fluctuation that occurs due to the temperature-dependent population fraction. The argon laser could be used to measure pressure fluctuations provided that the flow was isentropic and the corresponding temperature fluctuations could be calculated. However, most interesting nonsteady flowfields are not isentropic, such as the turbulent boundary layer or the wake of a bluff body, and this would be a severe limitation in such flows. Therefore, a tunable, narrow-bandwidth laser, notably a dye laser, must be used for quantitative measurements of pressure fluctuations. Amplification of such continuous-wave laser output by a high-powered pulsed laser and use of a sheet of laser radiation and an intensified analog or digital camera would enable the pressure distribution to be recorded instantaneously in an entire cross-sectional plane of a flowfield.

Measurement of pressure fluctuations using laser-induced iodine fluorescence can be made simultaneously with measurements of velocity fluctuations using laser anemometry. The inelastic fluorescence from seeded iodine molecules can be spectrally separated from the elastic Mie scattering due to seeded particles, permitting their simultaneous monitoring. The continuous fluorescence signal can be sampled whenever a particle passes through the coincident laser-anemometer scattering volume for a simultaneous measurement of pressure and velocity. Such a technique would allow the direct measurement of the pressure-velocity correlation at a point, a quantity of fundamental importance in statistical theories of turbulence. Other obvious applications include the correlation of the output of a nonintrusive fluorescence probe with that of a standard pressure probe placed at a location where it did not perturb the flowfield. It would be very interesting, for example, to correlate the pressure fluctuation in the nearfield of a jet or a bluff body flow with farfield microphone measurements of the radiated acoustic sound pressure. This would be a very valuable capability in aeroacoustic studies of the mechanisms of sound generation and propagation.

Summary

A method of quantitative, nonintrusive pressure measurement with both spatial and temporal resolution has been discussed. The method utilizes laser-induced fluorescence from iodine molecules, seeded into the flowfield without perturbation and induced to fluoresce by a tunable, narrow-bandwidth laser. The following important points were presented.

1) The results of a rate-equation model for the fluorescence, examined in the high-pressure limit, showed that the fluorescence excited at linecenter is inversely proportional to pressure and independent of other flowfield parameters, except for the population fraction.

2) Calculations of the temperature variation of the fractional population of the ground state excited by the laser showed that this factor can be essentially removed by proper choice of the iodine absorption line.

3) The pressure necessary for the high-pressure limit to apply was evaluated. It was shown that both the quenching and lineshape factors exhibit the high-pressure limit by about 150 Torr at room temperature. Decreasing the temperature was shown to extend this limit to lower pressures. At room temperature the fluorescence signal may be considered to be in the high-pressure limit above about 300 Torr and above about 100 Torr at 100 K (the temperature at Mach 3 from a room temperature reservoir).

4) The variation of the fluorescent signal with Mach number for a steady isentropic flow showed the range of Mach number over which the technique could be used for pressure measurement with different reservoir pressures. Increasing the reservoir pressure extends the Mach number range for pressure measurement to higher Mach numbers but decreases the fluorescent signal level.

5) Experimental data collected in a static cell at room temperature confirmed the pressure dependence of the linecenter fluorescent signal for the case of constant temperature and iodine density. Data collected along the centerline of an underexpanded-jet flowfield verified that the linecenter fluorescent signal is inversely proportional to pressure in an iodine-seeded flowfield as long as the pressure remains above approximately 150 Torr.

6) A direct print of a photographic negative, exposed by the fluorescence distribution in the underexpanded-jet flowfield, illustrated the capability for recording directly the pressure distribution in an entire cross-sectional plane.

7) The fluid velocity can affect the fluorescent signal through a Doppler frequency shift of the absorption line. A 407-m/s velocity component in the direction of the laser beam was shown to change the linecenter fluorescent signal by 10% at 0.5 atm and room temperature.

8) The results of a preliminary experiment establish that the technique can be used to resolve pressure fluctuations on the order of 1% at frequencies up to a few kilohertz with good spatial resolution.

Acknowledgements

This work was initiated while the author was a postdoctoral research associate at Stanford University under the support of AFOSR contract F49620-80-C-0091 and is supported at the University of Virginia by NASA grant NAG-1-373 and NSF grant MEA-8307317.

References

[1] Cantwell, B. J., "Organized Motion in Turbulent Flow," Ann. Rev. Fluid Mech., Vol. 13, 1981, pp. 457-515.

[2] Willmarth, W. W., Winkel, R. E., and Sharma, L. K., "Axially Symmetric Turbulent Boundary Layers on Cylinders: Mean Velocity Profiles and Wall Pressure Fluctuations," Jour. Fluid Mech., Vol. 76, Jan. 1976, pp. 35-64.

[3] Dinkelacker, A., Hessel, M., Meier, G. C. A. and Schewe, G., "Investigation of Pressure Fluctuations beneath a Turbulent Boundary Layer by means of an Optical Method," Phys. Fluids, Vol. 20, Oct. 1977, pp. 5216-5224.

[4] Goldstein, S., ed., Modern Developments in Fluid Dynamics, Vol. 1, Oxford University Press, Oxford, England, 1938.

[5] Miller, D. R., and Comings, E. W., "Force-Momentum Fields in a Dual-Jet Flow," Jour. Fluid Mech., Vol. 7, Feb. 1960, pp. 237-256.

[6] Hinze, J. O., Turbulence, 2nd ed., McGraw-Hill, New York, 1975.

[7] Fairbank, W. M. Jr., Hansch, T. W., and Schawlow, A. L., "Absolute Measurement of Very Low Sodium-Vapor Densities using Laser Resonance Fluorescence," Jour. Opt. Soc. Am., Vol. 65, Feb. 1975, pp. 199-204.

[8] Peterson, C. W., "A Survey of the Utilitarian Aspects of Advanced Flowfield Diagnostic Techniques," AIAA Journal, Vol. 17, Dec. 1979, pp. 1352-1360.

[9] Zimmerman, M. and Miles, R. B., "Hypersonic Helium Flow Field Measurements with the Resonant Doppler Velocimeter," Appl. Phys. Lett., Vol. 37, Nov. 1980, pp. 885-887.

[10] Epstein, A. H., "Quantitative Density Visualization in a Transonic Compressor Rotor," Engr. Power, Vol. 99, July 1977, pp. 460-475.

[11] Rapagnani, N. L. and Davis, S. J., "Laser-Induced Iodine Fluorescence Measurements in a Chemical Laser Flowfield," AIAA Journal, Vol. 17, Dec. 1979, pp. 1402-1404.

[12] Cenkner, A. A. and Driscoll, R. J., "Laser-Induced Fluorescent Visualization on Supersonic Mixing Nozzles that Employ Gas-Trips," AIAA Journal, Vol. 20, June 1982, pp. 812-819.

[13] McDaniel, J. C., Baganoff, D., and Byer, R. L., "Density Measurement in Compressible Flows using Off-Resonant Laser-Induced Fluorescence," Phys. Fluids, Vol. 25, July 1982, pp. 1105-1107.

[14] McDaniel, J. C., Hiller, B., and Hanson, R. K., "Simultaneous Multiple-Point Velocity Measurement using Laser-Induced Iodine Fluorescence," Opt. Letters, Vol. 8, Jan. 1983, pp. 51-53.

[15] McDaniel, J. C., "Investigation of Laser-Induced Iodine Fluorescence for the Measurement of Density in Compressible Flows," Ph.D. Thesis, Dept. of Aeronautics and Astronautics, Stanford University, Stanford, Calif., Dec. 1981.

[16] Goodman, J. W., Introduction to Fourier Optics, McGraw-Hill, New York, 1968, pp. 149-154.

[17] Simmons, J. D. and Hougen, J. T., "Atlas of the I_2 Spectrum from 19,000 to 18,000 cm^{-1}," Jour. Res. Nat. Bureau Standards, Vol. 81A, Jan.-Feb. 1977, pp. 25-80.

[18] Tellinghuisen, J., "Intensity Factors for the I_2 B-X Band System," J. Quant. Spect. Rad. Transfer, Vol. 19, 1978, pp. 149-161.

[19] Liepmann, H. W. and Roshko, A., Elements of Gasdynamics, Wiley and Sons, New York, 1957, p. 53.

[20] Levenson, M. D. and Schawlow, A. L., "Hyperfine Interaction in Molecular Iodine," Phys. Rev. A., Vol. 6, July 1972, pp. 10-20.

[21] Ackermann, U., Baganoff, D., and McDaniel, J. C., "Dependence of Laser-Induced Fluorescence on Fluctuations in Compressible Flows" (to be published).

Laser-Induced Schlieren Effect in Sodium-Nitrogen Mixtures

Jim W. L. Lewis* and Jim D. Selman†
The University of Tennessee Space Institute
Tullahoma, Tennesse

Abstract

The feasibility has been evaluated of using a laser-induced Schlieren technique (LIST) for the measurement of trace species concentrations in a host gas mixture. Using a resonantly absorbed laser pump beam tuned to a transition of the trace species, the gas heating and thermal lens produced by the pump beam are detected by a nonabsorbed probe laser whose magnitude of deflection is proportional to the local concentration of the trace species. The evaluation was performed for Na-N_2 mixtures and N_2 with trace quantities of Na and nonabsorbing H_2O vapor, and time dependent deflections were calculated for pulsed pump beams. The feasibility of LIST was demonstrated for the ternary mixture, and it is estimated that, for Mach numbers less than 0.1, the Na measurement sensitivity is on the order of 0.1 ppm for measurements using a 1-W laser source with 10^{-3}-pulse width.

Presented as Paper 83-1467 at the AIAA 18th Thermophysics Conference, Montreal, Canada, June 1-3, 1983. Copyright © American Institute of Aeronautics and Astronautics, Inc., 1983. All rights reserved.
*Professor of Physics, Applied Physics Research Group.
†Graduate Research Assistant, Applied Physics Research Group.

Introduction

The measurement of the local concentration of atomic and molecular trace gas species in a host gas environment is a recurring requirement in such application areas as combustion studies and the tracking and monitoring of tracer species in a flowfield of varying density and temperature. Both in situ and sample extraction techniques have been applied to this measurement problem. Among the in situ techniques, the use of coherent anti-Stokes Raman scattering suffers from the problem of scattering signal contributions from the nonresonant electronic susceptibility of the gas mixture.[1] The problem is no easier for other well-known techniques such as resonant fluorescent scattering and its radiation quenching corrections[2] and absorption spectroscopy whose results are modified by line of sight spatial averaging over density and temperature gradients. Mass spectrometric sampling probes fare no better in cases of general application due to the sometimes complicated mass spectral pattern and mass resolution problem. Although these specific problems are avoided in some cases by the use of matrix isolation spectroscopy, the difficulty of obtaining an accurate calibration of the sampling efficiency of the system[3] is not eliminated.

Resurgence of interest in the field of opto-acoustic spectroscopy[4] resulted from the need to measure the concentration of small quantities of a trace gas species would be accomplished. This goal has, in fact, been accomplished, but under rather restrictive conditions, for acoustic refraction effects can be a most serious problem for flowfields as well as static samples of nonuniform gas properties. The sensitivity of the phenomenon to inter- and intramolecular rate processes makes difficult the quantitative interpretation of results acquired for gas samples for which the constituency is both variable and unknown a priori.

A variation on the theme of opto acoustic spectroscopy has been proposed and used by Hermann and Pohl[5] for trace gas concentration measurements. The proposed approach is the use of thermal deflection spectroscopy which is based upon the absorption of radiative energy from a pump beam source by the trace species of interest, the heating of the gas sample as the excited species transfers its energy to the translational mode of motion of the gas mixture, and the deflection of a nonabsorbing probe beam by the lens action of the heated gas sample. For isobaric heating, the magnitude of the deflection of the probe beam is proportional to the concentration of the absorbing trace species over the volume of intersection of the pump and probe beams. The geometrical arrangement of the two beams is shown in Figure 1. For the work reported in Ref. 5, measurements were determined by the concentration of trace quantities of water vapor in laboratory air samples. The pump laser source was a pulsed CO laser of approximately 3-ms pulse width, and a He-Ne laser was used for the probe beam. An estimated H_2O vapor detection sensitivity of approximatley 1 ppm/$\sqrt{Hz}$ could be obtained for water vapor with a spatial resolution of about 1 mm^3. Although Hermann and Pohl demonstrated experimentally that detectable deflections of the probe beam could be obtained, the analysis presented in Ref. 5 did not demonstrate the critical parameters inherent in the process.

An application of the techniques to other trace gas species or gas mixtures or flowfields requires an examination of the

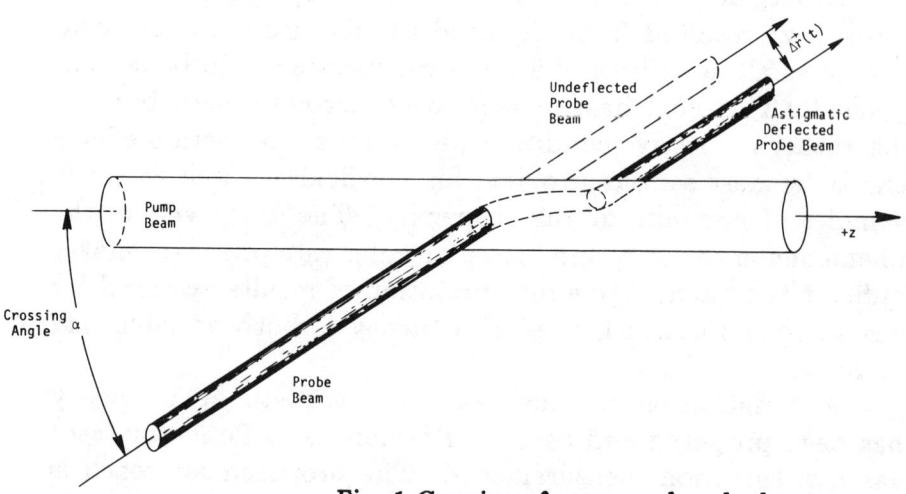

Fig. 1 Crossing of pump and probe beams.

laser pumping, radiative and collisional energy transfer mechanisms, and the diffusion thermal wave spreading of the heated gas mixture and the subsequent time dependent deflection of the probe beam. To address these questions, the Na-N_2 mixture was selected since resonant pumping of the Na 3P levels and the subsequent radiative and collisional transfer processes are comparatively well understood and the trace Na mixture is easily prepared for future experimental study. Further, detailed rate kinetics calculations are available[6] to predict and monitor the time dependent progress of energy transfer from the Na pumping process through the energy transfer mechanisms to the deposition of energy into the translational mode of the gas mixture. Using the Na-N_2 mixture and the kinetics model of Ref. 6, the critical parameters and regime of applicability of the LIST technique as proposed by Hermann and Pohl have been investigated, and the results of this study are presented in this work.

Theory and Results

Simple order-of-magnitude estimates of some of the critical parameters of LIST can be obtained before embarking on detailed numerical calculations. The deflection of the probe beam of optical path length s in a medium of varying index of refraction $\mu(\vec{r})$ is found by solution of the equation[7]

$$\frac{d}{ds}[\mu(\vec{r})\frac{d\vec{r}}{ds}] = \nabla\mu(\vec{r}) \qquad (1)$$

where $\vec{r}$ is the position vector of a point on the ray and s is the ray's path length. Further, a variation $\delta\mu$ of the index of refraction due to pump beam heating of the gas is

$$\delta\mu/(\mu - 1) = -\delta T/T \qquad (2)$$

where the energy transfer is assumed to be isobaric. For a crossing angle α and a pump beam of characteristic width w = 1 mm, one can show that astigmatic probe beam deflections of the measurable magnitude range of 10-100 μm are produced for $\delta T/T = 0.1$ and $|\delta\mu/\alpha^2| \ll 1$. The pump laser energy E_p required to produce $\delta T/T = 0.1$ can be estimated for isobaric binary gas mixture heating by the relation

$$\delta T/T = X_1(M_1/M_2) \cdot (\gamma - 1)$$
$$\cdot [\sigma_{abs}(1)/A_B] \cdot (E_p/k_B \cdot T) \tag{3}$$

where X_1 is the mole fraction of the trace species "1" which has the absorption cross section $\sigma_{abs}(1)$. The molecular weight of species i is M_i, γ is the specific heat ratio of the mixture (and $\gamma \simeq \gamma_2$) as shown and k_B is Boltzmann's constant. The pump laser is of energy E_p and cross-sectional area A_B. Using Eq. (3) and $\sigma_{abs}(1)$ for the broadened resonant absorption cross section for the Na $3S_{1/2} \mapsto 3P_{1/2}$ transition in an N_2 host gas at 300 K, one finds that approximately 1 ppm of Na will produce $\delta T/T = 0.1$ for a 1-mm-diam. pump beam of 1 mJ energy.

The use of pulsed pump sources is required to maximize the gas heating and to minimize convective heat loss for a gas sample of flow speed v_∞. If, relative to initiation of the laser pulse, τ_f is the time required to complete the deflection measurement during a single laser pulse cycle where the pulse width is τ_p, it is clear that τ_f and $\tau_p \ll \tau_c$, where τ_c is the characteristic convective flow time w/ v_∞. Further, since the isobaric thermal diffusion wave solution is chosen to represent the effects of heat addition to the gas, it is necessary that the acoustic wave solution be negligible. If, relative to initiation of the laser pulse, the deflection measurement is begun at time τ_i, the acoustic wave can be neglected if $\tau_i \geq 10\tau_a$ (Ref. 8), where the characteristic acoustic time $\tau_a = w/c_\infty$ and c_∞ is the mixture sound speed at the point of the measurement. Combining the above inequalities, one finds that the Mach number M_∞ of the flow must satisfy the approximate inequality $M_\infty \leq 0.1$ if the thermal diffusion wave solution is to accurately represent the time-space variation of the gas temperature $T(\vec{r}, t)$. Extension of the technique to supersonic flow regimes is, of course, possible if isobaric heating is not assumed and one recognizes the increased sensitivity of the results to collisional relaxation times. Also, it should be noted that no firm restriction applies to the Peclet number $Pe = \tau_D/\tau_c$, where τ_D is the characteristic conduction time $w^2/4\kappa$ and κ is the thermal diffusivity of the mixture. Finally, assuming the existence of a characteristic time T_v for transfer of the internal excitation energy to the translational mode of the mixture, it is clear that, to minimize convective heat loss, $\tau_v \ll \tau_c$.

Having defined the critical parameters of LIST and shown the basic feasibility for subsonic flows or static gas samples, detailed calculations were performed for the Na/N$_2$ mixture to evaluate the time-dependent effects on the probe beam deflection of thermal wave spreading and collisional energy transfer processes. For this evaluation, a spectral and spatial Gaussian-shaped laser pump beam was assumed to pump predominantly either the $3^2S_{1/2}$ to $3^2P_{1/2}$ or $3^2P_{3/2}$ transitions of Na at wavelengths 589.6 and 589.0 nm, respectively. Collisional energy transfer processes considered for Na included the exchange process $3P_{1/2} \rightleftarrows 3P_{3/2}$ and quenching of both the $P_{1/2}$ and $P_{3/2}$ levels. N$_2$ pumping resulted from the Na to N$_2$ electronic to vibrational (E-V) transfer, and the non-Boltzmann N$_2$ distribution relaxed to near-Boltzmann configurations by vibration-vibration (V-V) and vibration-translation (V-T) processes. Figure 2 shows the relevant energy levels of the system. A detailed discussion of the rate kinetics model, collisional rate constants, and method of numerical solution of the coupled rate equations for the level specific number density distributions is presented in Ref. 6. The primary differences between the rate kinetics models of this work and Ref. 6 are that in this work a Gaussian spectral shape was assumed for the pump laser and the absorption line shapes are represented by Voigt profiles. The collision broadening parameters for the absorption line shapes were calculated using the cross-section values given in Ref. 6. Finally, the temporal pulse shape for the laser source spectral energy density ρ_ν was of the form of the difference of two Heaviside functions

$$\rho_\nu(t) = \rho_{\nu o}\{\tanh(\tau_p t/\tau_p) - \tanh[W(t - \tau_p)]\} \qquad (4)$$

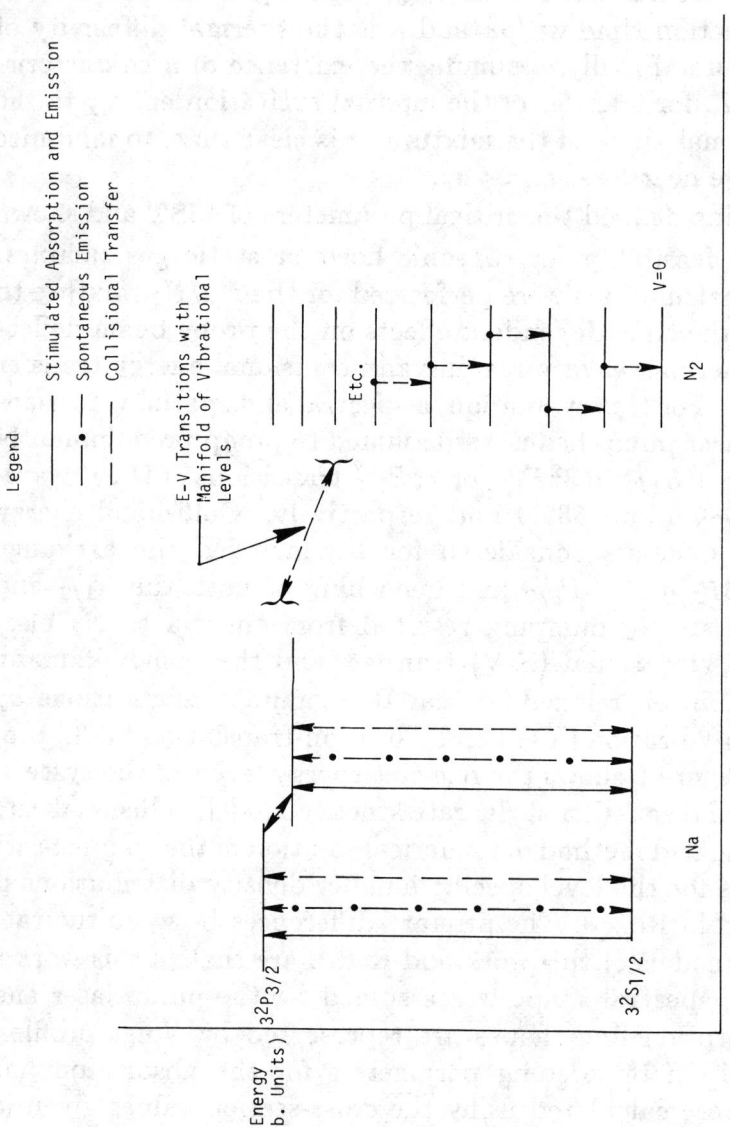

Fig. 2 Partial energy level diagram of Na and N_2.

where $\rho_{\nu o}$ characterizes the pulse height and τ_p the FWHM of the pulse duration. Variation of the parameter W provides widely ranging pulse shapes and CW sources are provided by increasing τ_p to very large values.

The attenuated laser pump beam intensity was assumed to be of the form $I_p(z) = I_{po} \exp(-\bar{\alpha}z)$, where z is the direction of propagation of the collimated beam, and a train of k pulses, each of FWHM pulse width τ_p equally spaced in time, irradiates the sample. The method of solution for the temperature increase δT of Bailey et al.[9] in their study of CO_2 laser thermal blooming was employed. Bailey's solution is for a single, delta-function pulse; modifications were made to use the normalized Heaviside Function, $\bar{\rho}_\nu(t_i^{(j)})$ as an envelope for a series of delta-function pulses. The temperature rise for the jth pulse at its nth interior time, $t_n^{(j)}$, is given by

$$\delta T_n^{(j)} = (1/\tilde{C}_p \tilde{\rho} \hat{\tau}_v) \cdot (\bar{\alpha}/2\pi w) \cdot \exp(-\bar{\alpha}z) \cdot E_{po}$$

$$\{\sum_{i=1}^{n} \bar{\rho}_\nu(\hat{t}_i^{(j)}) \hat{t}_i^{(j)} \exp\{\frac{-(\hat{t}_i^{(j)} - t_{i-1}^{(j)})}{\hat{\tau}_v}\}$$

$$\int_o^{\hat{t}_n^{(j)}} d\hat{t}' \cdot (\frac{1}{\hat{t}_n^{(j)} - \hat{t}' + 1})$$

$$\cdot \exp(\frac{-\hat{t}'}{\hat{\tau}_v}) \exp(\frac{-\hat{r}^2}{\hat{t}_n^{(j)} - \hat{t}' + 1})$$

where the nth interior time is simply the relation

$$t_n^{(j)} = t - (j-1)\tau_p$$

Finally, for pulses, such that $1 \leq j \leq k$, the total temperature rise is

$$\delta T(t) = \sum_{j=1}^{k} \delta T_n^{(j)} \qquad (5)$$

The incident pulse energy is E_{po} and the molar heat capacity and molar density are $\tilde{C}_p$ and $\tilde{\rho}$, respectively. The radial coordinate r is divided by w, the $1/e$-radius of the Gaussian-shaped pump beam, to yield $\hat{r}$, and the characteristic thermal conduction time τ_D is used to make all time parameters nondimensional; e.g., the time of the nth pulse, t_n, is given by $\hat{t}_n \cdot \tau_p$. Further, a characteristic vibrational relaxation time τ_v is assumed to exist, and $\tau_v = \hat{\tau}_v \cdot \tau_D$, and, as will be shown later, τ_v is obtained from the numerical solution of the coupled rate kinetics code. Finally, the time-dependent absorption coefficient α_{ij} for the $i \mapsto j$ transition in Na is obtained from the following equation

$$\alpha_{ij} = A_{ji} \cdot (\lambda_{ij}^2/8\pi) \cdot g(\nu \cdot [n_i - (\frac{\overline{g}_i}{\overline{g}_j}) \cdot n_j] \cdot I_\nu, \qquad (6)$$

where A_{ji} is the spontaneous emission coefficient for the absorption wavelength λ_{ij} and n_i and g_i are the number density and degeneracy, respectively, of level i. $g(\nu)$ is the line-shape factor and $I_\nu = c\rho_\nu$, where c is the speed of light. For Eq. (5), α was taken to be the sum of the absorption coefficients for excitation of the $^2P_{1/2}$ and $^2P_{3/2}$ levels.

With $\alpha(t)$ provided by the rate kinetics calculation and Eq. (6), $\delta T(\vec{r}, t)$ was obtained by numerical integration, and, for each time-step, the deflection of the probe beam was determined by numerical integration of the component equations of Eq. (1). For the deflection, as well as the rate kinetics model, the GEAR multistep method[10] was used.

For reasons best not discussed in this paper, the incident probe beam was elliptically-shaped with semi-major and minor diameters of 1.06 and 1.0 mm, respectively, and the angle of intersection with the pump beam was approximately 20°. The Gaussian diameter of the pump beam was 2 mm, the interaction volume was approximately 20 mm^3, the FWHM pump pulse duration was 1 msec with a 50% duty cycle. The incremental temperature field $\delta T(\vec{r}, t)$ and the probe beam deflection $\Delta \vec{r}(t)$ were calculated for a range of Na/N$_2$ number density values for $T = 300$ K. As expected, with n(Na) $\lesssim 10^{10}$ cc^{-1}, due to the extremely large values of τ_v for N_2, insignificant beam heating occurred with practical pump laser powers. Consequently, the mixture was assumed to consist of Na/N_2 and 10 ppm of nonabsorbing H_2O vapor which decreased τ_v of N_2 by a factor of 1000, and measurable deflections resulted. For this ternary mixture, calculations were performed for Na $3^2P_{1/2}$ pumping for $n(N_2) = 10^{19} cc^{-1}$, and $n(H_2O) = 10^{14} cc^{-1}$. The range of sodium number density was $10^7 \leq n(Na) \leq 10^{13} cc^{-1}$ and the power, P_p of the pump laser of 1 GHz spectral width was varied over the range of 10^{-2} - 10^2 W. Figure 3 shows the time-dependent values of $n_{Na}(3^2P_{1/2}, 3^2P_{3/2}$ and $3^2S_{1/2})$ for a sodium number density of $10^{10} cc^{-1}$ and $P_p = 1$ W. τ_v was determined using the vibrational relaxation equation for the vibrational energy $E_v(t)$

$$dE_v/dt = -(1/\tau_v)(E_v - E_{ve}), \qquad (7)$$

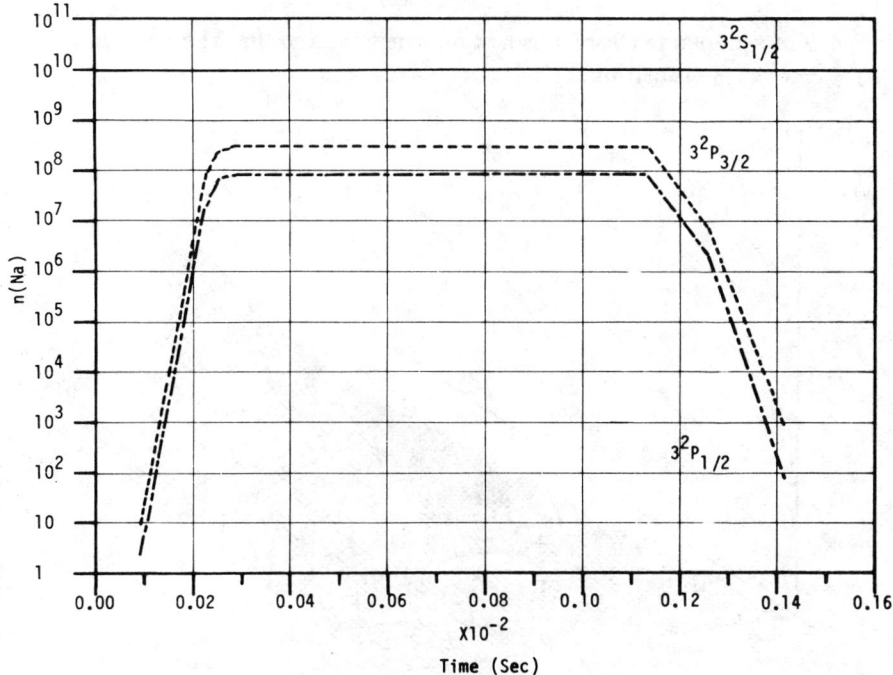

Fig. 3 Variation with time of sodium number density n values.

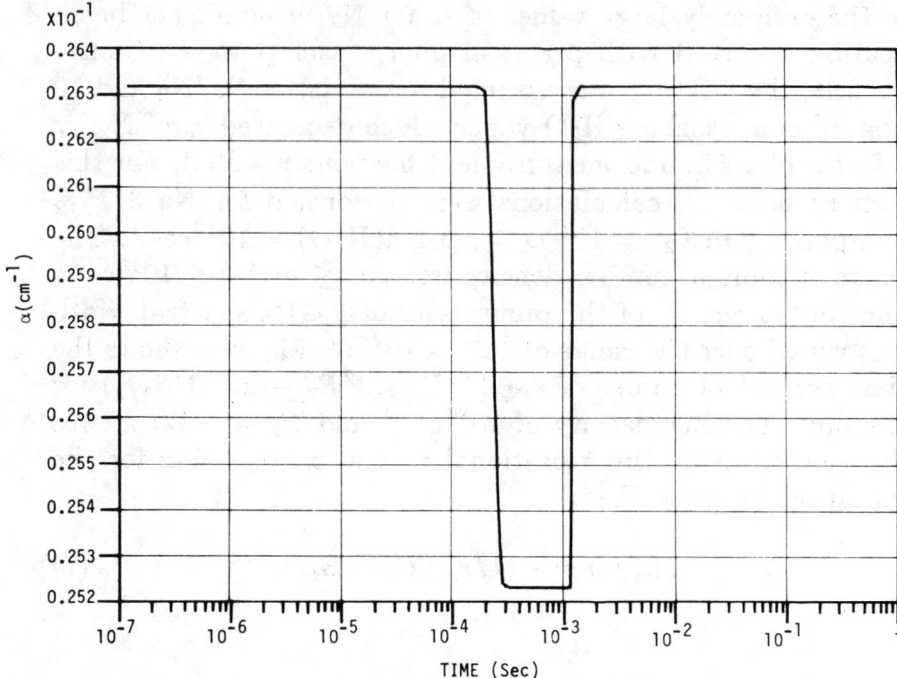

Fig. 4 Time dependence of absorption coefficient α for the sodium $3^2S_{1/2} \mapsto 3^2P_{1/2}$ transition.

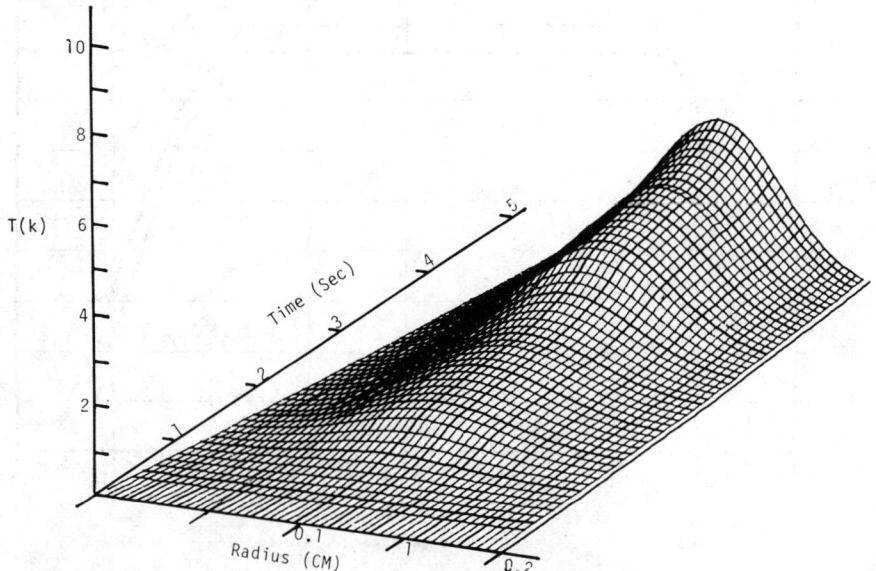

Fig. 5 Temperature field space-time profile for laser beam heating.

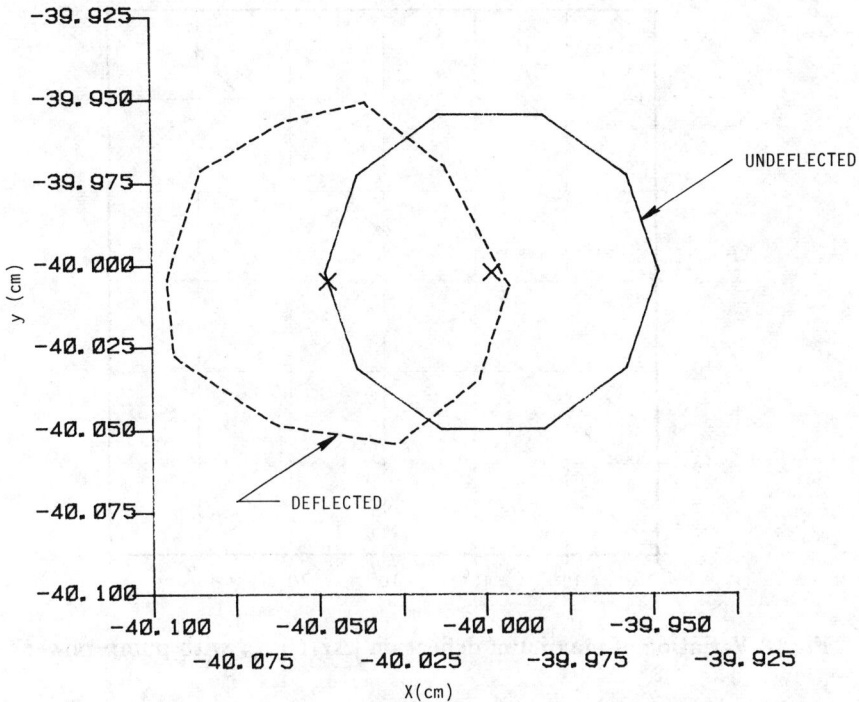

Fig. 6 Deflection lows of probe beam.

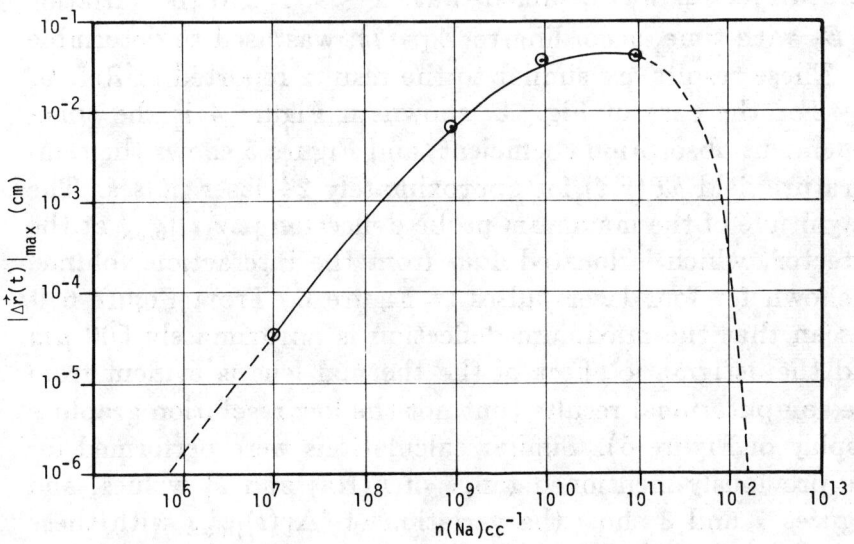

Fig. 7 Variation of maximum deflection $|\Delta \vec{r}(t)|_{max}$ with sodium number density n(Na).

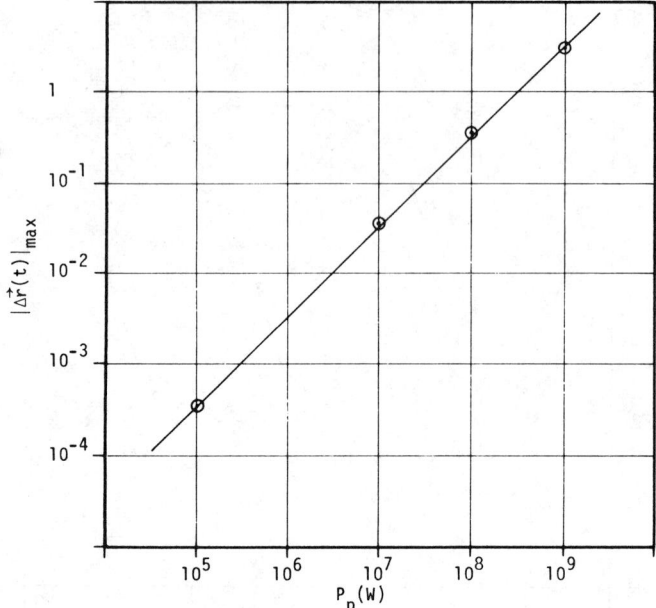

Fig. 8 Variation of maximum deflection $|\Delta \vec{r}(t)|_{max}$ with pump power.

where E_{ve} is the equilibrium vibrational energy at the gas temperature, T. It was assumed that $\delta T << T$ and the variation of E_v with time, according to Eq. (7), was used to determine τ_v. These results are similar to the results reported in Ref. 6.

For the case of Fig. 3, shown in Figure 4 is the time-dependent absorption coefficient, and Figure 5 shows the temperature field $\delta T(\vec{r}, t)$ for approximately $2\frac{1}{2}$ laser pulses. The magnitude of the maximum probe deflection $|\Delta \vec{r}(t)|_{max}$ at the detector, which is located 1 m from the interaction volume, is shown for four laser pulses in Figure 6. From Figure 6 it is seen that the maximum deflection is aproximately 500 μm and the astigmatic effect of the thermal lens is evident from the computational results (but not the low resolution graphics display of Figure 6). Similar calculations were performed for the previously-mentioned range of $n(\text{Na})$ and P_p values, and Figures 7 and 8 show the variation of $|\Delta r(\vec{t})|_{max}$ with these parameters, again, for four laser pulses. Figure 7 shows that, if 10 μm is selected to be the minimum measurable deflection, for a fixed power $P_p = 1$ W the minimum measurable n(Na) is approximately 10^8cc^{-1}. Above n(Na) = 10^{11}cc^{-1}, nonlinearity precludes measurements. Consequently, the measurement range of n(Na) for $P_p = 1$ W is approximately 10^8 to 10^{11}cc^{-1}. Similarly, from Figure 8 for $n(\text{Na}) = 10^{10}\text{cc}^{-1}$, it is seen that $|\Delta r(t)|_{\max}$ is linear with P_p for $0.01 \leq P_p \leq 10$ W. Higher values of P_p grossly violate the approximation $\delta T << T$.

Conclusions

From the results presented in Figures 7 and 8 and those of similar but unreported calculations for pumping of the $3^2P_{3/2}$ level, it has been shown that, for slowly flowing or static gas samples, sodium concentrations in the range of 10^8 to 10^{11}cc^{-1} in an N_2 environment of 10^{19}cc^{-1} with H_2O number density no less than 10^{14}cc^{-1} can be measured using LIST and a 1-W pump beam. Because of the extremely long τ_v value of N_2, trace quantities of H_2O were required to decrease the τ_v/τ_c ratio and to increase the heating of the gas. The restriction to subsonic flows can, in principle, be removed for LIST if the currently neglected acoustic wave solution is utilized, and this improvement is in progress.

Finally, although this work concerned the evaluation of LIST for measurement of trace gas concentrations, the approach and deflection results are of more general interest. Specifically, these results yield information regarding the deflection of probe beams of any technique which intersects pump beams in an absorbing medium, and the deflections, in addition to resulting in positional errors, can be of importance for high resolution, narrow slit width spectroscopic studies.

References

[1] Levenson, M. D., Introduction to Nonlinear Laser Spectroscopy, Academic Press, New York, 1982.

[2] Omenetto, N., Analytical Laser Spectroscopy, John Wiley and Sons, New York, 1979.

[3] French, B. J., "Continuum Source Molecular Beams," AIAA Journal, Vol. 3, March 1965, pp. 993-1000.

[4] Pao, Y-H, Optoacoustic Spectroscopy and Detection, Academic Press, New York, 1977.

[5] Hermann, W., and Pohl, D. W., "Trace Analysis in Gases by Laser-Induced Schlieren Technique," Infrared Physics, Vol. 18, August 1979, pp. 455-459.

[6] Campbell, D. H. and Lewis, J. W. L., "Detailed Temporal Behavior of Laser-Excited Sodium Tracer in Nitrogen and Application to Nitrogen Number Density Measurements at Low Densities," Applied Optics, Vol. 20, December 1981, pp. 4102-4109.

[7] Born, M. and Wolf, E., Principles of Optics, Pergamon Press, New York, 1975.

[8] Fader, W. J., "Density Perturbations Caused by Weak Absorption of a Laser Pulse", Journal of Applied Physics, Vol. 47, May 1976, pp. 1975-1978.

[9] Bailey, R. T., Cruickshank, F. R., Pugh, D., and Johnstone, W., "Pulsed Source Thermal Lens Part I - Theoretical Analysis" Journal of the Chemical Society Faraday Transactions II (GB), Vol. 76, December 1980, pp. 633-647.

[10] Gear, C. W., Numerical Initial Value Problems in Ordinary Differential Equations, Prentice-Hall, Englewood Cliffs, N.J., 1971.

Use of Laser-Induced Fluorescence for Fundamental Gas-Phase Kinetic Measurements

Arthur Fontijn*
Rensselaer Polytechnic Institute, Troy, New York

Abstract

The uses of laser-induced fluorescence (LIF) for rate measurements on reactants and products, as well as for product species identification and product state determination, are illustrated by the use of examples. LIF studies discussed are preferentially chosen from work on single reactions at elevated temperatures. LIF is compared to other techniques for observations on combustion intermediates. Reactions of refractory species are treated, followed by reactions of O and OH with hydrocarbon molecules and some reactions of other radicals formed in flames. Finally, some work on reactions of vibrationally and electronically excited species is reviewed. Extensive literature references are provided for detail.

Introduction

The basis of fluorescence measurements is the pumping of a lower state by a light source and observation of emission from the upper state thus produced. The advent of lasers as high intensity pump radiation sources is one major factor that has led to the current prominence of fluorescence measurements. The other major, concurrent, factor is the rapid development in deconvolution of the electrical signals from photomultiplier tubes, the common type of emission detection device used. Indeed in a very

---

Presented as Paper 83-1469 at the AIAA 18th Thermophysics Conference, Montreal, Canada, June 1-3, 1983. Copyright © American Institute of Aeronautics and Astronautics, Inc.,1984. All rights reserved.
*Professor, Department of Chemical Engineering and Environmental Engineering.

few years[1-3] laser-induced fluorescence (LIF) has developed into one of the major techniques for the measurement of concentration variation of neutral reaction intermediates (atoms and free radicals) in flames.† However, with few exceptions, even simple laboratory flames are too complicated to yield reaction kinetic data directly. Rather, reactors for observations on isolated elementary reactions have to be used, chiefly laminar fast-flow reactors and real-time photolysis reactors.[4] The use of LIF to obtain kinetic information on such systems is the subject of this paper. Together with species concentration profiles from flames these kinetic data and appropriate models are leading to quantitative descriptions of combustion chemistry.

The very high sensitivity which can be achieved with LIF is a major reason for its frequent use. In flames a species such as OH can be monitered at concentrations[1,5] down to $\approx 10^{10} - 10^{11}$ cm^{-3}, however in the atmosphere OH concentrations as low as $10^6 - 10^7$ cm^{-3} have been measured.[6] The majority of important flame intermediates are small molecules. LIF is particularly suited to their observation. Combustion intermediate species which have been observed using LIF include,[6-9] AlO, BCl, BO, BO_2, BaO, O, N, OH, CM, CN, C_2, C_2O, CH_2O, CH_3O, C_2H_3O, CO, NH, NH_2, NO, NO_2, NCO, NO, S_2, SH, SO, SO_2, C_2, CS_2, C_2O, C_3, and HCN. LIF is less sensitive for larger molecules since the greater number and closer spacing of energy levels results in more effective quenching and more widely spread spectra.[5,10] Consequently, other techniques especially mass spectrometry, which has a sensitivity limit of $\approx 10^{11}$ cm^{-3} may be preferred. However, mass spectrometry cannot readily distinguish between species of the same mass, nor can it give detailed information on internal excitation states; moreover it is an intrusive technique.

A fundamental limitation of LIF is that the excited state must emit, rather than dissociate or undergo radiationless transitions. As can be seen from the above list, most important flame intermediates are not subject to this problem. A significant exception is HO_2, which has, however, been studied using laser magnetic resonance in discharge-flow systems.[10]

---

†Short-pulsed lasers are most commonly used in LIF work. The development of data processing techniques referred to here pertains to the resulting short fluorescence pulses. Techniques for handling fluorescence data from continuous light sources well predate the advent of lasers, which also have found some use in kinetic studies. See further below.

Like other major techniques LIF has been used in so many investigations that an exhaustive review in a single paper is not practical. Instead some of the more prominent fundamental kinetic application methods of combustion interest are illustrated using somewhat arbitrarily chosen examples. These have been selected preferentially from work at elevated temperatures, which are of more direct importance to combustion problems than room-temperature studies. There is no principle difference between these techniques and those used at or nearer room temperature. Usually the lower temperature experiments are more facile; their literature is more extensive.

Refractory Species Oxidation Reactions

In many practical combustion systems refractory species are present either as a result of unwanted, but not economically removable, impurities in the fuel or as an integral part of it, such as in metalized solid propellants. Figure 1 shows a form of the flow apparatus developed for study of refractory species oxidation reactions over the 300 - 1900 K temperature range.[11,12] In the source section of this HTFFR (high-temperature fast-flow reactor) metal atoms Me are produced by vaporization. Refractory monoxide or monohalide radicals MeY can be produced by chemical reaction, usually also at elevated temperatures.[12-14] Rapid entrainment and transport by the

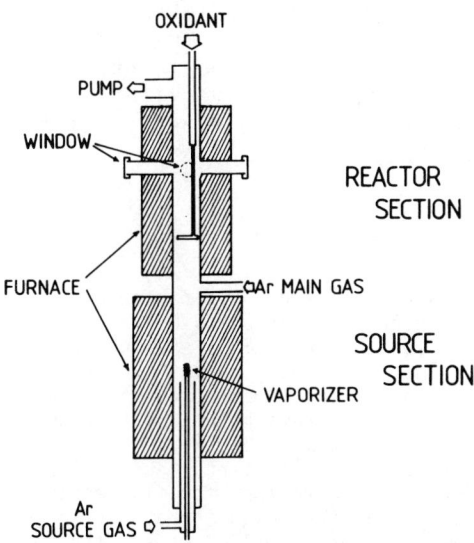

Fig. 1   Schematic of high-temperature fast-flow reactor.

bath gas Ar introduces these Me or MeY species into the reactor section, where the oxidant OX is introduced through a movable inlet under conditions such that [OX]>>[Me] or [MeY]. Downstream, at the window plane, the Me or MeY concentration, relative to that in the absence of oxidant, is measured as a function of time (proportional to distance), [OX], and total pressure. For atomic species a hollow cathode lamp of the element of interest often suffices for these relative concentration measurements, which are then made in absorption at the window opposite the lamp or in fluorescence at the window at 90 deg from the light source.[12] For measurements on radicals LIF is needed. Figure 2 gives a schematic presentation of the arrangement we are currently using for work on BCl. The $A^1\Pi-X^1\Sigma(0,0)$ transition is pumped with the frequency doubled radiation from an excimer pumped dye laser. With the aid of the energy meter the resulting fluorescence intensitites are normalized for variation in the laser intensity. The time-integrated fluorescence pulses‡ are recorded by the transient digitizer and sent to a microcomputer for kinetic analysis. In most fast-flow reactor LIF work to date boxcar integrators have been used for measurement of the time-integrated fluorescence pulses. The transient digitizer, a more recent development similar to storage oscilloscopes, simultaneously shows the intensity decrease with time, i.e., the pulse shape. This yields experimental lifetimes τ (e-folding times) of the emitting states and allows discrimination against accidentally excited species or states, which in general will have different e-folding times.[15,16] The experimental lifetimes can be used to obtain quenching rate coefficients k_Q and radiative lifetimes τ_{rad} as discussed in the section titled Kinetics of Excited States Produced by Laser Pumping.

---

‡The length of laser pulses used is on the order of 10^{-8}-10^{-6} s, depending on the type of laser used. Flash lamp pumped dye laser have longer pulses than Nd: YAG, excimer, or nitrogen laser pumped dye lasers. To discriminate against scattered light one should use a laser pulse of duration shorter than the life time of the resulting excited state. Pumping of upper states of free radicals with radiative lifetimes $\approx 10^{-5}$-10^{-4} s, i.e., metastable states, involves absorption coefficients too small to yield adequate excited radical con centrations for fluorescence to be measured. However, LIF has been used for atomic species with upper state lifetimes as long as $\approx 10^{-2}$ s, see the section entitled Kinetics of Excited States Produced by Laser Pumping.

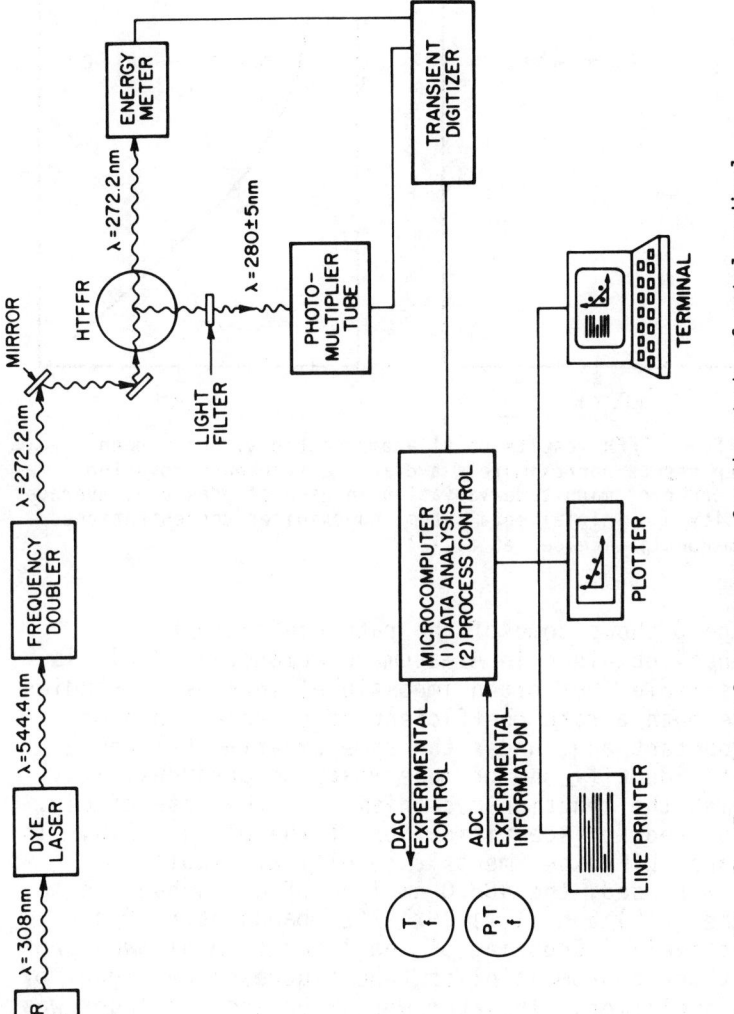

Fig. 2 Experimental arrangement for the study of metal radical reactions using laser-induced fluorescence. ⁓ radiation; —— electrical connections (T = temperature; f = volume flow rate; P = pressure; DAC = digital to analog converter; ADC = analog to digital converter).

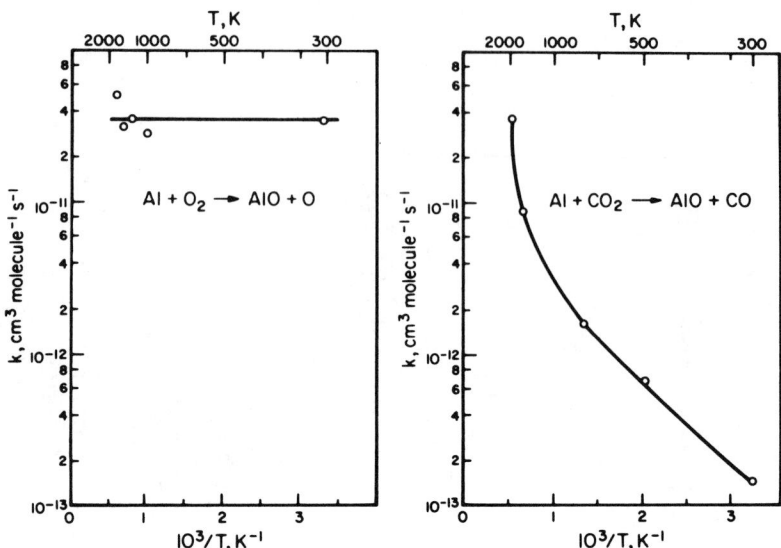

Fig. 3 Some HTFFR results on Al atom reactions. Each open circle represents approximately a dozen measurements covering about an order of magnitude variation in each of pressure, average gas velocity and initial metal atom and oxidizer concentrations. Data from Fontijn, Felder et al.[17,18]

Figure 3 shows some of the rate coefficient measurements§ obtained in Al-atom reactions.[17,18] It is highly desirable (but often impossible) in kinetic studies to measure both a rate coefficient of disappearance of initial reactant and, under the same experimental conditions, to identify one or more reaction products, i.e., to establish the reaction mechanism. In the case of Fig. 3, LIF was used for identification of the product AlO. In some of these LIF experiments a readily available Ar^+ laser was used, the 488.0-nm line of which happens to overlap the (0,0) and (1,1) $B^2\Sigma^+$-$X^2\Sigma^+$ bands at K=49 and 56, respectively. Chopping of the laser beam allowed processing of the photomultiplier tube fluorescence signal by a lock-in amplifier. In later work a pulsed dye laser was used. With it the pump radiation wavelength could be selected and a boxcar integrator was used for signal processing. As a boxcar or a transient digitizer typically registers only during the short time intervals of the

---

§It may be seen that these reactions show non-Arrhenius behavior, in common with many reactions with small intrinsic energy barriers when observed over wide temperature ranges. For an extensive discussion of these phenomena see Fontijn and Zellner.[4]

fluorescence decay, the pulsed laser experiment offers better background rejection.

Since in this work both the [Al] and [AlO] were from relative measurements, mere identification does not demonstrate quantitative conversion. However, such a 1:1 relation was further indicated by experiments in which it was shown that both reactions yielded the same ratios for the maximum AlO fluorescence intensity over the prereaction Al concentration.[19] Observation of the AlO fluorescence also allowed determination[17] of $k(T)=2\times10^{-13} \exp(+333/T)$ cm^3 molecule^{-1} s^{-1} for

$$AlO + O_2 \rightarrow AlO_2 + O$$

This much slower reaction could be observed by using higher [O_2] than used for AlO formation, see Fig. 4. In this case it was not possible to identify the product AlO$_2$, the electronic spectrum of which is not known. However, in a similar experiment at room temperature, we succeeded in identifying both BO and BO$_2$ in the reaction[13,20]

$$BO + O_2 \rightarrow BO_2 + O$$

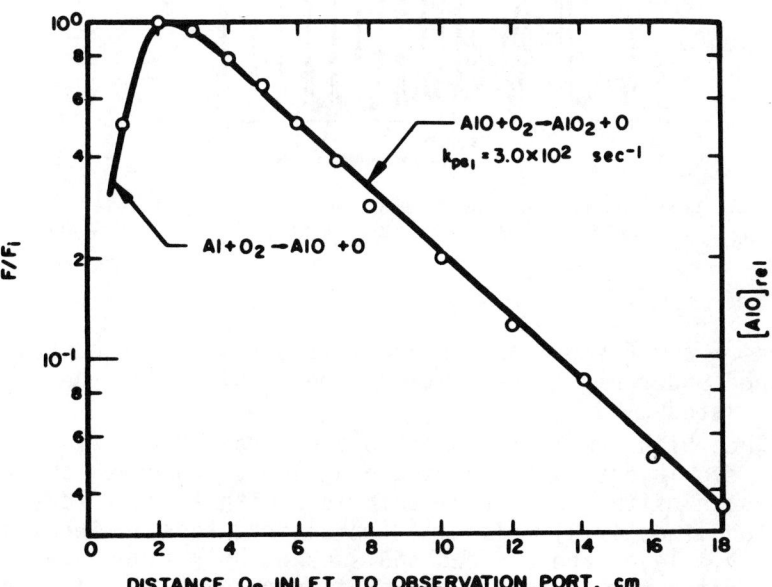

Fig. 4   AlO concentration profile. From Fontijn et al.[17] Copyright 1977, the Combustion Institute.

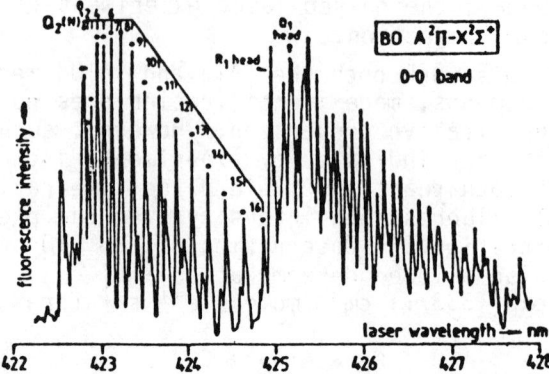

Fig. 5 Laser excitation spectrum of the 0-0 band of BO (A-X) at 10 pm laser bandwidth. From Clyne and Heaven.[20] Copyright 1980, North Holland Publishing Co.

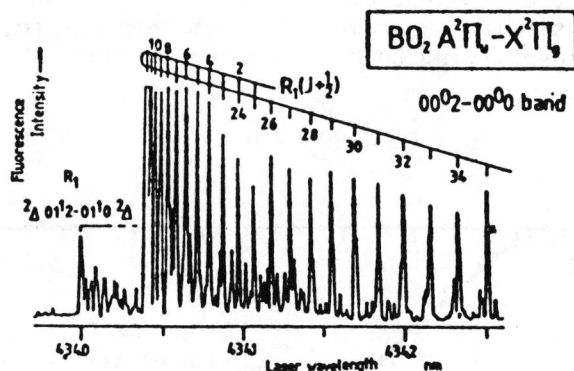

Fig. 6 Laser excitation spectrum of BO_2(A-X) at 1 pm laser bandwidth. From Clyne and Heaven.[20] Copyright 1980, North Holland Publishing Co.

Figures 5 and 6 show the respective fluorescence spectra, obtained under higher resolution then required for the kinetic study.[20]

Such high resolution spectra are obtained by using pulsed dye lasers in a slow scanning mode. Whenever the wavelength of the laser beam coincides with a specific rotation line, $(v,j) \rightarrow (v',j')$ transition, those molecules in the v,j level are excited and subsequently fluoresce. By measuring the intensity of the fluorescence signal as a function of laser wavelength it is possible to deduce the relative v,j population distribution from a knowledge of Frank-Condon and detection wavelength response factors.

If such work is carried out under sufficiently low-pressure conditions that no relaxation occurs then not only the product species can be identified but also their nascent distribution,π thus yielding important reaction dynamic information. This approach has been followed extensively by Zare's school in molecular beam scattering experiments,[15,21-23] Fig. 7. The highest v,j levels populated also allow determination of a lower bound to the dissociation energy of the product species. In their work, the LIF observations are particularly useful for electronic ground-state products or metastable electronically excited products. Short-lived electronically excited species produce chemiluminescence on the time scale of the experiment, which allows deduction of similar dynamic and thermochemical information.[12,21-23] Another interesting application of these beam-type experiments is that they have allowed characterization of metal clusters formed in supersonic free jet expansions[16,23,24] and observations on reactions of these clusters.[25] Under certain conditions, the work has also allowed determination of reaction cross sections[26] and Arrhenius type activation energies over narrow effective Boltzmann temperature ranges. Depending on the reaction studied these ranges were centered as low as about 800 K and as high as about 2300 K.[27,28]

Hydrocarbon Molecule Reactions with O and OH

Reactions of O atoms and OH radicals with hydrocarbons form the largest group of recent investigations using fluorescence detection. These reactions are both of combustion and atmospheric[29] interest and most of the work has been done at room temperature or within ≈ 100 K of room temperature. Here we will again emphasize the relatively few studies at temperatures of more direct combustion interest. In that work the real-time photolysis technique with fluorescence monitoring (RTPF) has played the major role. $[O]_{relative}$ and $[OH]_{relative}$ are measured in the presence of excess hydrocarbon. There are a number of variants of the RTPF technique depending on the photolysis and fluorescence excitation methods used, though the basic idea remains the same. Figure 8, from the recent work of Tully[30] on OH + C_2H_4, illustrates the prin-

---

π Nascent distribution: electronic, vibrational, rotational distribution resulting from the species formation reaction, i.e., the distribution before any intermolecular or intramolecular relaxation has taken place.

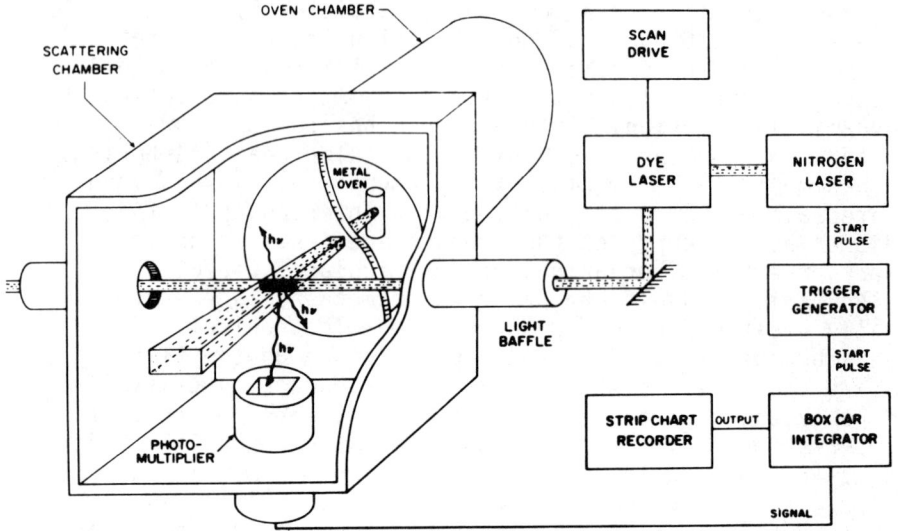

Fig. 7 Schematic of a beam+gas scattering apparatus with laser-induced fluorescence detection. From Pruett and Zare.[15] Copyright 1975, The American Institute of Physics.

ciple. A short photolysis pulse from the excimer laser (or in other work a flash lamp or pulsed lamp) results in a radical concentration [A] in the range of $10^{10} - 10^{11}$ cm^{-3}. The decrease of [A], on a timescale of $10^{-3} - 10^{-1}$ s, in essentially the same volume element of its formation, is monitored by LIF using quasi-cw laser radiation and registered on the multichannel analyzer as shown. Since the exciting radiation is essentially continuous, the rate of decrease of the fluorescence intensity with time in these experiments is proportional to that of [O] or [OH] and is not related to the lifetime of the upper state.** The compound to be photolyzed and the hydrocarbon reactant are premixed with the bath gas and flow slowly through the reactor to provide a fresh mixture for each photolysis pulse. Typically a few hundred pulses are

---

**Because of the relation of τ_{rad} to the Einstein transition moment and the absorption coefficient, τ_{rad} is one factor contributing to the absolute intensity. Other factors involved are k_Q, concentration of the absorber and intensity of the laser pulse.

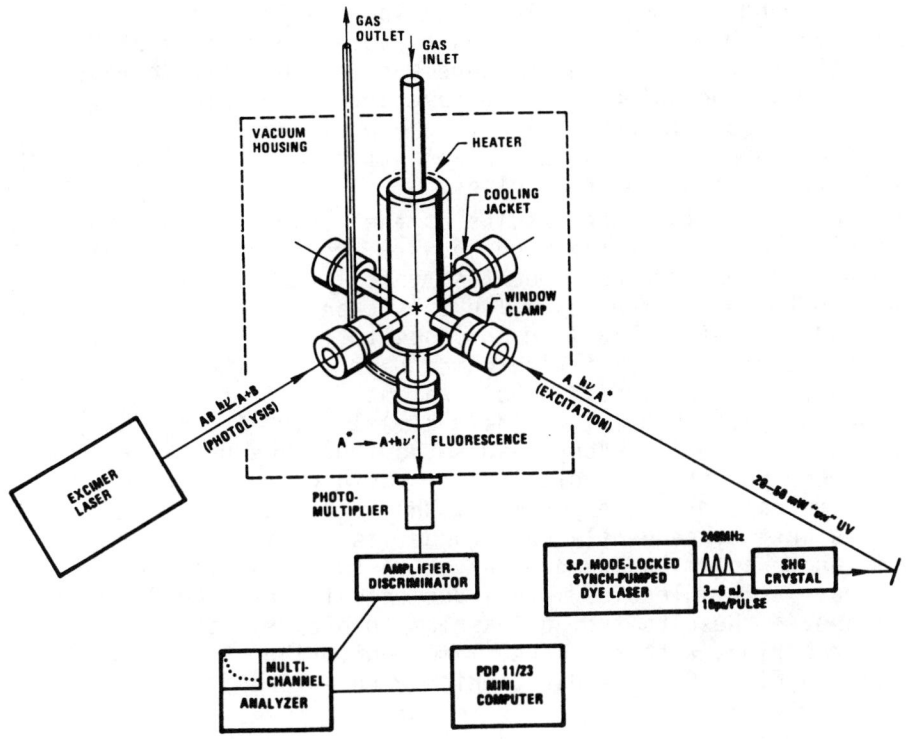

Fig. 8 RTPF apparatus. From Tully.[30] Copyright 1980, North Holland Publishing Co.

required to obtain an accurate decay plot; total pressures may vary in the range of 10^{-2} - 1 bar.

In other work, cw microwave discharge flow lamps have been used to provide the O or OH resonance radiation. Such lamps are excellent line sources for O and other atoms, such as N, H, Cl, etc.[31] However, while OH lamps have been used successfully up to ≈ 1150 K,[32] pulsed lasers are to be preferred for measurements on radicals, because of the resulting much higher intensities and better background rejection. This is especially true when working at temperatures above ≈ 1150 K, where radiation from reactor walls becomes an interference factor, even at wavelengths near 300 nm where OH is measured. But also at lower temperatures fluorescence detection of radicals can be subject to interference by the chemiluminescence accompanying hydrocarbon oxidation reactions.[33] Several groups[34,35] have used pulsed laser radiation to generate the LIF for radical decay measurements. Operating in that manner only one point on the decay curve is obtained per photolysis pulse. Generation of the whole curve then requires measurements with variable time delay between the photolysis and diagnostic pulses.

Figure 9 gives the results of the above mentioned OH + C_2H_4 study.[30] The reaction shows a pronounced negative temperature dependence. Such is typical for reactions leading to adduct formation (in this case HOC_2H_4), when the adduct preferentially decomposes to give the original reactants and has a lifetime which decreases with increasing temperature.[4] The straight part of the Arrhenius plot, Fig. 9, yields a $k(T)=1.7 \times 10^{-12} \exp(460/T)$ cm^3 molecule^{-1} s^{-1}. Above 440 K the [OH]-time profiles were nonexponential, indicative of rapid OH reformation on the timescale of the experiment; the corresponding k-values must consequently be regarded as estimates. Similar behavior has been observed in the reactions of OH with aromatic molecules, e.g., the reaction between OH and toluene.[32] Results from a flashlamp photolysis study of this reaction, with resonance lamp monitoring of OH, are given in Fig. 10. The data, which were extended to a

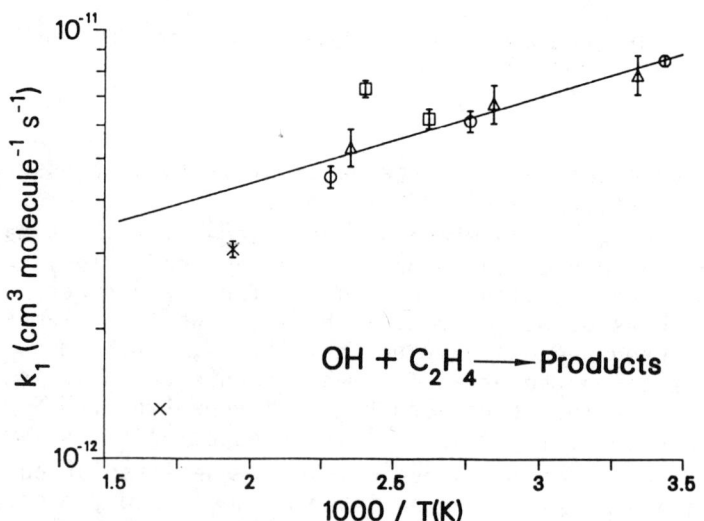

Fig. 9 Arrhenius plot for the reaction OH + C_2H_4 → Products. From Tully.[30] o, exponential [OH] decays; x, nonexponential [OH] decays; Δ, ☐ are from work quoted in Ref. 30. Copyright 1983, North Holland Publishing Co.

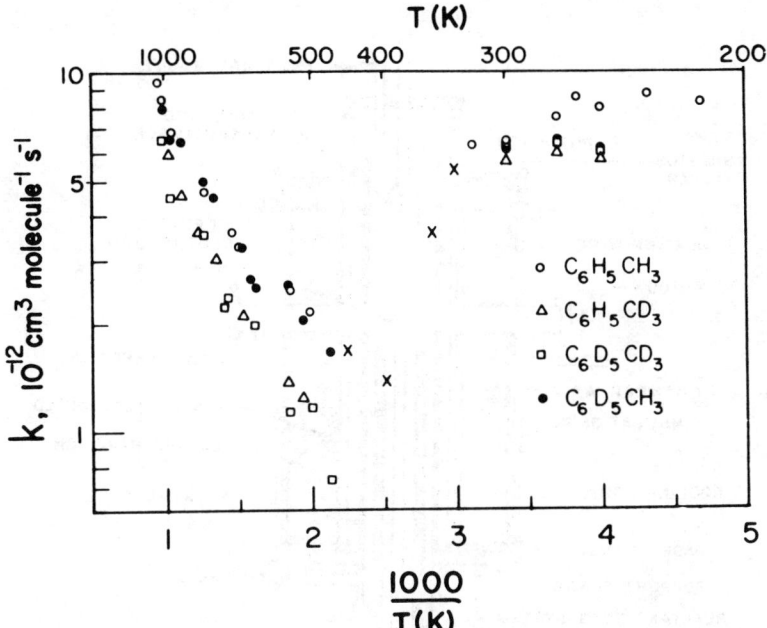

Fig. 10 Arrhenius plot for reactions of OH with four isotopically labeled toluenes. From Tully, Ravishankara et al.[32] Copyright 1981, The American Chemical Society.

higher temperature (1150 K) than those of the OH + C_2H_4 work, show initially the same behavior, i.e., a decrease in k with increasing T, followed by a region of nonexponential decays. However, above 500 K the rate coefficient increases again, indicative of another reaction, i.e., H-atom abstraction by the OH. Comparison of the various D substituted toluenes, Fig. 10, suggests that side-chain-atom abstraction dominates over ring abstraction, i.e., the dominant reaction above ≈ 500 K is

$$OH + C_6H_5CH_3 \rightarrow C_6H_5CH_2 + H_2O$$

Fast flow reactors have also been used for resonance fluorescence measurements on O-atom hydrocarbon reactions up to about 1150 K, especially in the work of Klemm, Michael et al.[36,37] Higher temperatures are not practical with such systems for this type of reaction because of rapid atom recombination on the walls[††] and hydrocarbon

---

[††]Similarly OH, H, etc., studies would not be practical in such reactors at higher temperatures.

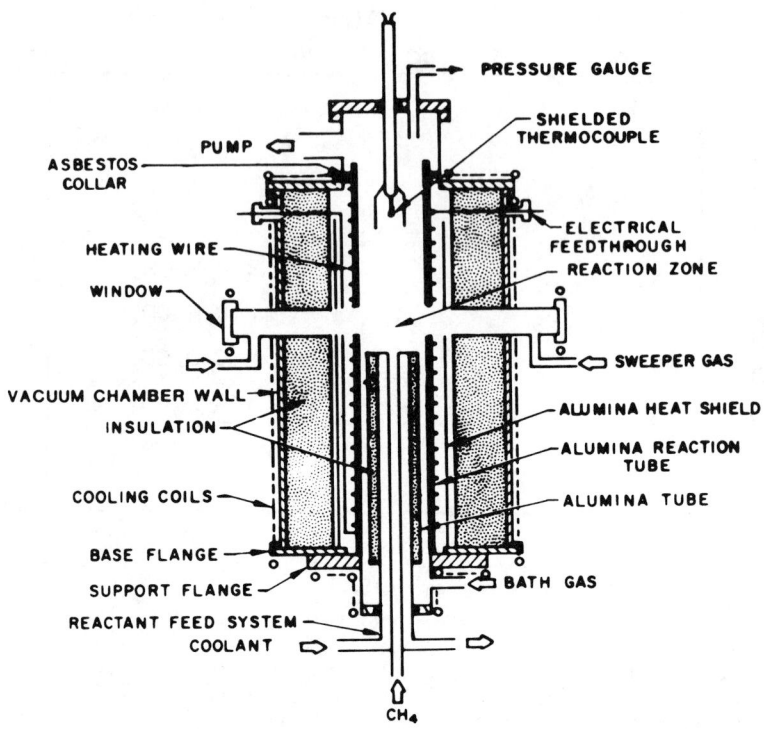

Fig. 11 Schematic of an HTP reactor, side view. The photolytic parent compound (O_2, CO_2) can be introduced with the bath gas or with the hydrocarbon reactant (CH_4 in the case shown). From Felder, Fontijn et al.[38] Copyright 1980, The American Institute of Physics.

wall pyrolysis. Since in RTFP the observed reaction zone, Fig. 8, can be kept well away from the walls, that technique is better adaptable to temperatures above ≈ 1150 K. By using apparatus similar in construction to the HTFFR, discussed in the section titled Refractory Species Oxidation Reactions, we have adapted the RTPF technique for work up to ≈ 1800 K,[38] and named this modification HTP (high-temperature photochemistry). The basic design of an HTP reactor is shown in Fig. 11. The hydrocarbon is introduced through the air-cooled inlet to adequately reduce the time it spends at high temperatures. The distance from this cooled reagent inlet to the reaction zone can be varied. Heating in the configuration shown is provided by Pt/Rh resistance wire. The reaction (observation) zone temperature is measured with a shielded thermocouple, which is withdrawn (as shown) during an actual experiment. Figure 12 summarizes the results obtained for the reaction[39]

$$O + CH_4 \rightarrow OH + CH_3 ,$$

using the HTP technique from 400-1700 K. For comparison are given the shock-tube absorption data of Roth and Just[40] (1500-2250 K) and the flow-tube resonance fluorescence data of Klemm et al[36] (500-1160 K), showing excellent agreement.

The reactions between O atoms and C_2 hydrocarbons form an important group of reactions. The O +

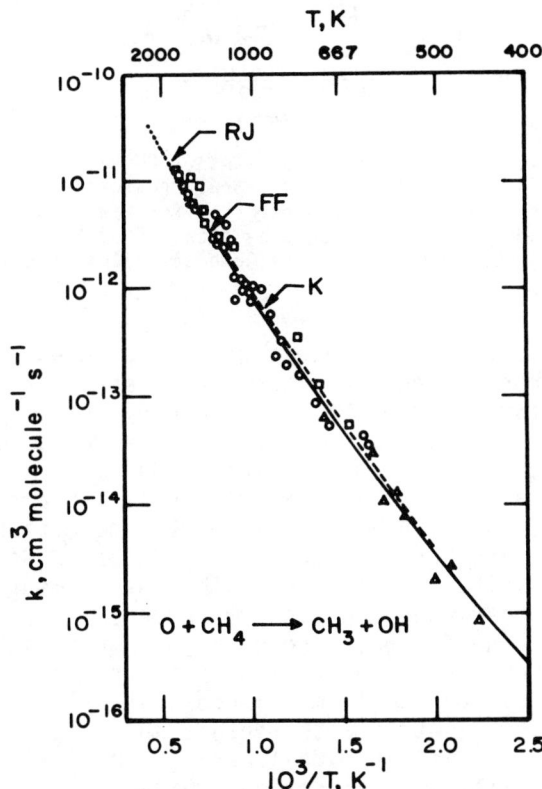

Fig. 12 Arrhenius plot of O + CH$_4$ rate coefficient measurements. ─────── FF and individual points: HTP data from Felder and Fontijn.[39] Individual HTP measurement points:
○ - O$_2$ 147 nm pulsed photolysis, resonance fluorescence
☐ - CO$_2$ flash photolysis, resonance fluorescence
△ - O$_2$ 147 nm pulsed photolysis, O/NO chemiluminescence
········ RJ: Fit to shock-tube absorption data of Roth and Just[40]
--------- K: Fit to flow-tube resonance-fluorescence data of Klemm et al.[36] Copyright 1980, The Combustion Institute.

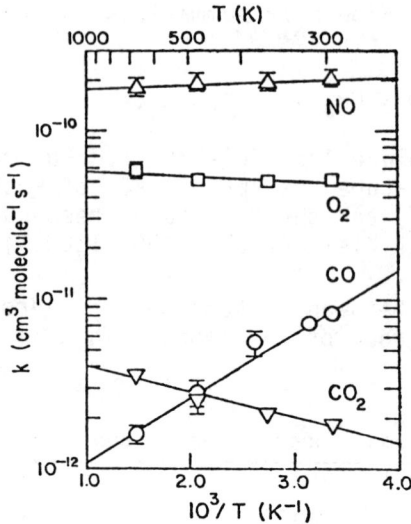

Fig. 13 Arrhenius plots for the reactions of CH radicals with NO (△), O_2 (□), CO_2 (▽) and CO(○). The total pressure was maintained at 0.13 bar by the addition of Ar buffer gas. The solid lines are weighted linear least-squares fits. From Berman, Fleming et al[34]. Copyright 1982, The Combustion Institute.

C_2H_4 reaction has been the subject of many studies in recent years. For a long time, based on Cvetanovic's work, the reaction was thought to proceed by initial adduct formation of the O atom to the double bond, leading principally to CH_3 and CHO as initial products.[41] However, Lee et al[42] found in a single collision crossed-molecular beam study with mass spectrometer detection that the only initial product was the vinoxy radical $H_2C=CHO$, indicating a substitution reaction. This assignment was confirmed by Kleinermanns and Luntz[43] under similar conditions using LIF detection. Inoue and Akimoto[44] in a bulk-kinetic microwave-discharge flow reactor study identified this radical by LIF from $O + C_2H_4$, as well as from reactions of F and Cl atoms with organic molecules. The work of Hunziker et al[45] in 0.05-1 bar N_2, where vinoxy was found in 36% yield, then provided what appeared a satisfactory picture, i.e., direct production of CH_2CHO as in the single collision studies and collision stabilized adduct formation leading mainly to CH_3 + CHO, but also some H + CH_2CHO and H_2 + CH_2CO in agreement with Cvetanovic. However, some preliminary reports from bulk studies in several laboratories, in which the H-atom yield is measured by fluorescence, suggest a 100% yield of these atoms, compatible only with

$$O + C_2H_4 \rightarrow H + CH_2CHO$$

as in the single collision studies. It will be some time before these questions of the room-temperature bulk reactions are settled; possibly the nature of the bath gas may be more important then considered thus far.

In a RTPF study from 300-950 K Nicovich and Ravishankara[46] noted a slight upward curvature of the rate coefficient with increasing T, similar to the $O + CH_4$ reaction discussed above. Using an HTP reactor we are currently extending the work to higher temperatures, where possibly the slightly endothermic (≈ 32 kJ mol^{-1})[47] abstraction reaction

$$O + H_2C=CH_2 \rightarrow OH + HC=CH_2$$

could become important.

Reactions of Some Other Combustion Intermediate Species

In the preceding section we have discussed reactions of O and OH with hydrocarbon molecules. Several groups have produced hydrocarbon and nitrogenous radical combustion intermediate species by laser photolysis (e.g., CH, CH_2, C_2, C_2O, CN, NH_2, etc.) and studied their reactions, again mainly at room temperature.[8,48,49] Figure 13 shows results from one such study where the temperature was varied.[34] CH was produced in a RTPF experiment by multiphoton dissociation of $CHBr_3$ at 266 nm, using the fourth harmonic of a Nd:YAG laser, and detected by LIF from a pulsed dye laser near 430 nm. Over the 300-675-K range studied the reactions of CH with NO and O_2 indicate no clear temperature dependence. The reaction with CO_2 has normal Arrhenius behavior and the CO reaction has a strong negative temperature dependence, indicative of the type of adduct formation process discussed above. Several reaction paths are possible in these reactions and the authors[34] are planning product identification experiments.

A negative temperature dependence was also observed in studies of the reaction of NH_2 with NO in a flash photolysis apparatus[50] and in a fast-flow reactors,[51,52] all with LIF detection. This reaction as well as other NH and NH_2 reactions are of major current interest because of the role they are thought to play in the thermal de-NO_x process. (In this process NH_3 is added to post-combustion gases, containing O_2, for removal of NO.)

Reactions of Excited Species

Knowledge of the kinetics of reactions of excited species is, for various reasons, desired in combustion work. In the LIF work discussed in the preceding sections only relative concentrations were required. To get an idea of absolute concentrations generally requires calibration against a known system. Even then differences in quenching rate coefficients of the upper electronic state have to be taken into account to achieve better than order-of-magnitude accuracy.[53] In actual flames the dependence of k_Q on T is required, but usually not known.[5] Quenching is not a problem when saturated absorption is achieved,[54] but at the resulting higher concentrations of electronically excited species their differences in reactivity[55-57] from ground-state species have to be considered.[58] Alternately, such differences may offer prospects for influencing the flame chemistry in a desired direction.

While equilibrium concentrations of electronically excited states in flames are low, at flame temperatures significant concentrations of vibrationally excited molecules are present in equilibrium. These molecules can react at much faster rates than the corresponding ground-state species.[4,59] LIF is a major tool to help elucidate the kinetics of both vibrationally and electronically excited species.

LIF experiments on excited species kinetics can be separated into two groups, which here are discussed consecutively. In the first, the excited species are produced by laser pumping and the intensity of the fluorescent emission from the pumped state yields information on the kinetics of species in this state. The pumped state in this type of reaction rate measurement must have a τ_{rad} long compared to its quenching τ in the medium investigated. In the second, the excited species are produced by any number of means (such as an electrical discharge, or again a laser) and the products of the reactions of these excited species are probed by LIF.

Kinetics of Excited States Produced by Laser Pumping

Breckenridge et al[55,60,61] have studied the reactions of Group II metastable excited atoms with H_2. They used a 5×10^{-7} -s, 457.1-nm pulse from a flash lamp pumped dye laser to pump ground-state $Mg(^1S_0)$ atoms to $Mg(^3P_1)$, τ_{rad} = 4.5 x 10^{-3} s. In the 0.4-bar He bath gas of the experiment these quickly equilibrated with $^3P_0$ and $^3P_2$ atoms, which also are metastable. Decay of the excited $Mg(^3P_J)$ population was then monitored in the presence of H_2 or D_2 by observing the $Mg(^3P_1)$ fluorescence, starting 2 x 10^{-5} s after the laser pulse. Measurements of the quenching rate coefficients from 600 to 800 K showed a strong positive temperature dependence. The quenching is predominantly chemical in nature:

$$Mg(^3P_J) + H_2 \rightarrow MgH + H.$$

This reaction is endoergic by 47 kJ mol^{-1}. By contrast, at lower temperatures inefficient physical (or energy transfer) quenching dominates:

$$Mg(^3P_J) + H_2 \rightarrow Mg(^1S_0) + H_2.$$

Clyne's group has determined quenching rate coefficients for a number of diatomic free radicals by laser excitation at pressures on the order of 1 x 10^{-6} bar. Radicals are produced in a microwave-discharge flow reactor and expanded through a pinhole into a low-pressure chamber, in the center of which they interact with the pulsed laser beam. For example,[62] ground-state SO radicals were formed by $O + O_2 + CS_2$ reaction and pumped to the $A^3\Pi$ state. At the initial chamber pressure of 0.5 mTorr (7 x 10^{-7} bar) the mean time between collisions of the $SO(A^3\Pi)$ excited radicals was on the order of 2 x 10^{-4} s, thus the probability of a quenching collision during the 1.6 x 10^{-5} -s radiative lifetime of $SO(A^3\Pi)$ was less than 0.1. Figure 14 shows the reciprocal of the experimental lifetimes of the $SO(A^3\Pi)$ as a function of O_2 added to the observation chamber. A quenching rate coefficient of (6.4 ± 1.6) x 10^{-11} cm^3 $molecule^{-1}$ s^{-1} at 300 K was determined from this Stern-Volmer plot.

Several groups are investigating the kinetics of vibrationally excited molecules. Heteronuclear diatomics can effectively be pumped with infrared lasers. Their reactions can then be monitored using the infrared fluorescence of the pumped molecules. Wolfrum et al. have followed this approach.[59,63] Figure 15 shows their results for the O + HCl system in the study of which, in addition to the HCl laser, mass spectrometry and atomic resonance line absorption were used for quantitative detection of all reactants and products. The reaction for ground-state HCl is 3.3 kJ mol^{-1} endoergic, while the excitation energy

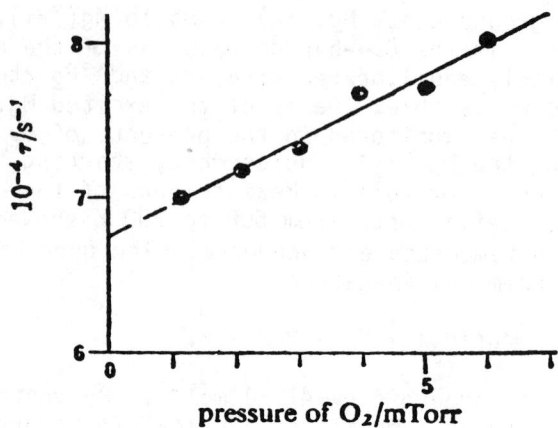

Fig. 14 Quenching of SO(A-X) fluorescence by O_2. From Clyne and McDermid[62] Copyright 1979, The Royal Society of Chemistry.

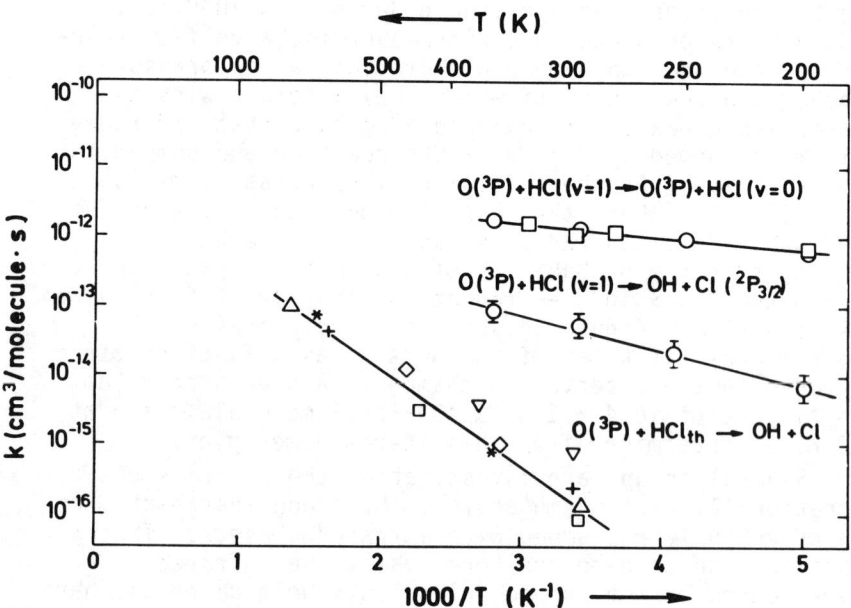

Fig. 15 Experimental data for the temperature dependence of the rates for vibrational relaxation and reaction of HCl (v = 1,0) with O atoms. The reaction leading to OH + Cl is 3.3 kJ mol^{-1} endoergic for v = 0 and 31 kJ mol^{-1} exoergic for v = 1. From Wolfrum.[59] Copyright 1983, Academic Press.

of the first vibrational quantum of HCl is 34.3 kJ mol^{-1}. It may be seen that the vibrational excitation leads to a major increase in the rate coefficient as compared to normal thermal HCl (bottom line). Thermal HCl contains increasing amounts of HCl(v=1) with increasing temperature, in accord with the Maxwell-Boltzmann equation. Most of the HCl(v=1) is physically quenched by the O atoms; only a small fraction reacts to form OH + Cl, as may be seen from the figure. The data indicate that at 200 K thermally excited HCl(v=1) contributes < 10^{-3}% to the total HCl consumption; at 2000 K the contribution is about 10%.

LIF Observation on Excited Product Molecules

Dagdigian[64] has used LIF apparatus similar to that of Fig. 7 to probe the MgO product states from the Mg($^3P_0$) reactions with O_2 and N_2O. The metastable $^3P_0$ atoms were produced by adding provisions for a mild electrical discharge to the apparatus. Vibrational-rotational

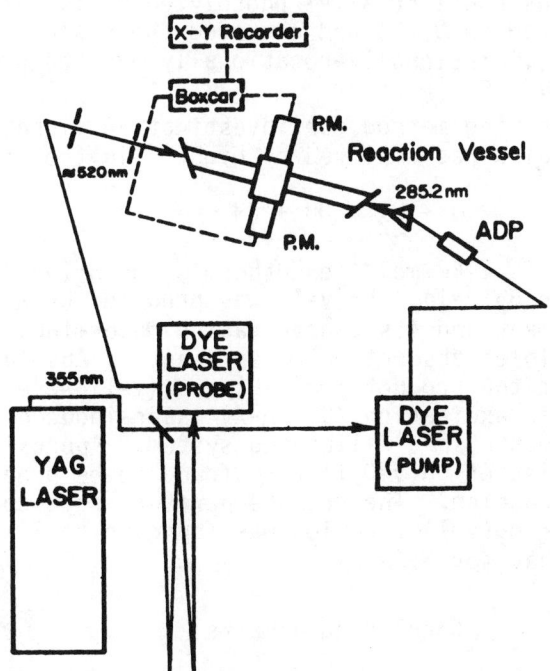

Fig. 16 A schematic diagram of the laser "pump-and-probe" apparatus. From Breckenridge and Umemoto.[66] Copyright 1982, American Institute of Physics.

distributions of the ground ($X^1\Sigma$) and excited electronic states ($a^3\Pi$ and $A'^1\Pi$) formed were measured.

Another interesting approach to the study of product state distributions of excited atom reactions is the pump-and-probe technique illustrated by Fig. 16. Two dye laser heads are pumped from the same Nd:YAG laser, the output of the third harmonic of which is split. One portion is used to excite atoms (premixed with a second reactant) with the frequency-doubled radiation from one dye laser. The remaining portion of the Nd:YAG radiation, which oscillates the second dye laser for probing of the reaction products, is spatially delayed. Delay times of $< 10^{-8}$ s are used which are short compared to the mean collision times of $\approx 10^{-7}$ s at the total reaction pressure of $\approx 1 \times 10^{-3}$ bar, thus nascent distributions of the product molecules are obtained. Breckenridge[55,65,66] developed this technique for short-lived (large absorption coefficient) species such as $Cd(^1P_1)$, $\tau_{rad} = 1.7 \times 10^{-9}$ s, and $Mg(^1P_1)$, $\tau_{rad} = 2.0 \times 10^{-9}$ s. Reactions of metastable excited species can however be similarly observed if they can be produced by short-pulse laser photolysis of an absorbent molecule. Thus Luntz et al.[67] photolyzed O_3 in $O_3/H_2/Ar$ mixtures to produce $O(^1D)$ and observed the nascent distribution of the vibrationally-rotationally excited product OH radicals by LIF.

An interesting method for investigating vibrationally excited molecules was employed by Light[53] in the study of

$$O + H_2 \rightarrow OH + OH$$

This reaction is 8 kJ mol^{-1} endothermic for $H_2(v=0)$, but exothermic for $H_2(v=1)$. $H_2(v=1)$ was produced by a heated tungsten filament and its concentration determined by vacuum ultraviolet absorption measurements. Absolute concentrations of the products $OH(v=0)$ and $(v=1)$ were deduced from electronic excitation (OH, A-X) laser-induced fluoresence measurements in a calibrated system. Approximately equal quantities of $OH(v=0,1)$ were found to be produced in the $H_2(v=1)$ reaction. The room-temperature rate coefficient of the $H_2(v=1)$ reaction was found to be 3×10^3 higher than that for $H_2(v=0)$.

Concluding Remarks

The various applications discussed give an idea of the variety of ways in which LIF has been used for fundamental kinetic measurements of combustion interest. No claim to completeness is made and new ways of using LIF continue to be reported. However, it is hoped that this short overview will leave no doubt that LIF has, in a short time, become an important tool and has led to major advances in the knowledge of reaction kinetics and dynamics.

Acknowledgments

I am grateful for support of our LIF work by the Air Force Office of Scientific Research, Air Force Systems Command, USAF, under Grant No. AFOSR-82-0073, Dr. Leonard H. Caveny, Technical Monitor. The United States Government is authorized to reproduce and distribute reprints for Governmental purposes notwithstanding any copyright notation hereon. I thank Dr. Judith A. Halstead, Andrew J. English, and Paul E. Foster for helpful comments and discussions.

References

[1] Laser Probes for Combustion Chemistry, edited by D. R. Crosley, ACS Symposium Symposium Series 134, American Chemical Society, Washington, D.C., 1980.

[2] Fontijn, A., "Concentration Measurements by Fluorescence," Combustion Measurements, edited by R. Goulard, Hemisphere Publishing Corporation, Washington,D.C., 1976, pp. 127-128.

[3] Muller,C. H., Schofield, K., Steinberg, M. and Broida, H. P., "Sulfur Chemistry in Flames," Seventeenth Symposium (International) on Combustion. The Combustion Institute, Pittsburgh, Pa., 1979, pp. 867-879.

[4] Fontijn, A. and Zellner, R., "Influence of Temperature on Rate Coefficients of Bimolecular Reactions," Reactions of Small Transient Species. Kinetics and Energetics, edited by A. Fontijn and M. A. A. Clyne, Academic Press, London, 1983, Chap. 1.

[5] Crosley, D. R., "Collisional Effects on Laser-Induced Fluorescence Flame Measurements," Optical Engineering, Vol. 20, July/Aug. 1981, pp. 511-521.

[6] Schofield, K., "Atomic and Molecular Fluorescence as a Stratospheric Species Monitor," Journal of Quantitative Spectroscopy and Radiative Transfer, Vol. 17, Jan. 1977, pp 13-51.

[7] Clyne, M. A. A. and McDermid, I. S., "Laser-Induced Fluorescence: Electronically Excited States of Small Molecules," Dynamics of the Excited State, edited by K. P. Lawley, John Wiley and Sons, New York, 1982, pp.1-104.

[8]Reisler, H., Mangir, M., and Wittig, C., "Laser Kinetic Spectroscopy of Elementary Processes," Chemical and Biochemical Applications of Lasers, Vol. 5, edited by C. B. Moore, Academic Press, New York, 1980, pp. 139-173.

[9]Lin, M. C. and McDonald, J. R., "Production and Detection of Reactive Species with Lasers in Static Systems," Reactive Intermediates in the Gas Phase. Generation and Montoring, edited by D.W. Setser, Academic Press, New York, 1979, pp. 233-304.

[10]Smith, I. W. M., "Lasers in the Study of Elementary Processes," Berichte der Bunsen-Gesellschaft für physikalische Chemie, Vol. 86, May 1982, pp. 395-401.

[11]Fontijn, A., Kurzius, S. C. and Houghton, J. J., "High-Temperature Fast-Flow Reactor Studies of Metal-Atom Oxidation Kinetics," Fourteenth Symposium (International) on Combustion, The Combustion Institute, Pittsburgh, Pa., 1973, pp. 167-174.

[12]Fontijn, A. and Felder, W., "High Temperature Flow Tubes. Generation and Measurement of Refractory Species," Reactive Intermediates in the Gas Phase. Generation and Monitoring, edited by D.W. Setser, Academic Press, New York, 1979, pp. 59-149.

[13]Llewellyn, I. P., Fontijn, A. and Clyne, M. A. A., "Kinetics of the Reaction $BO + O_2 \rightarrow BO_2 + O$," Chemical Physics Letters, Vol. 84, Dec. 15, 1981, pp. 504-508.

[14]Light, G. C., Herm, R. R. and Matsumoto, J. H., "Rate Coefficients for the Reactions of BF with O and O_2," Chemical Physics Letters, Vol. 70, March 1, 1980, pp. 366-370.

[15]Pruett, J. G. and Zare, R. N., "Lifetime-separated Spectroscopy: Observation and Rotational Analysis of the BaO $(A'^1\Pi)$ State," The Journal of Chemical Physics, Vol. 62, March 15, 1975, pp. 2050-2059.

[16]Heaven, M., Miller, T. A., English, J. H., and Bondybey, V. E., "Laser-Induced Fluorescence Spectra of YAG Laser Vaporized Se_2," Chemical Physics Letters, Vol. 91, Sept. 17, 1982, pp. 251-257.

[17]Fontijn, A., Felder, W. and Houghton, J. J., "HTFFR Kinetics Studies. Temperature Dependence of Al/O_2 and AlO/O_2 Kinetics from 300 to 1700/1400 K", Sixteenth Symposium (International) on Combustion, The Combustion Institute, Pittsburgh, Pa., 1977, pp. 871-879.

[18]Fontijn, A. and Felder, W., "HTFFR Kinetics Studies of $Al + CO_2 \rightarrow AlO + CO$ from 300 to 1900 K, a non-Arrhenius Reaction," The Journal of Chemical Physics, Vol. 67, Aug. 1977, pp. 1561-1569.

[19]Fontijn, A. and Felder, W., "HTFFR Kinetics Studies of the Al/SO_2 Reaction from 700-1600 K. Implications for D(Al-O)," The Journal of Chemical Physics, Vol. 71, Dec. 15, 1979, pp. 4854-4859.

[20] Clyne, M. A. A. and Heaven, M. C., "Laser-induced Fluorescence of the BO and BO$_2$ Free Radicals," Chemical Physics, Vol. 51, Oct. 1, 1980, pp. 299-304

[21] Zare, R. N. and Dagdigian, P. J., "Tunable Laser Fluorescence Method for Product State Analysis," Science, Vol. 185, Aug. 30, 1974, pp. 739-747.

[22] Zare, R. N., "Fluorescence of Free Radicals: A Method for Determining Dissociation Energy Limits," Berichte der Bunsen-Gesellschaft für physikalische Chemie, Vol. 78, Febr. 1974, pp. 153-157.

[23] Gole, J. L., "High Temperature Chemistry: Modern Research and New Frontiers," Annual Reviews of Physical Chemistry, Vol. 27, 1976, pp. 525-551.

[24] Sinha, M. P., Caldwell, C. D., and Zare, R. N., "Alignment of Molecules in Gaseous Transport: Alkali Dimers in Supersonic Nozzle Beams," The Journal of Chemical Physics, Vol. 61, July 15, 1974, pp. 491-503.

[25] Crumley, W. H., Gole, J. L. and Dixon, D. A., "Metal Cluster Oxidation: Chemiluminescence from the Reaction of Sodium Polymers (Na$_n$, n ⩾ 3) with Halogen Atoms (X = Cl, Br, I)," The Journal of Chemical Physics, Vol. 76, June 15, 1982, pp. 6439-6441.

[26] Dickson, C. R., George, S. M. and Zare, R. N., "Determination of Absolute Photon Yields under Single Collision Conditions," The Journal of Chemical Physics, Vol. 67, Aug. 1, 1977, pp 1024-1030.

[27] Preuss, D. R. and Gole, J. L., "The Temperature Dependence of "Single Collision" Bimolecular Beam-Gas Chemiluminescent Reactions. I. Theory," The Journal of Chemical Physics, Vol. 66, April 1, 1977, pp. 2994-2999.

[28] Gole, J. L. and Preuss, D. R., "The Temperature Dependence of "Single Collision" Bimolecular Beam-Gas Chemilumiscensent Reactions II. Experimental Studies," The Journal of Chemical Physics, Vol. 66, April 1, 1977 pp. 3000-3011.

[29] DeMore, W. B. et al., "Chemical Kinetics and Photochemical Data for Use In Stratospheric Modeling," Evaluation Number 5, JPL Publication 82-57, NASA, Jet Propulsion Laboratory, California Institute of Technology, Pasadena, CA, July 1982.

[30] Tully, F. P., "Laser Photolysis/Laser-induced Fluorescence Study of the Reaction of Hydroxyl Radical with Ethylene," Chemical Physics Letters, Vol. 96, April 1, 1983, pp. 148-153.

[31] Davis, D. and Braun, W., "Intense Vacuum Ultraviolet Atomic Line Sources," Applied Optics, Vol. 7, October 1968, pp. 2071-2074.

[32] Tully, F. P., Ravishankara, A.R. et al, "Kinetics of the Reactions of Hydroxyl Radical with Benzene and Toluene," The Journal of Physical Chemistry, Vol. 85, July 23, 1981, pp. 2262-2269.

[33] Gaydon, A. G., The Spectroscopy of Flames, 2nd Ed., Chapman and Hall, London, 1974.

[34] Berman, M. R., Fleming, J. W., Harvey, A. B. and Lin, M. C., "Temperature Dependence of CH Radical Reactions with O_2, NO, CO, and CO_2," Nineteenth Symposium (International) on Combustion, The Combustion Institute, Pittsburgh, Pa., 1982, pp. 73-79.

[35] Lam, L., Dugan, C. H. and Sadowski, C. M., "The Gas Phase Reactions of CN and NO," The Journal of Chemical Physics, Vol. 69, Sept. 15, 1978, pp. 2877-2881.

[36] Klemm, R. B., Tanzawa, T., Skolnik, E. G., and Michael, J. V., "A Resonance Fluorescence Kinetic Study of the $O(^3P)$+ CH_4 Reaction over the Temperature Range 474 to 1156 K," Eighteenth Symposium (International) on Combustion, The Combustion Institute, Pittsburgh, Pa., 1981, pp. 785-799.

[37] Michael, J. V., Keil, D. G., and Klemm, R. B., "A Resonance Fluorescence Kinetic Study of Oxygen Atoms + Hydrocarbon Reactions, V: $O(^3P)$ + Neopentane (412-922K)," Nineteenth Symposium (International) or Combustion, The Combustion Institute, Pittsburgh, Pa., 1982, pp. 39-50.

[38] Felder, W., Fontijn, A., Volltrauer, H. N. and Voorhees, D. R., "High-Temperature Photochemistry Reactor for Kinetic Studies of Isolated Elementary Gas-Phase Reactions," Review of Scientific Instruments, Vol. 51, Febr. 1980, pp. 195-200.

[39] Felder, W. and Fontijn, A., "High-Temperature Photochemistry (HTP), A New Technique for Rate Coefficient Measurements over Wide Temperature Ranges: Measurements on the O + CH_4 Reaction from 420-1670K," Technical Meeting Eastern Section, The Combustion Institute, Pittsburgh, Pa., 1980, Paper 59.

[40] Roth, P. and Just, Th., "Atomabsorptionsmessungen zur Kinetik der Reaktion CH_4 + O → CH_3 + OH in Temperaturbereich 1500< T < 2250 K," Berichte der Bunsen - Gesellschaft für physikalische Chemie, Vol. 81, June 1977, pp. 572-577.

[41] Cvetanovic, R. J., "Addition of Atoms to Olefins in the Gas Phase," Advances in Photochemistry, Vol. 1, 1963, pp. 115-182.

[42] Buss, R. J., Baseman, R. J., He, G., and Lee, Y. T., "Reaction of Oxygen Atoms with Ethylene and Vinyl Bromide," Journal of Photochemistry, Vol. 17, Nov./Dec. 1981, pp. 389-396.

[43] Kleinermanns, K. and Luntz, A. C., "Laser-Induced Fluorescence of CH_2CHO in the Crossed Molecular Beam Reactions of $O(^3P)$ with Olefins," The Journal of Physical Chemistry, Vol. 85, July 9, 1981, pp. 1966-1968.

[44] Inoue, G. and Akimoto, H., "Laser-Induced Fluorescence of the C_2H_3O Radical," The Journal of Chemical Physics, Vol. 74, Jan. 1, 1981, pp. 425-432.

45Hunziker, H. E. Kneppe, H., and Wendt, H. R., "Photochemical Modulation Spectroscopy of Oxygen Atom Reactions with Olefins," Journal of Photochemistry, Vol. 17, Nov./Dec. 1981, pp. 377-387.

46Nicovich, J. M. and Ravishankara, A. R., "A Study of the Reaction of $O(^3P)$ with Ethylene," Nineteenth Symposium (International) on Combustion, The Combustion Institute, Pittsburgh, Pa., 1982, pp. 23-30.

47McMillen, D. F. and Golden, D. M., "Hydrocarbon Bond Dissociation Energies," Annual Review of Physical Chemistry, Vol. 33, 1982, pp. 493-532.

48Pasternack, L., Nelson, H. H. and McDonald, J. R., "Reactions of Free Radicals Produced by Laser Photolysis," Journal of Chemical Education, Vol. 59, June 1982, 456-462.

49Baronavski, A., Umstead, M. E. and Lin, M. C., "Laser Diagnostics of Reaction Product Energy Distribution," Photoselective Chemistry, Part 2, edited by J. Jortner, R. D. Levine and S. A. Rice, John Wiley and Sons, New York, 1981, pp. 85-131.

50Stief, L. J., Brobst, W. D., Nava, D. F., Borkowski, R. P., and Michael, J.V., "Rate Constant for the Reaction NH_2 + NO from 216-480 K," Journal of the Chemical Society, Faraday Transactions 2, Vol. 78, 1982, pp. 1391-1401.

51Silver, J. A. and Kolb, C. E., "Kinetic Measurement for the Reaction of NH_2 + NO over the Temperature Range 294-1215 K," The Journal of Physical Chemistry, Vol. 86, Aug. 5, 1982, pp. 3240-3246.

52Andresen, P., Jacobs, A., Kleinermanns, C., and Wolfrum, J., "Direct Investigations of the NH_2 + NO Reaction by Laser Photolysis at Different Temperatures," Nineteenth Symposium (International) on Combustion, The Combustion Institute, Pittsburgh, Pa., 1982, pp. 11-22.

53Light, G. C., "The Effect of Vibrational Excitation on the Reaction of $O(^3P)$ with H_2 and the Distribution of Vibrational Energy in the Product OH," The Journal of Chemical Physics, Vol. 68, March 15, 1978, pp. 2831-2843.

54Daily, J. W. and Chan, C., "Laser-Induced Fluorescence Measurement of Sodium in Flames," Combustion and Flame, Vol. 33, Sept. 1978, pp. 47-53.

55Breckenridge, W. H., "Reactions of Electronically Excited Atoms," Reactions of Small Transient Species. Kinetics and Energetics, edited by A. Fontijn and M. A. A. Clyne, Academic Press, London, 1983, Chap. 4.

56Slanger, T. G., "Reactions of Electronically Excited Diatomic Molecules," ibid, Chap. 5.

[57] Schofield, K., "Critically Evaluated Rate Constants for Gaseous Reactions of Several Electronically Excited Species," Journal of Physical and Chemical Reference Data, Vol. 8, 1979, pp. 723-798.

[58] Muller, C. H. III, Schofield, K., and Steinberg, M., "Laser Induced Flame Chemistry of $Li(2^2P_{1/2,3/2})$ and $Na(3^2P_{1/2,3/2})$. Implications for Other Saturated Mode Measurements," The Journal of Chemical Physics, Vol. 72, June 15, 1980, pp. 6620-6631.

[59] Wolfrum, J., "Reactions of Vibrationally Excited Molecules," Reactions of Small Transient Species. Kinetics and Energetics," edited by A. Fontijn and M. A. A. Clyne, Academic Press, London, 1983, Chap. 3.

[60] Breckenridge, W. H. and Nikolai, W. L., "Temperature Dependence of the Quenching of $Mg(^3P_J)$ by H_2: Evidence for a Change from Physical to Chemical Exit-Channel Control," The Journal of Chemical Physics, Vol. 73, Sept. 15, 1980, pp. 2763-2766.

[61] Breckenridge, W. H. and Stewart, J., "The Temperature Dependence of the Quenching of $Mg(^3P_J)$ by H_2 and D_2. Endoergic Chemical Reaction as Rate Limiting," The Journal of Chemical Physics, Vol. 77, Nov. 1, 1982, pp. 4469-4473.

[62] Clyne, M. A. A., and McDermid, I. S., "$A^3\Pi_i$ and $B^3\Sigma^-$ Excited States of the SO Radical. Part 1. Laser-Induced Fluorescence Study of the A-X System; Excitation Spectra and Lifetimes," Journal of the Chemical Society, Faraday Transactions 2, Vol. 75, 1979, pp. 905-922.

[63] Kneba, M. and Wolfrum, J., "Absolute Rates for the Reaction of O-Atoms with Vibrationally Excited HCl-Molecules," Seventeenth Symposium (International) on Combustion, The Combustion Institute, Pittsburgh, Pa., 1979, pp. 497-504.

[64] Dagdigian, P. J., "Laser Fluorescence Study of MgO formed from $Mg(3s3p^3P^0) + O_2$ and N_2O under Single Collision Conditions," The Journal of Chemical Physics, Vol. 76, June 1, 1982, pp. 5375-5384.

[65] Breckenridge, W. H., Malmin, O. K., Nikolai, W. L., and Oba, D., "A Rapid "Pump-and-Probe" Laser Technique for Determining State-Resolved Product Distributions," Chemical Physics Letters, Vol. 59, Nov. 1, 1978, pp. 38-42.

[66] Breckenridge, W. H. and Umemoto, H., "Reaction of $Mg(3s3p^1P_1)$ with a variety of Alkyl C-H bonds: Identical Initial MgH ($X^2\Sigma^+$, v=0) Rotational Quantum State Distributions," The Journal of Chemical Physics, Vol. 77, Nov. 1, 1982, pp. 4464-4468.

[67] Luntz, A. C., Schinke, R., Lester, W. A. Jr., and Gunthard, Hs. H., "Product State Distribution in the Reaction $O(^1D_2) + H_2 \rightarrow OH + H$: Comparison of Experiment With Theory," The Journal of Chemical Physics, Vol. 70, June 15, 1979, pp. 5908 - 5909.

# Chapter III. Particle Diagnostics

Nonintrusive Laser-Based Particle Diagnostics

E. Dan Hirleman*
Arizona State University, Tempe, Arizona

Abstract

The evolution of nonintrusive optical techniques for particle size analysis has provided an array of powerful diagnostics. The techniques either probe the light scattering/attenuation properties of the aerosol particles or form photographic or holographic images. This paper discusses the theoretical basis for in situ particle sizing techniques and reviews some practical applications as well. A number of subtle considerations which affect the reliability and interpretation of data from optical particle sizing instruments are discussed.

Nomenclature

C_{sc}	= partial light scattering cross section
d	= particle diameter
D_{32}	= volume-to-surface area mean diameter
F	= differential light scattering cross section
i_1, i_2	= scattering intensity functions
I	= intensity or time-averaged radiant energy per unit area normal to the propagation direction
I_{sc}	= scattered intensity
I_{inc}	= intensity incident upon a particle
J_1	= spherical Bessel function of first kind and first order
k	= proportionality constant in Eq. (7)
n	= complex refractive index
$n(\alpha)$	= particle number distribution function
N	= exponent parameter for Rosin-Rammler particle size distribution

---

Presented as Paper 83-1514 at the AIAA 18th Thermophysics Conference, Montreal, Canada, June 1-3, 1983. Copyright © American Institute of Aeronautics and Astronautics, Inc., 1983. All rights reserved.
*Associate Professor, Mechanical and Aerospace Engineering.

P_{sc} = scattered optical power
r = distance from origin to observation point in particle centered light scattering coordinate system
S = light scattering signal amplitude
x = mean diameter in Rosin-Rammler particle size distribution
α = particle size parameter $\pi d/\lambda$
δ = fringe spacing
λ = wavelength
θ = scattering angle measured from the incident beam propagation vector
φ = azimuthal scattering angle

Introduction

There are many instances when conventional batch sampling methods for particle size analysis are either impractical or impossible to implement. Further, it is often the case that the intrusive nature of sampling methods introduce unacceptable levels of interference into the aerosol flow of interest. For these reasons the development of nonintrusive optical diagnostics for particle size and concentration measurements has been the objective of a significant amount of research and development. Successful applications of this technology are being reported with increasing frequency.

Optical techniques for particle measurements can be divided into three broad areas. First, photographic and holographic methods analyze simultaneously recorded images of a number of individual particles to build a discrete particle size histogram. Secondly, ensemble or multi-particle analyzing methods utilize aggregate light scattering or extinction properties of a large number of particles to determine parameters of the particle size distribution. Finally, single particle counters (SPC) size and count individual particles traversing a relatively small optical sample volume, and a sequence of particles are sampled in order to build up a discrete size distribution. The three approaches are complementary in the sense that they are optimized for different types of applications.

Single particle counters are the optimum choice for analyzing particles greater than about 0.3 μm in applications demanding high specificity and the potential for simultaneous velocity measurements. The existing commercial technology of imaging techniques is generally limited to particles larger than a few micrometers with time response longer than a few seconds. Imaging techniques can provide information on particle morphology not retrievable with light scattering methods. Ensemble methods generally require less sophisticated optical systems for implementation but inherently provide less information as the optical characteristics of the individual particles are superimposed and can never be totally recovered.

This paper first presents a brief discussion of the fundamental principles of light scattering which underlie laser-based particle sizing technology. Then details of some of the techniques for nonintrusive particle diagnostics are reviewed. For the purposes of this paper an instrument is considered to be nonintrusive if no sampling probes are involved and the working space between optical elements and the optical measurement volume is on the order of 10 cm or greater.

Light Scattering by Particles

An infinite, planar electromagnetic wave can propagate through a homogeneous, nonabsorbing medium undisturbed. This propagation is rigorously described by Maxwell's equations.[1] However it is also useful to consider Huygens' principle[2] which states that each point on a wavefront (surface of constant phase in the electromagnetic wave field) serves as the source of spherical secondary wavelets such that the wavefront at some later time is determined by the envelope of these wavelets. The secondary wavelets propagate with the same frequency and speed as the primary wave would at each point in space. The fact that an infinite planar wavefront in a homogeneous medium propagates as a plane wave is readily visualized with Huygens' construction.[2]

If we consider the homogeneous medium to be a gas, then the secondary wavelets derive from electrons in the molecules comprising the gas which are harmonically accelerated by the time-varying E-field in the electromagnetic wave. This occurs because each accelerating electron, by virtue of Ampere's and Faraday's Laws,[2] produces its own secondary electromagnetic wave (i.e. a scattered wavelet) which propagates spherically outward. The superposition of these scattered wavelets with the unscattered incident wave define the entire electromagnetic field. From a quantum point of view, the gas molecule absorbs a photon which causes an electron to be excited into a virtual (unstable or disallowed) state for a very short (< psec) time. In elastic scattering events of interest here the electron then drops back to its original state emitting a second photon of the same frequency as the incident photon. This emission or scattering process is random in the sense that the photon can propagate with equal probability in any direction (at least in the plane normal to the polarization vector of the incident E-field). The memory of the molecule retains only the phase and polarization of the incident photon and not the direction of incidence.

It is also possible for the energy coupled into the electron from the incident photon to be dissipated by collisions of the excited electron with other nuclei or electrons. In that situation the photon energy would have been absorbed and converted into thermal (internal kinetic) energy. Both the scattering and absorption processes are included in rigorous light scattering theory.

Individual Spherical Particles

The parameters controlling the scattering of planar electromagnetic radiation by isolated spherical particles are the size parameter α, the complex refractive index n of the particle relative to the surroundings, and the polarization state of the incident radiation. The three scattering regimes of importance can be delineated as Rayleigh scattering for $\alpha \ll 1$, geometric optics for $\alpha \gg 1$, and Lorenz-Mie scattering for $\alpha \sim 1$. For visible radiation Rayleigh scattering approximations are valid for particle diameters $d < 0.05$ μm, and geometric optics approximations for roughly $d > 5$ μm. In the Rayleigh regime all of the electrons (or charge dipoles) in a particle are subjected to the same E-field by virtue of their close proximity (relative to the wavelength) and therefore oscillate in phase. The properties of the scattered radiation are then given in a very simple form applicable to the harmonic oscillation of a charge dipole. In the geometric optics limit, the wavelength is much smaller than the particle dimensions and the incident radiation can be considered to be a bundle of rays. The scattered field at any point distant from the particle (far field) can be calculated by coherent superposition of the refracted and reflected rays with the diffracted field.

In contrast with the Rayleigh scattering and geometric optics regimes, no approximations are possible for particle sizes on the order of the wavelength and the complete set of Maxwell's equations must be solved for the particle and the surroundings. The theoretical difficulties here arise from the fact that the E-fields experienced by the various electrons or charge dipoles distributed throughout the particle depend on position, and therefore these electrons emit secondary wavelets which are out of phase. The formulation for this intermediate case, known as Lorenz-Mie theory, is the general solution for all particle sizes. Exhaustive treatises of light scattering are given by van de Hulst[3] and Kerker.[4] Computer codes for calculating the

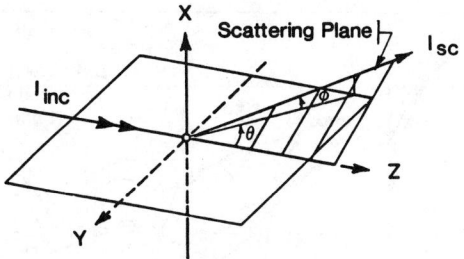

Fig. 1 Light scattering coordinate system. The functions i_1 and i_2 are for scattered light polarized perpendicular and parallel to the scattering plane respectively.

scattering characteristics of spherical particles of arbitrary size are readily available.

Consider the scattering geometry in Fig. 1 with a particle at the origin illuminated by linearly polarized electromagnetic radiation propagating in the +z direction with incident intensity I_{inc}. The scattered intensity I_{sc} at some point a distance r from the origin is given by

$$I_{sc} = \frac{I_{inc}\lambda^2}{4\pi^2 r^2} [i_1(\alpha,n,\theta)\sin^2\varphi + i_2(\alpha,n,\theta)\cos^2\varphi] \qquad (1)$$

where i_1 and i_2 are dimensionless intensity functions for scattered light polarized perpendicular and parallel to the scattering plane, respectively. The functions i_1 and i_2 are composed of spherical Bessel and associated Legendre functions and their first derivatives, and are integral parts of Lorenz-Mie theory.[3,4] It is convenient to normalize Eq. (1) by the incident intensity and other constants and define the differential scattering cross section F:

$$F \equiv i_1(\alpha,n,\theta)\sin^2\varphi + i_2(\alpha,n,\theta)\cos^2\varphi \qquad (2)$$

Some computations of F are shown[5] in Figs. 2-4. Figure 2 indicates the angular dependence of the scattered light for particle diameters of 0.1, 0.5, and 1.0 μm, and Fig. 3 for 5.0 and 10.0 μm particles as well. Note the lobe structure which becomes a dominant factor as particle size increases. Figure 4 indicates the dependence of F on particle size. In the Rayleigh regime F increases as diameter to the sixth power, and then gradually changes to a diameter-squared dependence in the geometric optics regime. The oscillations present for θ = 45 and 90 deg in Fig. 4 are typical for off-axis scattering of nonabsorbing (no imaginary component of the refractive index) particles. Forward scattering (small

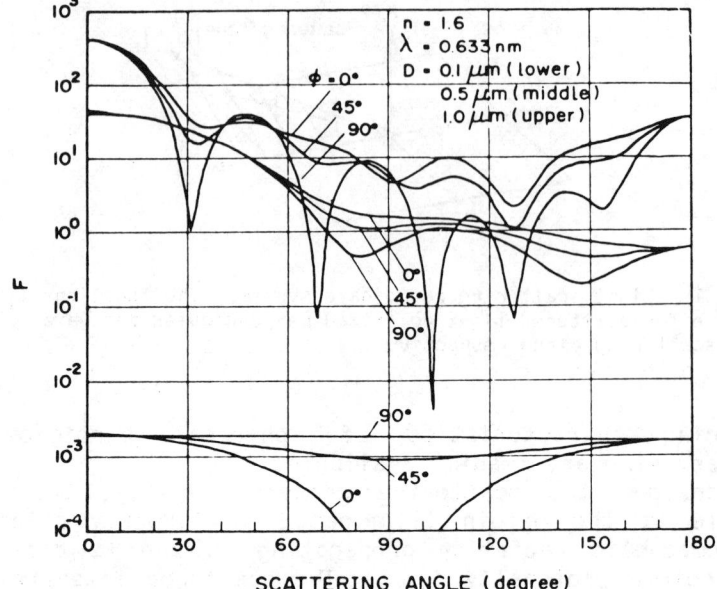

Fig. 2   Lorenz-Mie theory calculations of differential scattering cross-section F as a function of scattering angle θ for various particle diameters after Handa et al.[5]

θ) properties generally display much less structure as is also evident in Fig. 4.

The radiant power P_{sc} scattered into a detector with a finite collection aperture is obtained by integrating the scattered intensity over the solid angle subtended by the detector

$$P_{sc} = \frac{I_{inc}\lambda^2}{4\pi^2} \iint F(\alpha,n,\theta,\varphi) \sin\theta \, d\theta \, d\varphi \qquad (3)$$

The partial scattering cross section for a particular detector is defined as the scattered power divided by the incident intensity

$$C_{sc} = \frac{\lambda^2}{4\pi^2} \iint F(\alpha,n,\theta,\varphi) \sin\theta \, d\theta \, d\varphi \qquad (4)$$

Note that Eqs. (3) and (4) apply in a practical measuring system only if the scattered light wave experiences negligible distortion due to secondary scattering off of other particles in the field before reaching the detector. In other words Eqs. (3) and (4) are applicable in single

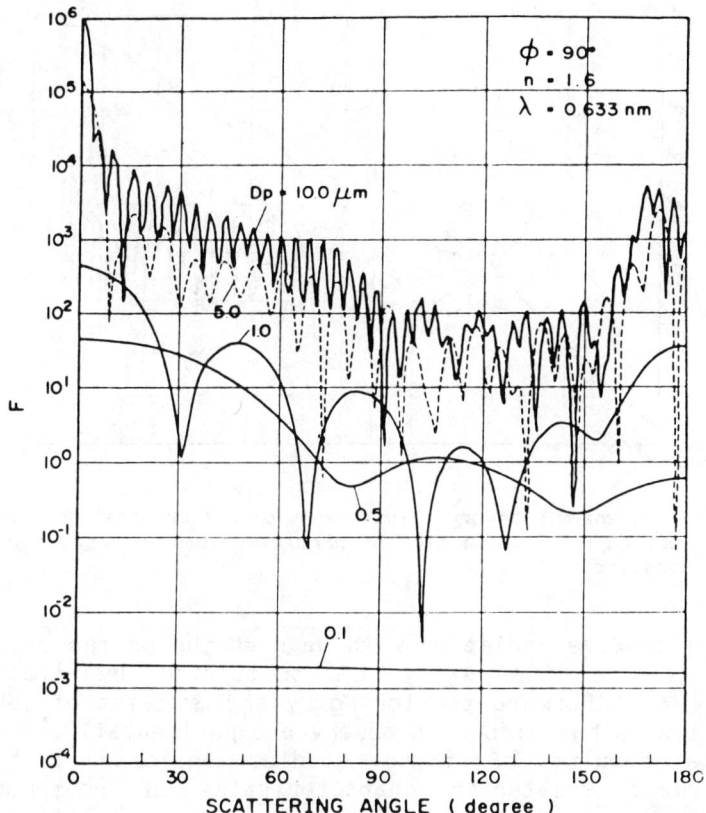

Fig. 3 Lorenz-Mie theory calculations of differential scattering cross-section F as a function of scattering angle θ for various particle diameters after Handa et al.[5]

scattering aerosols and must be altered when multiple scattering (secondary scattering events) is significant.

Individual Nonspherical Particles

It is not possible at present to calculate the scattering and absorption characteristics of particles of arbitrary shape and refractive index. There has, however, been some progress on theoretical models and calculations for certain nonspherical shapes such as ellipsoids,[4] spheroids,[6,7] clusters of spheres,[8] and cylinders.[9] The calculations are often valid for only limited values of refractive index.

Some experimental work on the scattering characteristics of nonspherical particles has been performed. The

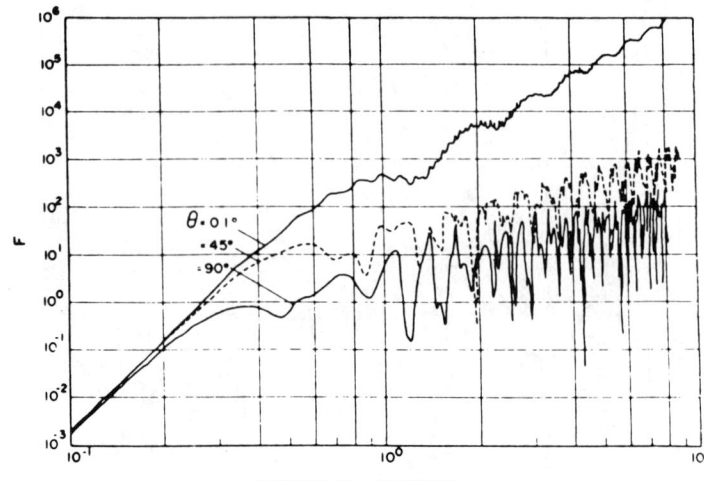

Fig. 4 Lorenz-Mie theory calculations of differential scattering cross-section F as a function of particle diameter for various θ after Handa et al.[5]

use of microwave radiation with wavelengths on the order of 1 cm permits the study of scattering by arbitrary shapes.[10,11] Forward scattering by agglomerates of spherical particles has also been observed experimentally.[12]

The results of these studies indicate that the near-forward scattering characteristics of nonspherical particles are predicted reasonably well by calculations for spherical particles of equal cross-sectional area. The off-axis scattering characteristics however are strongly dependent on the detailed particle shape. Concerning extinction (scattering plus absorption) spheres of equal volume or surface area have been used to approximate these optical properties of nonspherical particles.[13]

Scattering by an Ensemble of Particles

Often in particle diagnostics experiments it is either undesirable or impossible to define an optical sample volume small enough to ensure that less than one particle on average is in the volume. In that case the aggregate scattering properties of a number of particles are measured. Interpretation of the resulting ensemble or multiparticle scattering measurements (not to be confused with multiple scattering) is straightforward if the detected light has undergone only one scattering event. That is, if single scattering is predominant then the presence of other particles in the aerosol cloud has a negligible effect. In that situation light scattering in the far field is just the superposition of isolated single scattering contributions from each particle in the scattering volume. If there are a very large number of randomly positioned particles in the optical sample volume then superposition of the scattered intensity contributions (incoherent scatter) describes ensemble scattering properties. Conversely, if there are either relatively few particles or the particles are positioned in regular or quasi-ordered fashion then interference phenomena become important and the superposition must use scattered E-fields (coherent scatter) rather than intensities.

There are practical situations where single scattering approximations are not valid. For example two particles spaced closer than a few diameters apart will scatter as a single entity and Lorenz-Mie theory would not apply. Fortunately this situation occurs rather infrequently. Of more practical concern is the case where the particles on average are well separated but particles adjacent to the sample volume distort the scattered wave before it reaches the detector.

Multiple Scattering

As the physical size of an aerosol cloud increases, the probability that a scattered photon or ray will encounter another particle and be scattered again before leaving the aerosol increases as well. This phenomenon, termed multiple scattering, will clearly alter the characteristics of the scattered light which finally reaches the detector of a diagnostic instrument. Therefore the presence of multiple scattering significantly complicates the interpretation of light scattering measurements. The level of multiple scattering can be ascertained from the level of attenuation of the incident beam. For an axisymmetric aerosol with a centered optical sample volume the fraction of detected scattered light which has undergone only one scattering event is approximately equal to the square root of the fraction transmission of the incident beam. Some degree of multiple scattering is inherent in all measurements and the significance of this effect depends on the application. In particular, the anisotropy of the single scattering signature of the aerosol of interest plays a significant role in determining the sensitivity of measurements to multiple scattering. For example, Felton[14] performed a series of laser diffraction particle size measurements on 45 μm polystyrene latex spheres in a water flow cell. The ensemble scattering method assumed that the particle size distribution was Rosin-Rammler and determined the mean diameter and a width parameter. Measurements were taken for a series of latex particle concentrations which gave transmission fractions from 0.92 to 0.03. With increasing concentrations (decreasing optical transmission) the mean diameter decreased as expected since the secondary scattering events further diffused the forward scattered light giving the appearance of smaller particles. At 50% transmission Felton[14] observed approximately a 5% decrease in apparent mean diameter relative to the high transmission (single scattering) limit.

Ensemble (Multiparticle) Sizing Techniques

Optical techniques which analyze the light scattering and extinction properties of an ensemble of particles are invaluable in some applications. For measurements of particles smaller than about 0.1 μm, ensemble methods are the only viable options since SPC and imaging techniques generally cannot distinguish these particles. The lower size limit of a typical SPC is determined by one of two factors. First, scattering signals from individual small particles become rapidly indistinguishable from detector shot noise since the scattering cross sections decrease as d^6 in the Rayleigh regime. Second, typical particle number densities increase as d^{-4} (Junge distribution) making it eventually impossible to maintain the presence of only one particle in the optical sample volume. Imaging techniques are useless for particles smaller than several wavelengths, and since visible or in some cases near ultraviolet radiation is generally used imaging methods are limited to particles several μm and above.

Ensemble measurements inherently contain less information than SPC and imaging data as the scattering or extinction is averaged over all particle sizes in the aerosol. In some situations it is possible to mathematically invert the set of ensemble measurements and reconstruct or estimate the size distribution. The maximum resolution possible for the reconstructed size distribution is determined by the number of optical property measurements (e.g. the number of scattering angles), but practical considerations often limit here. It is often advantageous to estimate average parameters of the aerosol such as a mean diameter rather than perform the complete inversion. Similarly the form of the size distribution can be assumed and the measurements used to estimate the best fit parameters for the assumed size distribution.

Several ensemble-averaged optical properties of aerosols can be used in size analysis. These include spectral extinction, the angular dependence of scattered light, and finally for very small particles the spectral properties of the scattered light as Doppler-shifted by the Brownian motions of the particles. The following paragraphs discuss in further detail these ensemble methods.

Extinction Methods

The amount of light removed from a beam passing through an aerosol directly indicates the extinction cross sections of the particles along the beam path. If the refractive index and the volume concentration of the particles are known, then the volume-to-surface area mean diameter D_{32} (or Sauter Mean Diameter, SMD) can be determined from a single transmission measurement.[15] Further, the authors[15] studied the ratio of the transmittance at two probe wavelengths and found that it exhibited monotonic behavior when plotted as a function of D_{32} for nonabsorbing particles in the range $\lambda_1/3 < D_{32} < \lambda_2$. Ariessohn et al.[16] also studied this two-wavelength approach and found that the specific form of the particle size distribution, if it was not very narrow, had little influence on the measurement. The authors[16] considered measurements on coal ash particles which are weakly absorbing and found a compressed but useful sizing range of roughly $\lambda_2/10 < D_{32} < 1.3\lambda_2$ for $\lambda_1 = 0.325$ μm and $\lambda_2 = 3.39$ μm. Lester and Wittig[17] and Bro[18] utilized a similar method in shock tube studies of soot formation. Powell et al.[19] used spectral transmission data coupled with scattering measurements to study smoke particle sizes. Although in the works referenced above only mean diameters are determined, there have also been a number of studies on the use of spectral transmission measurements to determine the size distribution as well.[20] For optimum sensitivity the wavelengths used must roughly bracket the particle sizes of interest, so these techniques are in general useful for intermediate particle sizes near practical wavelengths.

Multiangle Scattering Measurements

It is clear from Figs. 2-4 that the angular scattering characteristics of an ensemble of particles will contain information on the particle size distribution. For small particles, say several μm and below, it is necessary to measure scattering characteristics over a large range of scattering angles. This can be accomplished for θ from 2 deg to 178 deg using a polar nephlometer as discussed by Hansen and Evans.[21] Hansen then used this technique[22] to estimate size distributions and refractive indices of an atmospheric aerosol. In some situations it is impractical to traverse a detector around the aerosol to measure angular

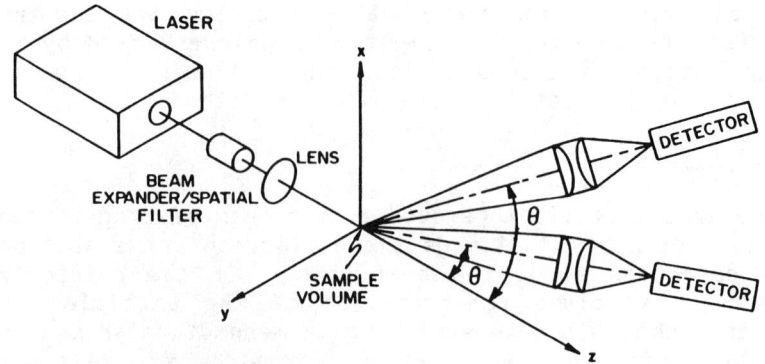

Fig. 5 Generalized schematic of a laser-based single particle counter.

scattering characteristics, and a few detectors at selected scattering angles are used.

Multiangle ensemble scattering techniques are utilized in some situations where SPC and imaging methods are not applicable. Measurements in solid-propellant rocket exhausts where the particle velocities are very high and the run times very short have been made by McCay et al.[23] using multiangle scattering and extinction. Measurements of soot particle sizes in flames require ensemble methods because of the small sizes (<100 nm). Recent studies on soot by Santoro and Semerjian[24] and Chang and Penner[25] have been completed although the presence of nonspherical agglomerates complicate interpretation of the data. The authors[23-25] used an optical system similar to that in Fig. 5 but with some detectors oriented in the backscatter direction because of the small particle sizes. Measurement of the polarization state of the scattered radiation is also useful in particle size analysis by ensemble multiangle scattering.

One problem for all multiwavelength or multiangle diagnostics for particle sizes of several micrometers and below is that the scattering characteristics can be strongly influenced by the refractive index which is in general not known. By increasing the number of measurements and assuming that the size distribution is monodisperse or of some particular form it is possible in theory to determine the refractive index along with the size distribution.[22,24,26]

As particle size increases it can be seen from Fig. 3 that the energy is scattered predominantly into the near-forward directions. Further, for particles greater than several um the dominant contributor to the forward lobe is diffractive scatter as opposed to refraction or reflection.

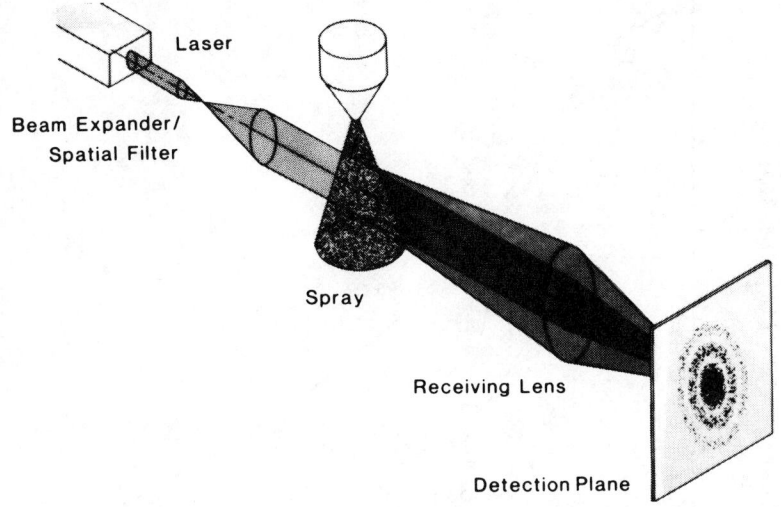

Fig. 6 Schematic of laser diffraction particle sizing instrument.

Analysis of the forward diffraction lobe has become a common diagnostic for particles and droplets larger than several micrometers in diameter.

The generalized schematic of a laser diffraction particle sizing apparatus is shown in Fig. 6. The beam from a laser, typically a several mW He-Ne model, is spatially filtered, expanded, and collimated to several mm diameter at the $1/e^2$ intensity points. This collimated probe beam is directed through the aerosol of interest and the transmitted (unscattered) portion is focused on-axis to a spot at the back focal plane of the receiving lens. Light scattered by particles in the probe beam which passes through the aperture of the receiving lens is directed to off-axis points on the observation or detection plane. A monodisperse ensemble of spherical particles large compared to the wavelength would produce the characteristic Airy diffraction pattern shown in Fig. 6 as described by Fraunhofer diffraction theory

$$I(\theta) = I_{inc} \frac{\alpha^4 \lambda^2}{16\pi^2} \left(\frac{2J_1(\alpha\theta)}{\alpha\theta} \right)^2 \qquad (5)$$

where J_1 is the first-order Bessel function of first kind. The obliquity correction $(1 + \cos^2\theta)/2$ has been neglected in Eq. (5) and the small angle approximation of $\sin\theta = \theta$ has been made.

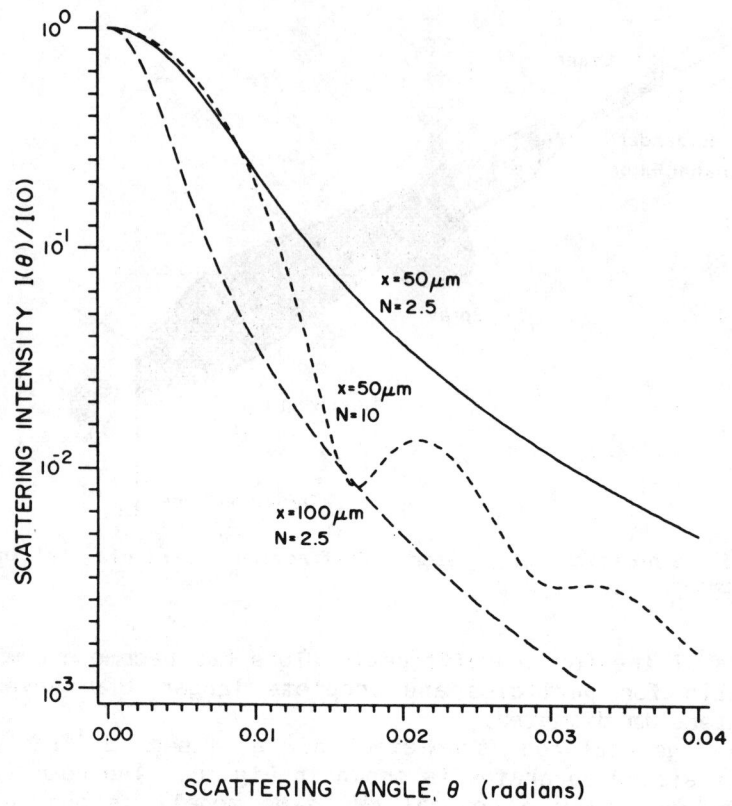

Fig. 7 Forward scattering signatures calculated using Fraunhofer diffraction theory for Rosin-Rammler particle size distributions. $\lambda = 0.6328$ µm.

In practical systems a distribution of particle sizes or a polydispersion is generally encountered. The composite scattered intensity profile is a linear combination of the characteristic profile of each droplet size with a weighting coefficient equal to the number of particles of that size in the sample volume. The diffraction signature of a polydisperse spray is given by

$$I(\theta) = I_{inc} \int_0^\infty \frac{\alpha^4 \lambda^2}{16\pi^2} \left(\frac{2J_1(\alpha\theta)}{\alpha\theta} \right)^2 n(\alpha) \, d\alpha \qquad (6)$$

where $n(\alpha)d\alpha$ is the number of particles in the laser beam with sizes between α and $\alpha + d\alpha$ and truncation of light diffracted at large angles by the receiving lens has been neglected.[27] A primary effect of broadened size distribu-

Fig. 8   Reproduction of the photosensitive elements of a monolithic P on N photodiode array detector after Hirleman.[27]

tions is elimination of the contrast in the diffraction pattern as shown in the diffraction signatures calculated for several Rosin-Rammler particle size distributions in Fig. 7. The two parameters in a Rosin-Rammler distribution are the mean diameter x and the exponent N. The width of the distribution increases with decreasing N, and as N approaches infinity the distribution becomes monodisperse.

The basic task in laser diffraction particle sizing is to detect and analyze the diffraction signature $I(\theta)$, and then mathematically invert Eq. (6) to determine parameters of the particle size distribution. Chin et al. in 1955[28] proposed several detection techniques, one of which was to traverse a pinhole/photodetector assembly across the diffraction pattern. Due to the mechanical traverse this detection approach requires a significant amount of time to cover the entire diffraction pattern. Further, the large dynamic range of the diffraction signature given by Eqs. (5) and (6) is another difficulty for such systems.

The advantages of real time analysis of the entire diffraction signature as opposed to traversing a detector across either the diffraction pattern itself or a photographic image thereof are obvious. Developments in monolithic solid state multielement detector arrays in the 1970's improved the situation by allowing the entire diffraction signature to be analyzed instantaneously. A monolithic detector designed for forward scattering measurements is shown in Fig. 8. Note the increasing thickness of the annular detector elements which, when coupled with increasing length (circumference), result in a significant increase in detector area as radius increases. This effect compresses the dynamic range of the scattering measurements. A detector similar to that in Fig. 8 designed for parts recognition applications[29] is utilized in a commercial laser diffraction particle sizing instrument[30] based on the work of Swithenbank et al.[31]

A number of data processing methods have been used to extract particle size information from measured diffraction patterns. Chin et al.[28] utilized the integral transform derivation of Titschmarsh[32] to analytically invert Eq. (6) to obtain $n(\alpha)$. Dobbins et al.[33] somewhat paradoxically observed that the diffraction signatures were relatively independent of the form of the droplet size distribution and depended primarily on D_{32}. The authors[33] utilized a single parameter of the diffraction pattern, the angle at which the scattered light intensity is down to 10% of the on-axis value, to determine D_{32}. Others[34,35] have since modified slightly this approach and it is still in use today.

Swithenbank et al.[31] analyzed the diffraction pattern with the annular ring detector discussed above and subsequently did a numerical inversion (as opposed to integral transform) of a discretized form of Eq. (6) to obtain the volume distribution in 7 discrete size bins. The inversion problem is ill-conditioned and as a second approach the authors[31] assumed that the size distribution was of the Rosin-Rammler form with two independent parameters. Recent data processing developments do not require an assumption of the form of the size distribution.[30,36]

Diffusion Broadening Spectroscopy

One problem with spectral extinction and multiangle scattering measurements of small particles is the dependence on refractive index which is generally unknown and might even vary between particles. One diagnostic which for certain applications does not require knowledge of the refractive index is diffusion broadening spectroscopy. Light scattered by molecules or particles is Doppler shifted due to Brownian motion. The magnitude of the frequency shift depends on the velocity of the particle and the angle at which the scattered radiation is collected. Light scattered from a large number of particles undergoing Brownian motion in a medium with a mass mean velocity of zero contains a distribution of frequencies centered around the incident laser frequency. If the light scattered by these particles is collected and mixed on a single detector (homodyne detection) then the frequency differences between waves scattered from the various particles will be present in the detector output with a resulting spectrum centered around zero frequency. The theoretical analysis for predicting the power spectrum and autocorrelation function of the homodyne scattered light signal for particles suspended in a stagnant or laminar flow is well known.[37]

The predictions depend on the scattering angle, the particle diameter, and the diffusion coefficient which in turn depends on temperature and viscosity. By measuring the half-width of the power spectra[38] after Penner et al. or the correlation time[39] from photon correlation after King et al. the diffusion coefficient of the particles can be determined. Introduction of some assumptions concerning the diffusion coefficient then allows the particle size of a monodisperse aerosol to be determined.

The optical system required for diffusion broadening spectroscopy is rather simple as shown in Fig. 5. The laser focus diameter is selected to minimize broadening effects due to finite particle residence time.[39] The output from the detector would then go to a spectrum analyzer or a digital photon correlator.

Diffusion broadening spectroscopy has been used successfully in flames[37-39] and other particle systems. It is only useful for particle diameters less than about 100 nm because the frequency shifts become very small as the Brownian diffusion velocities decrease for larger particles. Further, this technique is only independent of refractive index for monodisperse aerosols, and successful application in polydisperse systems seems unlikely.

Laser/Optical Single Particle Counters (SPC)

A generalized schematic of an optical SPC is presented in Fig. 5. The output beam from a laser or other source of radiation is directed (and typically focused) into the optical sample volume. This sample or probe volume can be thought of as that region of space where a single particle can generate a sufficient detector signal to be discriminated or "seen" over the background noise. As individual particles pass through the sample volume they interact with the incident radiation beam (i.e., scatter, absorb, and/or fluoresce light) and are observed by detection optics oriented at some angle(s) θ with respect to the beam propagation direction. The single particle signals obtained at the photodetector(s) are processed to provide information on the size and possibly the velocity of each particle. The various SPC approaches to particle sizing are discussed below.

Light Scattering Cross-section Measuring Techniques

The most common approach to particle sizing involves the principle that the amount of the light scattered by a particle is a nominally monotonic increasing function of

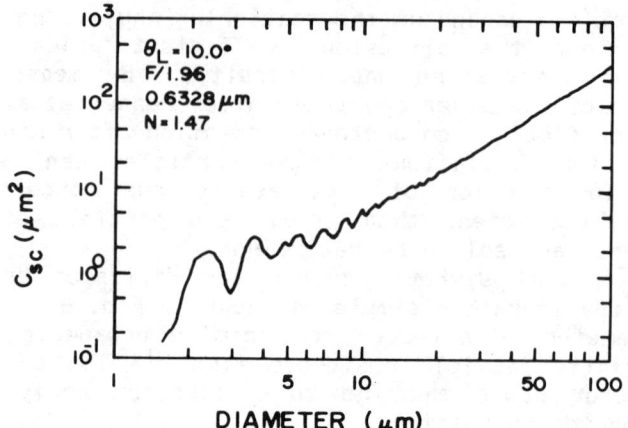

Fig. 9 Partial light scattering cross sections for spherical particles with refractive index n=1.47 for f/1.96 receiving optics oriented for 10 deg off-axis collection in the plane normal to the direction of polarization of the incident beam. The Lorenz-Mie theory calculations used λ = 0.6328 μm.

particle size. It follows that measurement of a scattering or extinction cross section can be used to infer particle size. The SPC scattering signal response S to a particle in an incident radiation field (uniform over the particle) of intensity I_{inc} is given by

$$S = k\, I_{inc}\, C_{sc} \qquad (7)$$

where k is the system gain in transducing radiant energy to voltage using a photodetector and C_{sc} is the partial light scattering cross section as determined from Eq. (4). The partial cross sections, as opposed to total cross sections, depend on the specific finite aperture detector configuration in use. A response function S(d) relating measured signal levels to the diameters of spherical particles of known refractive index passing through a SPC sample volume of known incident intensity I_{inc} can be determined from theoretical calculations of $C_{sc}(d)$. Here the factor k must be determined by calibration.

A plot of partial light scattering cross section for spherical particles illuminated by a coherent uniphase wave calculated using a Lorenz-Mie theory computer code[3] is given in Fig. 9. The calculations are for an off-axis f/1.96 collection lens centered at θ = 10 deg from the incident radiation propagation direction (forward scattering). The oscillatory behavior is due to resonance interactions in the scattering process and results in ambiguities in particle size determination from SPC scattering measurements. Another problem inherent in using the laser as a SPC radiation source is the nonuniform intensity profile across the beam.[12,40] An ambiguity in signal levels arises for in situ SPC since the particles are free to traverse the sample volume at any position. Thus, particles will experience different peak incident intensities I_{inc} depending on the trajectory and even a monodisperse (uniform size) aerosol will generate a broad distribution of signal amplitudes S.

A number of methods have been devised to eliminate the unknown incident intensity effect in cross-section measuring techniques. The basic approaches include: 1) analysis of only those particles which pass through a selected portion of the beam of known and constant intensity, 2) analysis of all particles and later correction for the known distribution of particle trajectories and corresponding incident intensities, 3) use of the ratio of scattering signals at two or more angles to cancel the incident intensity effect.

For in situ measurements various optical methods of discriminating those particles which pass through a control portion of the beam have been used, including coincidence detectors at 90 deg by Ungut et al.[41] and in the forward direction by Knollenberg.[42] It has also been suggested that a pointer laser beam tightly focussed within a larger probe beam be used to discriminate those particles which pass through the center of the probe beam.[43] This latter approach does not eliminate the ambiguity, but rather shifts the problem to the pointer beam where the effect is less significant. It is also possible to change the intensity profile across the laser beam from Gaussian to something approximating a tophat using specially designed filters. However any beam degradation due to windows or refractive index fluctuations would spread the profile and reintroduce the intensity ambiguity. It appears that no definitive studies on the use of tophat profiles have been reported.

Another somewhat similar technique proposed by Hirleman[44] involves the use of signals generated by particles traversing two adjacent laser beams. The dual peak signature is used to determine two velocity components and the trajectory of each particle. Given known laser beam properties the incident intensity history for a particle is then completely determined which permits a real time correction for the intensity ambiguity. After I_{inc} in Eq. (7) is determined a calibrated response function prediction such as Fig. 9 would be used to relate signal amplitudes to particle size. This technique[44] has been proposed for light scattering, extinction, and fluorescence cross-section measurements although experiments to date have used only light scattering.

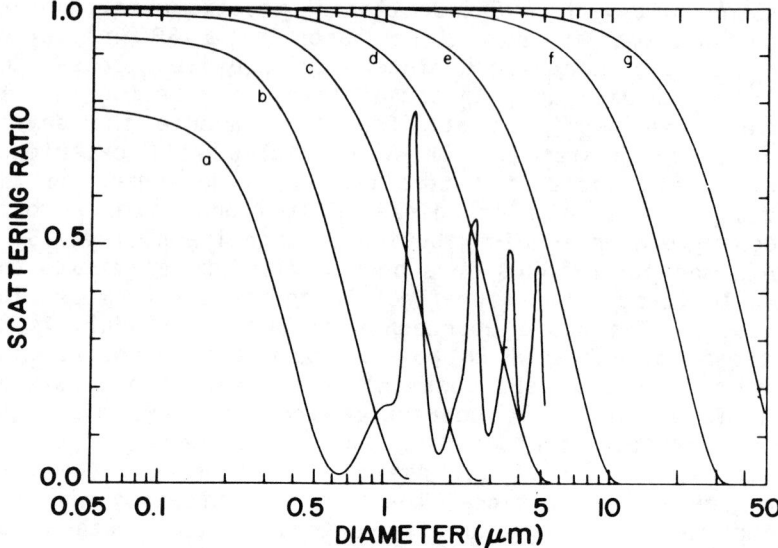

Fig. 10 Response functions for ratio-type SPC. The data apply to spherical particles with n=1.56-0.47i (soot) and λ = 0.6328 μm. The scattering angle pairs are a)48/24 deg, b)24/12 deg, c)12/6 deg, d)6/3 deg, e)3/1.5 deg, f)1/0.5 deg, g)0.5/0.25 deg. All but the 48/24 deg curve were truncated after the first minimum.

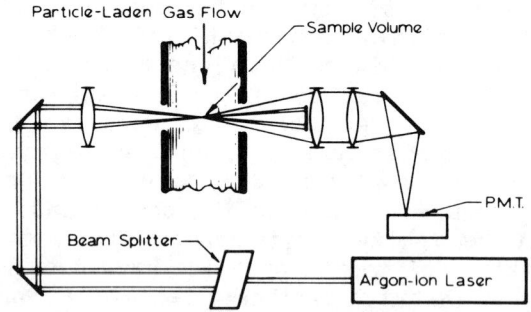

Fig. 11 Schematic of optical system for particle sizing interferometer after Houser.[51]

A second general approach to the ambiguous incident intensity problem is to correct after the fact. One implementation of this approach proposed by Holve and Self[45] is to first consider the distribution of scattering signal pulse heights generated by particles of one size passing with equal probability through all portions of the laser beam focus region. The optical system required again is like Fig. 5 using a single near-forward off-axis detector. The signal height distribution from a polydispersion is then

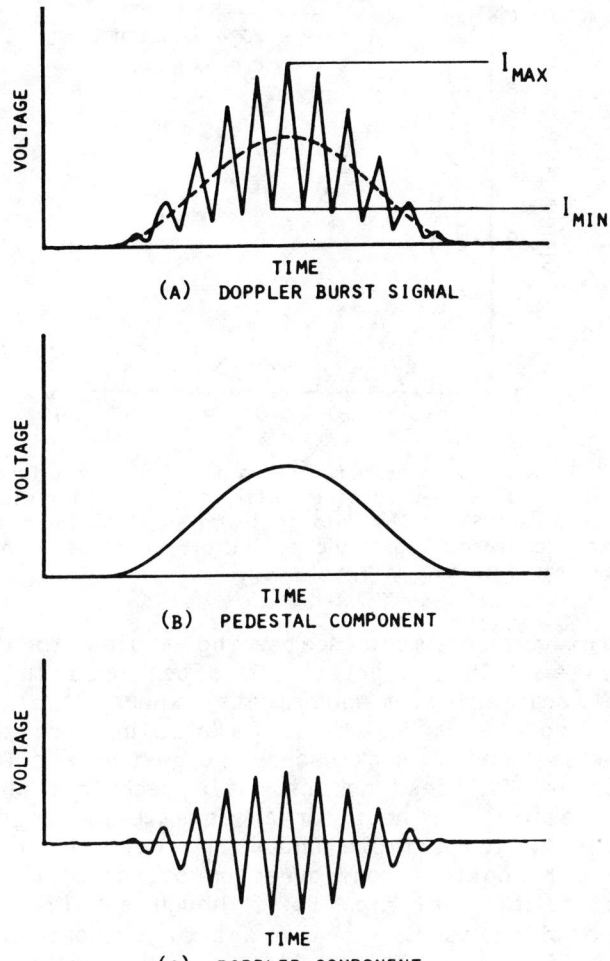

Fig. 12 Signals from particle sizing interferometer after Bachalo.[58]

a linear combination of the monodisperse particle response distributions. A numerical scheme was developed[45] to invert the resulting system of equations and solve for the linear coefficients which are proportional to concentrations in the discretized particle size intervals. This approach[45] has been successfully used for sizing burning droplets and particulates emitted from a coal combustor.

Scattering Intensity Ratio Techniques

The final method to eliminate the incident intensity ambiguity in SPC is to utilize the ratio of scattered light

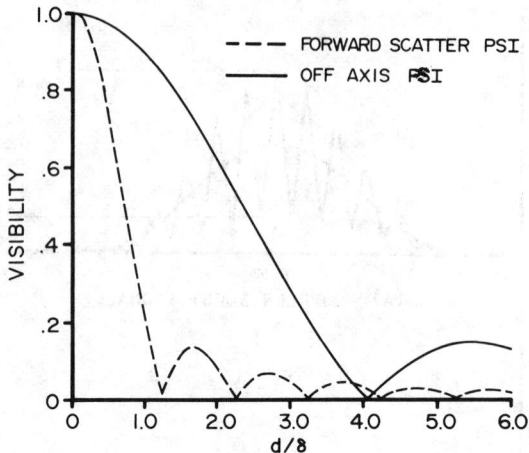

Fig. 13 Calculations for the fringe visibility V as a function of particle diam to fringe spacing ratio d/δ for particle sizing intereferometers (PSI). The data apply to a PSI collecting all of the forward scattered light and to an off-axis PSI with an f/2 collection lens oriented at θ = 20 deg.

signals from two or more scattering angles to determine particle size. This approach is often used in ensemble multiangle scattering measurements where the relative scattering profile rather than the absolute scattering at some angle is used. Hodkinson[46] suggested and Gravatt[47] implemented an SPC based on the ratio technique which used scattering ratios from near-forward scattering angles where the sensitivity to particle shape and refractive index is minimized. The optical configuration of ratio counters can be similar to that in Fig. 5, although annular detection schemes are often used.[12,48] A set of response functions for a ratio SPC is plotted in Fig. 10. One problem evident from Fig. 10 is the multivalued response function plotted for the largest angle pair. Outsize particles, or those larger than the first minimum in the ratio response functions in Fig. 10, will be incorrectly sized by ratio instruments which utilize only a single pair of scattering angles. The multiple ratio concept (MRSPC) developed by Hirleman and coworkers[12,48] was designed to eliminate this ambiguity problem.

Ratio counters still have an optical sample volume which depends on particle size and corrections for this effect must be considered.[12] Also, since forward scattering is generally used, ratio counters are relatively insensitive to particle shape and refractive index.[12]

A possible advance for ratio schemes may be to integrate photodiode array detectors to allow more scattering

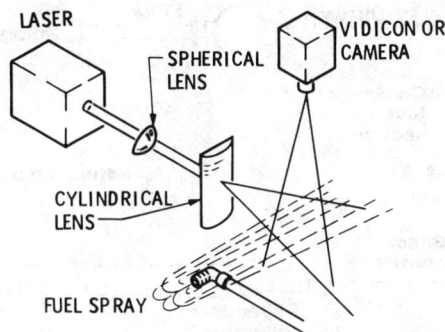

Fig. 14 Schematic of imaging particle sizing system after Fleeter et al.[60]

data to be collected without simply adding photomultiplier tubes. Bartholdi et al.[49] used a linear photodiode array in an SPC application and we are studying the use of intensified versions of the detector in Fig. 8.

Ratio SPC are applicable in the nominal size range of 0.3 - 10.0 µm for practical laser sources. They have been successfully applied in engine exhausts,[48] flame studies,[50] fluidized bed off-gas,[51] and in several other applications.

Particle Sizing Interferometry

Another approach which can provide particle size information independent of incident intensity is particle sizing interferometry (PSI). A schematic is shown in Fig. 11. As a single particle passes through the intersection region of two nonparallel laser beams, Doppler-shifted scattered light waves from each beam emanate from the particle. Heterodyning the two contributions of scattered light at a detector will produce the Doppler-difference frequency which is directly related to the particle velocity and the angle between the laser beam propagation vectors. This principle underlies the laser Doppler velocimeter (LDV). A particle crossing the LDV beam intersection region will produce an approximately Gaussian signal (pedestal) with the modulated Doppler-difference component written on the pedestal[52] as shown in Fig. 12. The ratio of the modulated signal amplitude to the pedestal amplitude, which is termed the visibility, provides a measure of particle size as shown by Farmer[52] and others[53,54] who used a scalar description of the process. For large apertures which collect all of the forward scattered (diffracted) light the visibility V as a function of particle diameter d and fringe

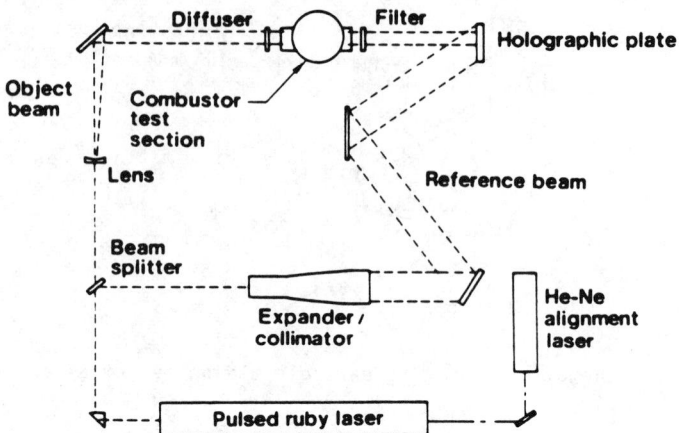

Fig. 15 Schematic of holographic particle sizing system after Chigler.[62]

spacing δ was shown by Robinson and Chu[54] to be

$$V = \frac{2J_1(\pi d/\delta)}{\pi d/\delta} \qquad (8)$$

where J_1 is a first-order Bessel function of first kind. A plot of V is given in Fig. 13.

Calculations considering the complete problem of scattering by a sphere simultaneously in two coherent, collimated laser beams[55] predicted a strong dependence of the visibility on particle refractive index, the detector aperture, and detector position relative to the beams. A number of experimental studies have confirmed the importance of careful receiving optics design[55,56] although conflicting observations have also been made.[57]

Another related approach is the off-axis PSI proposed by Bachalo[58] which utilizes the interference of refracted or reflected light scattering contributions rather than the diffractive scatter of a conventional PSI.[52] This method is applicable to particles significantly larger than the wavelength and is based on the difference in optical path length traveled by refracted rays from the two crossed beams which pass through the particle and arrive coincidently at the detector. The visibility response function for a typical off-axis PSI collection angle[57] of 20 deg is also shown in Fig. 13, and the expanded d/δ sizing range for this concept is apparent.

Although the visibility is a relative measurement, absolute light scattering cross sections and incident laser beam intensity distributions still control the PSI. Only those particles which scatter enough light to be detected above the background noise level will be sized. Thus a PSI will "size" the particles using a relative measurement but the frequency at which particles are "seen" or counted is biased toward large particles.

To correct some serious problems in sizing particles traversing the edge of off-axis PSI probe volumes, it has been suggested that the amplitude of the Doppler bursts from PSI instruments be used to size particles. The incident intensity ambiguity is reintroduced and a correction must be made. Those particles traversing the center of the intersection region can be discriminated using coincidence detection with small aperture detectors or using an additional, tightly focused pointer beam. Unfortunately the latter approach merely shifts the trajectory ambiguity problem from the PSI beams to the Gaussian pointer beam.

Photographic and Holographic Methods

Several different imaging methods have been used for particle and droplet sizing. These rely on a short light pulse to "freeze" the particle images so that direct measurements of size may be made. In the case of double flash photography two closely spaced light pulses are used to obtain double images of each droplet so that velocity can also be determined. Single and double pulse holography have been used as well, with the advantage that a volume of the aerosol can be captured rather than the limited depth of field afforded by photographic methods. The problem with both photographic and holographic methods is the tedious and expensive post processing needed to extract the data. Also, quantitative measurements of particle size distributions with imaging techniques are realistic only for particle sizes greater than 5 μm at best.

Automated data processing for particle photography has been reported by Simmons and Lapera[59] and Fleeter et al.[60] In the first system[59] a strobe light was used to form the image on a vidicon tube. The image is scanned to obtain drop size information and the cycle repeated roughly 10 times per second. Mean diameters and size distributions, were obtained at each point in the spray.[59] Fleeter et al.[60] utilize a pulsed ruby laser as shown in Fig. 14 to illuminate the particles which are imaged onto a 512x512 diode array camera. The image is then digitized[60] and transferred to a computer memory for processing. Knollenberg[42] analyzes individual particles by projecting images onto a linear photodiode array.

One correction factor required in the data analysis of incoherent imaging techniques is the effective depth of field vs droplet size. (Large particles are visible over a larger axial distance from the exact object plane than small particles.) This correction is analagous to sample volume corrections required with SPC and is mandatory before useful data can be obtained.

Photographic image analysis is a very convenient method of particle and droplet sizing under cold flow conditions. One limitation is the typical resolution limit of about five micrometers. In hot flows one would expect substantially poorer results due to image distortion by refractive index fluctuations in the flow. Performance also suffers in applications where windows must be located between the spray and the camera, particularly when the optical aperture is limited. In a recent study of optical methods for Diesel engine research, the threshold of size detection was 35 μm for high-speed photography and 8 μm for holography.[61]

Pulsed holography eliminates the sample volume correction required for photographic methods since the holograms, which contain three-dimensional information, can be observed in two dimensions while the third is scanned. A schematic diagram of a holographic system is shown in Fig. 15. Holographic methods for particle and droplet size analysis have apparently been used to observe particles down to about 5 μm.[63,64] Note however that the resolution of a holographic system is typically several micrometers so that the accuracy in sizing such small particles is very poor. Another problem encountered in particle holography is performance degradation when the laser beam transmission drops below about 10%.[65]

Conclusions

Laser-based techniques for nonintrusive diagnostics of particle size and concentration distributions have been reviewed. The most common diagnostics are imaging and light scattering techniques, and each instrument has its own unique set of limitations and range of applicability. It is imperative that the subtle factors which control the accuracy and reliability of data obtained with laser/optical instruments be understood by the user.

Acknowledgments

The author's research in optical particle diagnostics has been generously supported over the past five years by the National Science Foundation, Particulate and Multiphase Processes Group, Dr. Morris Ojalvo, Program Director, and by the Office of Naval Research, Propulsion Group, Dr. Albert D. Wood, Project Director.

References

[1] Born, M. and Wolf, E., *Principles of Optics*, Sixth edition, Permagon Press, New York, 1980.

[2] Hecht, E. and Zajac, A., *Optics*, Addison-Wesley, New York, 1974.

[3] van de Hulst, H. C., *Light Scattering by Small Particles*, John Wiley and Sons, New York, 1957.

[4] Kerker, M., *The Scattering of Light and Other Electromagnetic Radiation*, Academic Press, New York, 1969.

[5] Handa, T., Suda, K., Nagashima, T., Kaneko, K., Yamamura, T., Takahashi, Y., and Suzuki, H., "Size Determination of Submicron Particulates by Optical Counter," *Fire Research*, Vol. 1, 1978, pp. 255-263.

[6] Asano, S., "Light Scattering Properties of Spheroidal Particles," *Applied Optics*, Vol. 18, 1979, pp. 712-723.

[7] Latimer, P. and Wamble, F., "Light Scattering by Aggregates of Large Colloidal Particles," *Applied Optics*, Vol. 21, 1982, pp. 2447-2455.

[8] Borghese, F., Denti, P., Toscano, G., and Sinden, O. I., "Electromagnetic Scattering by a Cluster of Spheres," *Applied Optics*, Vol. 18, 1979, pp. 116-120.

[9] Albini, F. A. and Nagelberg, E. R., "Scattering of a Plane Wave by an Infinite Inhomogeneous, Dielectric Cylinder - An application of the Bern Approximation," *Journal of Applied Physics*, Vol. 33, 1962, pp. 1706-1713.

[10] Schuerman, D. W., Wang, R. T., Gustafson, B. A. S., and Schaefer, R. W., "Systematic Studies of Light Scattering. 1. Particle Shape," *Applied Optics*, Vol. 20, 1981, pp. 4039-4050.

[11] Zerull, R. H. and Giese, R. H., in *Planets, Stars and Nebulae Studied with Photopolarimetry*, T. Gehrels, ed., University of Arizona Press, Tucson, Ariz., 1974, p. 901.

[12] Hirleman, E. D. and Moon, H. K., "Response Characteristics of the Multiple Ratio Single Particle Counter," *Journal of Colloid and Interface Science*, Vol. 87, 1982, pp. 124-139.

[13] Chylek, P. and Ramaswamy, V., "Lower and Upper Bounds on Extinction Cross Sections of Arbitrarily Shaped Strongly Absorbing or Strongly Reflecting Nonspherical Particles," *Applied Optics*, Vol. 21, 1982, pp. 4339-4344.

[14] Felton, P.G., "Measurement of Particle/Droplet Size Distributions by a Laser Diffraction Technique," 2nd European Symposium on Particle Characterization PARTEC, Nurnberg, West Germany, September 24, 1979, Report HIC 326 Dept. of Chem. Engr., Univ. of Sheffield, Sheffield, England.

[15] Dobbins, R. A. and Jizmagian, G. S., "Particle Size Measurements Based on Use of Mean Scattering Cross Sections," *Journal of the Optical Society of America*, Vol. 56, 1966, pp. 1351-1354.

[16] Ariessohn, P. C., Self, S. A., and Eustis, R. H., "Two-wavelength Laser Transmissometer for Measurements of Mean Size and Concentration of Coal Ash Droplets in Combustion Flows," *Applied Optics*, Vol. 19, 1981, pp. 3775-3781.

[17] Lester, T. W., and Wittig, S. L. K., "Particle Growth and Concentration Measurements in Sooting Homogeneous Hydrocarbon Combustion Systems," in *Proceedings of the Tenth International Shock Tube Symposium*, G. Kamimoto, ed., The Shock Tube Research Society, Kyoto, Japan, 1975.

[18] Bro, K., "The Optical Dispersion Quotient Method for Sizing of Soot in Shock-induced Combustion," Ph.D. Thesis, Purdue University, Lafayette, Ind., 1978.

[19] Powell, E. A., Cassonova, R. A., Bankston, C. P., and Zinn, B. T., "Combustion-Generated Smoke Diagnostics by Means of Optical Measurement Techniques," *Experimental Diagnostics in Gas Phase Combustion: AIAA Progress in Aeronautics and Astronautics*, Vol. 53, B. T. Zinn, ed., AIAA, New York, 1977, pp. 449-463.

[20] Fymat, A. L., "Analytical Inversions in Remote Sensing of Particle Size Distributions: Multispectral Extinctions in the Anomalous Diffraction Approximation," *Applied Optics*, Vol. 17, 1978, pp. 1675-1676.

[21] Hansen, M. A. and Evans, W. H., "Polar Nephlometer for Atmospheric Particulate Studies," *Applied Optics*, Vol. 19, 1980, pp. 3389-3395.

[22] Hansen, M. Z., "Atmospheric Particulate Analysis Using Angular Light Scattering," *Applied Optics*, Vol. 19, 1980, pp. 3441-3448.

[23] McCay, T.D., Mundy, W.C., Mann, D.M., and Meserve, G.S., "Laser Mie Scattering Measurements of Particle Size in Solid Rocket Motor Exhaust," JANNAF 12th Plume Technology Meeting, CPIA Publication #332, Dec. 1980, p. 145.

[24] Santoro, R. J. and Semerjian, H. G., "Interpretation of Optical Measurements of Soot in Flames," AIAA Paper 83-1516, 18th Thermophysics Conference, Montreal, Canada, June 1983.

[25] Chang, P. H. P. and Penner, S. S., "Particle Size Measurements in Flames using Light Scattering: Comparison with Diffusion Broadening Spectroscopy," *Journal of Quantum Spectroscopy and Radiation Transfer*, Vol. 25, 1981, pp. 97-110.

[26] Powell, E. A. and Zinn, B. T. "In-situ Measurements of Complex Refractive Index of Combustion Generated Particulates," AIAA Paper 83-1518, 18th Thermophysics Conference, Montreal, Canada, June 1983.

[27] Hirleman, E. D., "On-line Calibration Technique for Laser Diffraction Droplet Sizing Instruments," ASME Paper 83-GT-232, 20th International Gas Turbine Conference, Phoenix, Ariz., March 1983.

[28] Chin, J. H., Sliepcevich, C. M., and Tribus, M., "Determination of Particle Size Distributions in Polydisperse Systems by Means of Measurements of Angular Variation of Intensity of Forward Scattered Light at Very Small Angles," Journal of Physical Chemistry Ithaca, Vol. 5, 1955, p. 841.

[29] Recognition Systems Inc., Van Nuys, Calif.

[30] Malvern Instruments Ltd., Malvern, Worcestershire, England.

[31] Swithenbank, J., Beer, J., Taylor, D. S., Abbot, D., and McCreath, C. G.,"A Laser Diagnostic Technique for the Measurement of Droplet and Particle Size Distribution," Experimental Diagnostics in Gas-Phase Combustion Systems: AIAA Progress in Astronautics and Aeronautics, Vol. 53, B. T. Zinn, ed., AIAA, New York, 1977, pp. 421-447.

[32] Titchmarsh, E. C., "Extensions of Fourier's Integral Formula to Formulae involving Bessel Functions," Proceedings London Mathematical Society, Vol. 23, 1924, p. xxii.

[33] Dobbins, R. A., Crocco, L., and Glassmann, I., "Measurement of Mean Particle Sizes of Sprays from Diffractively Scattered Light," AIAA Journal, Vol. 1, 1963, pp. 1882-1806.

[34] Roberts, J. H. and Webb, M. J., "Measurement of Droplet Size for Wide Range Particle Distributions," AIAA Journal, Vol. 2, 1964, pp. 583-585.

[35] Dieck, R. H. and Roberts, R. L., "The Determination of the Sauter Mean Droplet Diameter in Fuel Spray Nozzles," Applied Optics, Vol. 9, 1970, pp. 2007-2014.

[36] Alger, T. W. "Polydisperse-particle Size Distribution Function Determined from Intensity Profile of Angularly Scattered Light," Applied Optics Vol. 18, 1979, p. 3494.

[37] Benedek, G. B., Polarization, Matter, and Radiation, Presses Universitaires de France, Paris, 1969.

[38] Penner, S. S., Bernard, J. M., and Jerskey, T., "Laser Scattering from Moving Polydisperse Particles in a Flame II: Preliminary Experiments," Acta Astronautica, Vol. 3, 1976, p. 93.

[39] King, G. B., Sorenson, C. M., Lester, T. W., and Merklin, J. F., "Photon Correlation Spectroscopy used as a Particle Size Diagnostic in Sooting Flames," Applied Optics, Vol. 21, 1982, pp. 976-978.

[40] Hirleman, E. D., "Laser-based Single Particle Counters for in-situ Particulate Diagnostics," Optical Engineering, Vol. 19, 1980, pp. 854-860.

[41] Ungut, A., Yule, A. J., Taylor, D. S., and Chigier, N. A., "Simultaneous Velocity and Particle Size Measurements in Two Phase Flows by Laser Anomometry," AIAA Paper 78-74, 16th Aerospace Sciences Meeting, Huntsville, Ala., Jan. 1978.

[42] Knollenberg, R. G., "The Use of Low Power Lasers in Particle Size Spectroscopy," *Practical Applications of Low Power Lasers*, SPIE, Vol. 92, 1977, pp. 137-152.

[43] Hess, C., Spectron Development Laboratories, Private Communication, May 1983.

[44] Hirleman, E. D., "Laser Technique for Simultaneous Particle Size and Velocity Measurements," *Optics Letters*, Vol. 3, 1978, pp. 19-21.

[45] Holve, D. and Self, S. A., "Optical Particle Sizing for *in situ* Measurements, Parts I and II,' *Applied Optics*, Vol. 18, 1979, pp. 1632-1652.

[46] Hodkinson, J. R., "Particle Sizing by Means for the Forward Scattering Lobe,' *Applied Optics*, Vol. 5, 1966, p. 839.

[47] Gravatt, C. C., "Real Time Measurement of the Size Distribution of Particulate Matter by a Light Scattering Method," *APCA Journal*, Vol. 23, 1973, p. 1035.

[48] Hirleman, E. D., "*Optical Technique for Particulate Characterization in Combustion Environments: The Multiple Ratio Single Particle Counter*," Ph.D. Thesis, Purdue University, West Lafayette, Ind., 1977.

[49] Bartholdi, M., Salzman, G. C., Hiebert, R. D., and Seger, G., "Single Particle Light Scattering Measurements with a Photodiode Array," *Optics Letters*, Vol. 1, 1977, pp. 223-225.

[50] Samuelson, G. S., Hack, R. L., Poon, C. C., and Bachalo, W. D., "Study of Soot Formation in Premixed and Nonpremixed Flows with Complex Aerodynamics," Paper WSSCI-80-10, presented at the Western States Section Combustion Institute Meeting, Irvine, Calif., April 22, 1980.

[51] Houser, M. J. "Particle Field Diagnostics: Applications of Intensity Ratioing, Interferometry, and Holography," *Optical Engineering*, Vol. 19, 1980, pp. 873-877.

[52] Farmer, W. M., "Measurement of Particle Size, Number Density, and Velocity Using a Laser Interferometer," *Applied Optics*, Vol. 11, 1972, p. 2603.

[53] Fristrom, R. M., Jones, A. R., Schwar, M. S. R., and Weinberg, F. S., "Particle Sizing by Interference Fringes and Signal Coherence in Doppler Velocimetry," *Faraday Symposia of the Chemical Society*, Vol. 1, 1973, p. 183.

[54] Robinson, D. M. and Chu, W. P. "Diffraction Analysis of Doppler Signal Characteristics for a Cross-beam Laser Doppler Anemometer," *Applied Optics*, Vol. 14, 1975, p. 2177.

[55] Chu, W. P. and Robinson, D. M., "Scattering from a Moving Spherical Particle by Two Crossed Coherent Plane Waves," *Applied Optics*, Vol. 16, 1977, p. 619.

[56] Yule, A. J. Chigier, N. A., Atakan, S., and Ungut, A., "Particle Size and Velocity Measurement by Laser Anemometry," AIAA Paper 77-214, 15th Aerospace Sciences Meeting, Los Angeles, Calif., January 1977.

[57] Farmer, W. M., "Measurement of Particle Size and Concentrations Using LDV Techniques," *Proceedings of The Dynamic Flow Conference*, Dynamic Flow Conference, DK 2740, Skovlunde, Denmark, 1978.

[58] Bachalo, W. D., "Method for Measuring the Size and Velocity of Spheres by Dual-Beam Light-Scatter Interferometry," *Applied Optics*, Vol. 19, 1980, pp. 363-370.

[59] Simmons, H. C. "The Correlation of Drop-size Distributions in Fuel Nozzle Sprays," *Journal of Engineering for Power*, Vol. 99, 1977, pp. 309-314.

[60] Fleeter, R., Toaz, R., and Sarohia, V., "Application of Digital Image Analysis Techniques to Antimisting Fuel Spray Characterization," ASME Paper 82-WA/HT-23, American Society of Mechanical Engineers, New York, 1982.

[61] Lennert, A. E., Sowls, R. E., Belz, R. A., Goethert, W. H., and Bentley, H. T., "Electro-optical Techniques for Diesel Engine Research," *Experimental Diagnostics in Gas Phase Combustion: AIAA Progress is Astronautics and Aeronautics*, Vol. 53, B. T. Zinn, ed., AIAA, New York, 1977, pp. 629-656.

[62] Chigier, N., "Spray Combustion Processes: A Review," ASME Paper 82-WA/HT-86, American Society of Mechanical Engineers, New York, 1982.

[63] Trolinger, J. D. "Particle Field Diagnostics by Holography," Paper 80-0018, 18th Aerospace Sciences Meeting, Pasadena, Calif., January, 1980, American Institute of Aeronautics and Astronautics, New York, 1980.

[64] Thompson, B. J., "Holographic Particle Sizing Techniques," *Journal of Physics E*, Vol. 7, 1974, pp. 781-788.

[65] Jones, A. R., "Error Contour Charts Relevant to Particle Sizing by Forward-Scattered Lobe Methods," *Journal of Physics D*, Vol. 10, 1977, pp. L163-L165.

Interpretation of Optical Measurements of Soot in Flames

R. A. Dobbins,* R. J. Santoro,† and H. G. Semerjian‡
National Bureau of Standards, Washington, D.C.

Abstract

The mean cross sections for directional scattering and extinction are calculated for absorbing spheres obeying the log normal size distribution function using Mie theory. These properties are used to calculate the dissymmetry ratios, the scattering-extinction ratios, and the depolarization ratios for polydispersions of specified complex refractive index. The use of this information to deduce particle volume fraction, the various mean sizes and the width of the distribution, and the particle number concentration is discussed. Optical observations of agglomerated soot in flames in our laboratory and elsewhere are reviewed. The incompatability of these observations with the Mie theory for polydispersions of absorbing spheres noted by D'Alessio et al. is confirmed. It is concluded that this conflict arises because the loosely packed, low density agglomerates have an effective refractive index that is significantly reduced below that of the particulate material. The downward scaling of the refractive index in the manner suggested in the past for macroscopic aggregates of soot material with distributed finite void spaces alleviates the incompatabilities. When the particles display the characteristics of Mie scattering, it is possible to determine the soot volume fraction, the width of the distribution function and various mean diameters, the agglomerate

---

Presented as Paper 83-1516 at the AIAA 18th Thermophysics Conference, Montreal, Canada, June 1-3, 1983. This paper is declared a work of the U.S. Government and therefore is in the public domain.
 *Professor, Division of Engineering, Brown University, Providence, R.I.
 †Research Physicist, Center for Chemical Engineering, Chemical Process Metrology Division.
 ‡Group Leader, Center for Chemical Engineering, Chemical Process Metrology Division.

number concentration, and the effective refractive index of the soot agglomerates from certain optical observations. The solution for the soot properties is recovered from the observed data by the minimization of an aggregate relative error of the observations using a method for the least squares minimization of nonlinear functions. Illustrative examples based on recent observations in a laminar ethene/air diffusion flame are provided.

Introduction

Considerable interest has been generated in the use of laser light absorption and scattering by soot particles to follow their formation, transport, and oxidation in hydrocarbon flames[1-4]. The observations usually have been interpreted by means of the Rayleigh theory for absorbing spheres of uniform diameter. This method of data reduction may be acceptable if the size of the particles is small ($D<\lambda/10$) and their number concentration is sufficiently low ($\bar{N}<10^{10}$ cm^{-3}) so as to reduce effects due to agglomeration. These requirements are often not fulfilled in flames where measurements of the angular distribution of scattered light shows the Rayleigh theory to be inadequate or where high particle concentrations favor particle growth by agglomeration.

Earlier studies[5] have reported that the soot particle sizes found from dissymmetry ratios from Mie theory for polydisperse spheres were 60% higher than those yielded by the scattering-extinction ratio. The latter was considered to be a more reliable indicator of particle size because of its lower sensitivity to the standard deviation σg of the particle size distribution in the smaller size range. A second independent attempt to use extinction, angular dissymmetry, and polarization ratios for polydispersions obeying a logarithmic normal distribution was made for acetylene pyrolysis experiments.[6] Departures from sphericity and uncertainties of refractive index were considered to be the impediments to reliable interpretation of the measurements in terms of the properties of the soot aerosol.

These difficulties in interpreting the optical measurements of soot aerosols were recently reviewed.[7] It was pointed out that values of $Q_{hv}(90\ deg)/Q_{vv}(90\ deg)$ equal to 1.0×10^{-2} were found in a premixed flame where angular dissymmetry is evident while Mie theory predicts $Q_{hv}(\theta) = 0$ for all angles. It was also stated that the shift from 90 deg of the angle θ_m at which $Q_{hh}(\theta)$ is a minimum, which should accompany the observed dissymmetry of the angular scattering produced by large particles, could not be detected experimentally. Finally, it was pointed out that no optical model exists for loose clusters or chains formed by agglomeration.

Many of these trends have been confirmed in our own measurements in an ethene/air diffusion flame. In this paper, we will present a systematic approach to the interpretation of light scattering measurements based on the Mie theory of light scattering for polydispersed absorbing spheres. The proposed interpretation will be applied to measurements made in our laboratory on soot particles.

Optical Properties of a Polydispersion of Absorbing Spheres

Particle Size Distribution Function

The relative frequency of occurrence of the various sizes in a population is given by the size distribution function $P(D)$ and is defined by the relationship

$$\int_{D_1}^{D_2} P(D) dD = P(D_1 < D < D_2) \qquad (1)$$

where the righthand side represents the probability of occurrence of diameters greater than D_1 and less than D_2. From Eq. (1) it follows that the value of the integral over all sizes is unity.

Various moment ratios of the particle size distribution function, or their closely related mean diameters, are useful in describing a wide variety of physical phenomena. The general definition of the pth to q^{th} moment of the size distribution is

$$D_{pq}^{p-q} = \int_0^\infty (P(D) D^p dD) / \int_0^\infty (P(D) D^q dD) \qquad (2)$$

while the corresponding mean diameters are

$$D_{pq} = (D_{pq}^{p-q})^{1/(p-q)} \qquad (3)$$

In terms of a generalized mean diameter, the volume fraction of the particulate phase, for example, involves the third moment of the particle size distribution function and can be expressed as

$$\phi = \frac{\pi}{6} N \int_0^\infty P(D) D^3 dD = \frac{\pi}{6} N D_{30}^3 \qquad (4)$$

where N is the number of particles of all sizes per unit volume and D^3_{30} is third moment of the distribution function or the cube of the volume mean diameter.

A widely used expression for $P(D)$ is the logarithmic normal distribution that is given by

$$P(D) = \exp[-(\frac{\ln D/D_g}{\sqrt{2} \; \sigma_g})^2] / (\sqrt{2\pi} \; \sigma_g \; D) \qquad (5)$$

where D_g is the geometric mean diameter and σ_g is the geometric mean standard deviation. The generalized moment ratio expressed by Eq.(2) can be evaluated for the logarithmic normal distribution, Eq.(5), for specified D_g and σ_g. This results in the following expression for the mean diameter

$$\frac{D_{pq}}{D_g} = \exp(\frac{p+q}{2} \sigma_g^2) \qquad (6)$$

From Eq. (6) it is clear that if D_g and σ_g are known, then all mean diameters are known. Similarly any two mean diameters are also adequate to specify the values of both D_g and σ_g.

Another parameter that is used to describe the width of a size distribution function is the quantity

$$f_N = \frac{D_{60}^6}{(D_{30}^3)^2} = (\frac{D_{63}}{D_{30}})^3 \qquad (7)$$

From Eq.(6) we find that for a logarithmic normal distribution

$$f_N = \exp(9 \sigma_g^2) \qquad (8)$$

The quantity f_N plays a major role in determining the particle number concentration from the optical observations.

The size distribution function of the particles of a coagulating aerosol is evolutionary in that both σ_g and D_g change as a result of particle growth caused by interparticle collisions induced by Brownian motion. Under some circumstances the size distribution function may reach a limiting (dimensionless) form called the self-preserving size distribution.[8] The necessary conditions for the ultimate achievement of this size distribution are the absence of mechanisms of particle depletion, formation or growth, as well as the absence of other mechanisms to induce collisions. The properties of the self-preserving particle size distribution function that is achieved when the particles are small compared with the mean free path of the gas molecules are provided in Refs. 8 and 9. The value of f_N from the improved tabulation of Ref. 9 is 2.0788. If this number is inserted into Eq.(8), we find σ_g equals 0.28515 which is an important reference value defining the logarithmic normal distribution that duplicates the value of f_N for the self-preserving distribution function.

Some information on the particle size distribution function applicable for soot particles is available through several studies in which electron microscopy was used to examine soot withdrawn from flames by sampling techniques. An important study of this type[10] was performed on premixed acetylene/air flames at 20-Torr pressure in which it was concluded that the size distribution of the volume equivalent diameters was Gaussian near the burner mouth and logarithmic normal at greater heights. A second study using a slot burner supplied with premixed acetylene/air as well as a propane fired soot generator[11] found that the volume equivalent diameters obeyed the logarithmic normal distribution. Other tests[12] conducted on soot formed by the thermal decomposition of benzene in nitrogen also showed a transition from a Gaussian distribution to a logarithmic normal distribution as distance from the initial formation zone increased.

Recently a theoretical study[13] has been performed on the time-dependant behavior of an aerosol subject to the combined actions of free molecular agglomeration and sustained particle formation at rates representative of diffusion flames. An approximate integral solution was obtained in which the logarithmic normal size distribution function was assumed to be applicable. The important conclusion of this study is that the width of the distribution function increases at a dramatic rate during the period of active particle formation. The calculations show that the self-preserving distribution function is never reached in the lower part of the flame where particle formation is present. The previously mentioned necessary conditions for the ultimate achievement of this size distribution are not fulfilled in the lower part of the flame, where gas-to-particle conversion is active, nor in the upper portion of the flame, where oxidation of the soot particles occurs. For these reasons the width of the size distribution function should be treated as an unknown that is to be determined, where possible, by the optical experiments.

Light Scattering Properties of Polydisperse Absorbing Spheres

For the case of single scattering events, two optical properties are of principle interest when plane polarized light is incident upon a volume element containing many small particles. The first property is the mean (differential) scattering cross section $\bar{C}_{ij}(\theta)$ for the production of scattered light at a specified direction. This quantity is defined as the intensity of light scattered by the particles of various sizes in the illuminated volume at the angle θ measured from the direction of propagation, per unit of incident intensity, per particle. The subscripts i and j assume the letters v or h according to whether the state of polarization of the scattered (i) and incident (j) radiation is perpendicular or parallel, respectively, to the plane of observation. For spherical particles of isotropic materials only the case where i = j is of interest because cross polarization effects are absent. The mean scattering cross section is given by an integration over the size distribution function

$$\bar{C}_{ii}(\theta) = \int C_{ii}[\theta, \frac{\pi D}{\lambda}, \tilde{m}(\lambda)] \, P(D) \, dD \qquad (9)$$

where C_{ii}, the cross section for a specified particle size, is provided by electromagnetic theory. The closely related volumetric scattering cross section $Q_{ii}(\theta)$ is given by

$$Q_{ii}(\theta) = N \, \bar{C}_{ii}(\theta) \qquad (10)$$

The second property of interest is the mean (total) cross section for extinction which represents the sum of the scattering in all directions and absorption by the particles of various sizes upon which the transmitted beam is incident. This quantity is formulated in terms of the efficiency factor for extinction Q_{ext}, the particle cross section, and the distribution function as

$$\bar{C}_{ext} = \frac{\pi}{4} \int_0^\infty Q_{ext}[\frac{\pi D}{\lambda}, \tilde{m}(\lambda)] \, P(D) \, D^2 \, dD \qquad (11)$$

where Q_{ext} is given by electromagnetic theory.
The extinction coefficient is then given by

$$K_{ext} = N \, \bar{C}_{ext} \qquad (12)$$

and determines the optical transmission that is expressed by

$$\frac{I}{I_0} = \exp\left[-\int_0^L K_{ext}(s)\,ds\right] \qquad (13)$$

For polydispersions of larger spherical particles, we employ the appropriate expressions from Mie theory to find C_{ii} (i = v or h) and Q_{ext} for use in evaluating the integrals in Eqs. (9) and (11). These are represented by

$$C_{vv}(\theta,X,\tilde{m}) = \frac{i_1}{k^2} = \frac{1}{k^2}|S_1(\theta)|^2 \qquad (14)$$

$$C_{hh}(\theta,X,\tilde{m}) = \frac{i_2}{k^2} = \frac{1}{k^2}|S_2(\theta)|^2 \qquad (15)$$

where $X = \pi D/\lambda$, $k = 2\pi/\lambda$ and i_1, i_2, S_1, and S_2 are the intensity and complex amplitude functions in the customary notation.[14,15] The efficiency factor for extinction is given by

$$Q_{ext}(X,\tilde{m}) = 4\left(\frac{\lambda}{\pi D}\right)^2 \mathrm{Re}[S(0)] \qquad (16)$$

where $S(0)$ is $S_1 = S_2$ evaluated at $\theta = 0$. Unfortunately, Eqs. (14-16) are series expressions and no algebraically explicit relationships are available.

The general expressions for the mean differential and total cross-sections C_{ii} and C_{ext} applicable to polydispersions are found by inserting Eqs. (5), (14) or (15), and (16) into Eqs. (9) and (11). Because D_{63} is yielded directly for small size populations to which the Rayleigh limit of Mie theory applies (see next section), we use the quantity $X_{63} = \pi D_{63}/\lambda$ as the independent size variable. From Eq. (6) it follows that

$$X_{63} = X_g \exp\left(\frac{9}{2}\sigma_g^2\right) \qquad (17)$$

The properties defined by Eqs. (9) and (11) are used directly and also as ratios such as the scattering-extinction ratio $Q_{vv}(\theta)/K_{ext}$, where θ is a specified angle. Also of interest are the scattering ratios that describe the angular distribution of scattered light

$$R_{k1} = \frac{Q_{vv}(\theta_k)}{Q_{vv}(\theta_1)} \qquad (18)$$

and the polarization ratio

$$\rho_k(\theta) = \frac{Q_{hh}(\theta_k)}{Q_{vv}(\theta_k)} \qquad (19)$$

The use of these ratios to determine the properties of dispersed particulate phase of an aerosol is discussed in the following section.

Optical Properties of Small Absorbing Spheres in the Rayleigh Regime

When all sizes of the particle population are small compared with the wavelength of light ($D<\lambda/10$) all terms but the first in the Mie series are small and Rayleigh theory is recovered. The simple algebraic results that are yielded provide explicit relations involving the particle field properties of interest. The mean scattering cross section for a polydispersion of small spheres is given from Rayleigh theory[14,15] as

$$\bar{C}_{vv}^{Ray} = \frac{1}{k^2} \left(\frac{\pi D_{60}}{\lambda}\right)^6 F(\tilde{m}) \qquad (20)$$

and

$$\bar{C}_{hh}^{Ray} = \bar{C}_{vv}^{Ray} \cos^2\theta \qquad (21)$$

where

$$F(\tilde{m}) = |(\tilde{m}^2 - 1)/(\tilde{m}^2 + 2)| \qquad (22)$$

For small absorbing spheres, absorption dominates over scattering, and extinction, normally the sum of these two effects, is then simply equal to the absorption. The mean extinction cross section is then given by[14,15]

$$\bar{C}_{ext}^{Ray} = \frac{\lambda^2}{\pi} \left(\frac{\pi D_{30}}{\lambda}\right)^3 E(\tilde{m}) \qquad (23)$$

where

$$E(\tilde{m}) = -\text{Im}\left(\frac{\tilde{m}^2 - 1}{\tilde{m}^2 + 2}\right) \tag{24}$$

Several important features are displayed by Eqs. (20), (21), and (23). Both Eqs. (20) and (23) are expressed in terms of two moment ratios that apply for all size distribution functions. Secondly, the angular distribution of scattered light is independent of size ($R_{kl} = 1.00$ for all θ_k and θ_l; $\rho_k = \cos^2\theta_k$ for all θ_k). Finally the high powers associated with the mean diameters in Eqs. (20) and (23) implies that the smaller sizes make a small contribution to both scattering and extinction relative to the contribution made by the largest sizes in the tail of the distribution function.

For small absorbing spheres, Rayleigh theory can be applied to provide the particle field properties from optical measurements. For example, since from Eqs. (12) and (23), ND^3_{30} is directly measured by K_{ext} equals $N C_{ext}$, we may combine Eqs. (4), (12) and (23) to obtain an equation for particle volume fraction as

$$\phi = \frac{\lambda K_{ext} f_\phi}{6\pi E(\tilde{m})} \tag{25}$$

where f_ϕ, defined by

$$f_\phi = \frac{\bar{C}^{Ray}_{ext}}{\bar{C}_{ext}(X_{63}, \sigma_g, \tilde{m})} \tag{26}$$

is presently unity and is introduced for later purposes. Because K_{ext} is proportional to ND^3_{30} for any distribution function, a knowledge of the latter quantity is unnecessary to find ϕ whose determination rests solely on the measurement of the local extinction coefficient and on the knowledge of $E(\tilde{m})$.

A measure of the particle size can be obtained from the ratio $Q_{vv}(90 \text{ deg})/K_{ext}$ which is equal to $C_{vv}(90 \text{ deg})/C_{ext}$. From Eqs. (20) and (23) we find the mean diameter D_{63} is given by

$$D_{63} = \frac{\lambda}{\pi}\left[4\pi \frac{E(\tilde{m}) Q_{vv}(90 \text{ deg})}{F(\tilde{m}) K_{ext}}\right]^{1/3} \tag{27}$$

This Rayleigh limit result is again applicable for any particle size distribution function.

By combining Eqs. (4) and (7) we obtain an expression for the number concentration N in terms of ϕ and D_{63}

$$N = (6\phi\, f_N/\pi\, D_{63}^3) \qquad (28)$$

Clearly N can be calculated only if f_N is known. Two limiting cases are the monodispersion ($f_N = 1$) and the self-preserving distribution ($f_N \approx 2.1$), but a wide polydispersion can have a value of $f_N \approx 80$ if $\sigma_g \approx 0.7$. The determination of N requires a knowledge of the width of the distribution function that is characterized by f_N or σ_g. Rayleigh theory does not yield any additional information on the width of the distribution function. If independent information indicates a monodispersion is present, then $D_{63} = D_{30} = D$ and $f_N = 1.00$. Observations of $Q_{vv}(90 \text{ deg})$ and K_{ext} can be used with Eqs. (25), (26), and (28) to determine ϕ, D, and N.

We summarize the results for purely optical analysis of polydispersions of absorbing Rayleigh particles. Both ϕ and D_{63} are directly determined for a polydispersion obeying any size distribution function if Rayleigh scattering is observed and the refractive index is known. Measurements of the light scattered by small absorbing polydispersed spheres does not yield the information relating to the width the distribution function which is necessary to determine f_N and N. A test for the applicability of Rayleigh theory is afforded by the measurement of R_{ij} which must be approximately unity as will occur if $D \leq \lambda/10$.

Use of Optical Properties to Determine Particle Properties in the Mie Regime

Optical properties in the Mie regime are more diverse and therefore contain greater information about the scattering field. The consequence is that both mean size and the width of the distribution function can be deduced if the particle sizes lie in the Mie regime.

For particles of arbitrarily large size of specified refractive index we seek the values of X_{63} and σ_g which satisfy the observed optical observations, e.g., $Q(90 \text{ deg})/K_{ext}$ and R_{12}. Since the quantities are expressed as ratios they are independent of particle concentration and are solely properties of the local X_{63} and σ_g. This can be accomplished in a trial and error manner by searching through the X_{63}-σ_g space until

Table 1 $E(\tilde{m})$ and $F(\tilde{m})$ for selected values of $\tilde{m}(\lambda)$

Ref.	$\tilde{m}(\lambda)$	$E(\tilde{m})$	$F(\tilde{m})$	$[E(\tilde{m})/F(\tilde{m})]^{1/3}$
19	1.57-0.56i	0.2595	0.2173	1.061
20	1.90-0.55i	0.1927	0.2979	0.8649
21	1.70-0.68i	0.2780	0.2959	0.9794

Table 2 The scattering ratio R_{12} as a function of X_{63} for selected values of refractive index ($\sigma_g = 0.28515$)

λ=514.5 nm		R_{12}		
X_{63}	D_{30} (nm)	$\tilde{m}_1$ 1.57-0.56i	$\tilde{m}_2$ 1.90-0.55i	$\tilde{m}_3$ 1.70-0.68i
0.2	25.7	1.020	1.023	1.020
0.4	51.3	1.081	1.095	1.083
0.6	77.0	1.187	1.221	1.192
1.0	128	1.571	1.646	1.568
1.6	205	2.844	2.939	2.793
2.0	257	4.435	4.526	4.313

suitable agreement with observed quantities is achieved. A more systematic method is affforded by the solution to the nonlinear least-squares problem.[16,17] The Levenberg-Marquardt algorithm[17,18] has been found advantageous for this purpose.

Interpretation of the Optical Properties of Polydispersed Spheres

Optical Properties Based on the Refractive Index of Soot Particulate Material

The optical properties of a polydispersion of spheres are dependent on the refractive index, the mean diameter, and the width of the size distribution. In this section we will examine the sensitivity of the various scattered light quantities to these properties. In the Rayleigh regime these quantities depend solely on $E(\tilde{m})$, $F(\tilde{m})$, and the cube root of their ratio and can be calculated once λ is specified. At this time there is some disagreement on the value of $\tilde{m}(\lambda)$ for soot material and we list quantities of interest in Table 1 for three more recent recommendations[19-21] (see also discussion in Ref. 22). The

Table 3 The scattering ratio R_{12} for selected values of X_{63} and σ_g ($\tilde{m}$ = 1.70-0.68i)

X_{63} \ σ_g	0.10	0.28515	0.50	0.70
0.2	1.014	1.020	1.047	1.109
0.4	1.058	1.083	1.163	1.265
0.8	1.264	1.350	1.475	1.553
1.20	1.822	1.865	1.854	1.818
1.40	2.417	2.264	2.067	1.944
1.80	5.666	3.474	2.529	2.188
2.00	9.826	4.313	2.775	2.306

Table 4 The scattering ratio R_{13} for selected values of X_{63} and σ_g ($\tilde{m}$ = 1.70-0.68i)

X_{63} \ σ_g	0.10	0.28515	0.50	0.70
0.20	1.029	1.041	1.094	1.210
0.40	1.119	1.172	1.330	1.520
0.80	1.608	1.796	2.015	2.127
1.20	3.515	3.249	2.924	2.718
1.40	6.556	4.527	3.457	3.010
1.80	31.91	8.633	4.662	3.586
2.00	30.87	11.24	5.318	3.870

relatively modest variation of $E(\tilde{m})$, $F(\tilde{m})$ and $[E(\tilde{m})/F(\tilde{m})]^{1/3}$ for the three values of $\tilde{m}$ is noteworthy.

For our calculations we chose $\tilde{m}$ = 1.70 - 0.68i, a value obtained by observing the angular distribution of light scattered by a single electrostatically levitated particle. We chose the angles θ_1 equal to 45 deg, θ_2 equal to 90 deg, and θ_3 equal to 135 deg to allow for comparison with experiments to be described later in this paper.

The ratios R_{12} equal to $C_{vv}(45\ deg)/C_{vv}(90\ deg)$ and R_{13} equal to $C_{vv}(45\ deg)/C_{vv}(135\ deg)$ are primarily dependent on the mean size of the polydispersion and possess only a secondary dependence on the refractive index. This fact is illustrated in Table 2 where R_{12} is tabulated for selected values of X_{63} for the three values of $\tilde{m}$ listed in Table 1.

Tables 3 and 4 display R_{12} and R_{13} for polydispersions of various mean size X_{63} and width σ_g. The sensitivity of R_{12} and R_{13} to the width of the

Table 5 The scattering/extinction ratio, $10^4 \times Q_{vv}(90)/K_{ext}$ for selected values of X_{63} and σ_g ($\bar{m} = 1.70-0.68i$)

X_{63} \ σ_g	0.10	0.28515	0.50	0.70
0.2	6.596	6.628	6.609	6.064
0.4	47.73	47.43	42.72	32.84
0.6	132.7	124.8	99.38	70.08
1.00	313.9	263.4	193.6	135.6
1.40	317.0	287.4	233.9	176.0
1.80	171.0	234.5	238.3	197.2
2.00	103.3	200.1	233.2	203.0

Table 6 The polarization ratio ρ_2 for selected values of X_{63} and σ_g ($\bar{m} = 1.70-0.68i$)

X_{63} \ σ_g	0.10	0.28515	0.50	0.70
0.20	0.182(-4)	0.505(-4)	0.559(-3)	0.402(-2)
0.40	0.302(-3)	0.902(-3)	0.599(-2)	0.161(-1)
0.80	0.680(-2)	0.173(-1)	0.352(-1)	0.454(-1)
1.20	0.654(-1)	0.794(-1)	0.790(-1)	0.746(-1)
1.40	0.158	0.132	0.103	0.886(-1)
1.80	0.627	0.282	0.156	0.115
2.00	1.087	0.372	0.182	0.128

(n) Power of ten multiplier.

distribution function is somewhat more complex but for an intermediate range of σ_g the particle size continues to exert a dominant effect. The ability to determine both X_{63} and σ_g from measured values of R_{12} and R_{13} is apparent by examining these tables. The scattering-extinction ratio $Q_{vv}(90\ \text{deg})/K_{ext}$ is a particularly sensitive function of X_{63} for small size. However, as indicated by the values presented in Table 5, the sensitivity of this ratio to size decreases as X_{63} increases. A maximum occurs for a value of X_{63} between 1.0 and 2.0, depending on the value of σ_g. The large sensitivity of this ratio to size below its maximum makes the ratio an attractive parameter, provided the local value of K_{ext} is measurable to the desired accuracy.

The polarization ratio $\rho_2 = \bar{C}_{hh}(90\ \text{deg})/\bar{C}_{vv}(90\ \text{deg})$ proves to be the variable which possesses a primary sensitivity to σ_g as is illustrated in Table 6. Furthermore the sensitivity of ρ_2 to σ_g exists not only in the Mie regime but also for small values of X_{63} where Rayleigh theory for polydisperse particles predicts $\rho_2 = 0$. We explain the sensitivity of ρ_2 to σ_g at small X_{63} to be the result of the very strong influence of largest sizes in the tail of the particle size distribution function. An important observation is that ρ_2 is sensitive to σ_g even well into the size range where other properties [e.g. R_{12}, $Q_{vv}(90\ deg)/K_{ext}$] show little variation with σ_g.

Finally, Table 7 shows the sensitivity of f_ϕ to X_{63} and σ_g. The factor of f_ϕ is important because it extends the usefulness of the Rayleigh expression for volume fraction far beyond its normal range of validity. Even an approximate determination of X_{63} and σ_g provides a value of f_ϕ which possesses ample accuracy for most purposes.

A Self-Consistent Interpretation of Optical Data for Agglomerates

Studies of the tinting strength of carbon black first lead to the conclusion that the complex refractive index $\tilde{m}_s$ of a macroscopic aggregate of this material containing void spaces is less than the refractive index $\tilde{m}_c$ of the particulate material of which it is composed.[23] This idea was subsequently developed further[24] where the Lorentz-Lorenz formula was used to show the relationship between $\tilde{m}_s$ and $\tilde{m}_c$ for macroscopic aggregates to be given by

$$\tilde{G}(\tilde{m}_s) = \eta_v \tilde{G}(\tilde{m}_c) \qquad (29)$$

Table 7 The factor f_ϕ for selected values of X_{63} and σ_g ($\tilde{m} = 1.70-0.68i$)

σ_g X_{63}	0.10	0.28515	0.50	0.70
0.20	0.9628	0.9652	0.9703	0.9769
0.40	0.8605	0.8698	0.8927	0.9210
0.80	0.6366	0.6749	0.7435	0.8125
1.20	0.5909	0.6202	0.6809	0.7516
1.40	0.6242	0.6316	0.6737	0.7359
1.80	0.7291	0.6957	0.6884	0.7242
2.00	0.7898	0.7406	0.7058	0.7255

where $\widetilde{G}$ is a complex function of the complex refractive index $\widetilde{m}$ given by

$$\widetilde{G}(\widetilde{m}_i) = (\widetilde{m}_i^2 - 1) / (\widetilde{m}_i^2 + 2) \qquad (30)$$

and n_v is the volume fraction of the particulate phase within the macroscopic soot aggregate. The validity of Eq. (32) is subject to the inequality

$$2\pi a |\widetilde{m}|/\lambda \ll 1 \qquad (31)$$

where a is the radius of the void space within the macroscopic aggregate.

From the argument presented in Ref. 23 it is apparent that a spherical particle of diameter D containing numerous small voids of radius $a \ll \lambda$, but otherwise consisting of a homogenous isotropic material, will rigorously comply with the inequality [Eq. (31)] and with Eq. (29). This will be true for $D \gg \lambda$ if $a \ll \lambda$ and also for $D \sim$ or $< \lambda$ since $a \ll \lambda$ will then necessarily follow. The above relations also hold for the inverse geometry - a collection of numerous spheres which occupy a spherical envelope - provided the interparticle spacing $a \ll \lambda$. The latter configuration may be thought of as an agglomerate whose envelope is of spherical shape and, although such agglomerates may be infrequently observed, Eq. (29) provides the effective refractive index $\widetilde{m}_s$ of the agglomerate in this limiting case for all ratios of D/λ.

The sampling volume used in the laser scattering experiment contains about 10^{+5} to 10^{+7} soot agglomer-

Table 8 Optical properties of a polydispersion of loosely packed agglomerates with $\widetilde{m}_c = 1.7 - 0.68i$ for $X_{63} = 1.4$, $\sigma_g = 0.28515$ and $\lambda = 514.5$nm

$\widetilde{m}_s$ (n_v)	$10^4 \times \dfrac{Q_{vv}(90 \text{ deg})}{K_{ext}}$	R_{12}	θ_m	ρ_2	f_ϕ
1.70−0.68i (1.00)	287.3	2.264	97.3°	0.133	0.6316
1.35−0.242i (0.4840)	216.0	2.328	95.4°	0.0332	0.6968
1.175−0.110i (0.2451)	143.4	2.227	92.6°	0.00695	0.8072
1.0875−0.0533i (0.1236)	83.9	2.161	91.3°	0.00151	0.8904

ates and is viewed over a finite time period. The individual soot agglomerates are each randomly oriented and subject to rotary Brownian motion. Each particle can be visualized as occupying an effective optical volume which represents the average of its positions in the sampling time interval. We now propose to apply Eq. (29) to the individual, loosely packed, low density agglomerates. In this interpretation $\tilde{m}_c$ continues to represent the refractive index of the particulate material, while $\tilde{m}_s$ now represents the refractive index of the effective optical volume occupied by the agglomerate and η_v is the fraction of the optical mean volume $\pi D^3_{30}/6$ that is occupied by the particulate material.

By the definition of η_v the mean volume equivalent diameter D_v of the local agglomerate population is related to the optical diameter D_{30} by

$$D_v^3 = \eta_v D_{30}^3 \qquad (32)$$

From Eq. (6) we find D_v in terms of the primary unknowns X_{63} and σ_g by

$$D_v = \eta_v^{1/3} \frac{\lambda}{\pi} X_{63} \exp(-3\sigma_g^2) \qquad (33)$$

In this view, the various diameters D_{pq} and D_g apply to the optical sizes of the loosely packed agglomerates whose volume fraction based on optical sizes is ϕ_s. The quantity D_v is the diameter that would result if all loosely packed agglomerates were to coalesce into spheres of uniform density and the particulate material were to be redistributed to make all particles equal in diameter with no change in particle number concentration. Multiplying both sides of Eq. (32) by N, we find

$$\phi_c = \eta_v \phi_s \qquad (34)$$

where ϕ_c designates the volume fraction of particulate material. Therefore the particle number concentration is given by either

$$N = \frac{6}{\pi} \phi_c / D_v^3 \qquad (35)$$

or

$$N = \frac{6}{\pi} \phi_s / D_{30}^3 \qquad (36)$$

From Eq. (29) it follows that

$$n_v = \frac{\text{Re}(\tilde{G}_s)[1 + i\Gamma(\tilde{m}_s)]}{\text{Re}(\tilde{G}_c)[1 + i\Gamma(\tilde{m}_c)]} \quad (37)$$

where

$$\Gamma(\tilde{m}) = \frac{\text{Im}[\tilde{G}(\tilde{m})]}{\text{Re}[\tilde{G}(\tilde{m})]} \quad (38)$$

and Re and Im designate the real and imaginary portions of the complex function $\tilde{G}(\tilde{m})$. In order for n_v to be a scalar, it is necessary that $\Gamma(\tilde{m}_s) = \Gamma(\tilde{m}_c)$ and it follows from Eqs. (37) - (38) that

$$n_v = \frac{\text{Re}[\tilde{G}(\tilde{m}_s)]}{\text{Re}[\tilde{G}(\tilde{m}_c)]} = \frac{\text{Im}[\tilde{G}(\tilde{m}_s)]}{\text{Im}[\tilde{G}(\tilde{m}_c)]} \quad (39)$$

Most of the optical properties of the agglomerate are altered as its effective density $n_v \rho_p$ is reduced (where ρ_p is the density of the particulate material). This is illustrated in Table 8 where $n_s - 1 = \text{Re}(\tilde{m}_s - 1)$ is systematically lowered for specified values of $\tilde{m}_s$, X_{63}, and σ_g.

The optical properties shown in Table 8 demonstrate that downward scaling of $\tilde{m}_s$ has the effect of 1) reducing $Q_{vv}(90 \text{ deg})/K_{ext}$ while keeping R_{12} essentially constant 2) reducing θ_m, the angle at which $Q_{hh}(\theta)$ is a minimum toward its small particle limit of 90 deg; and 3) reducing the value of $\rho_2 = Q_{hh}(90 \text{ deg})/Q_{vv}(90 \text{ deg})$.

The Influence of Scaling of $\tilde{m}_s$ on Optical Properties of Agglomerates

The influence of the scaling procedure on the optical properties of the agglomerates is of paramount interest. From the application of Rayleigh theory to polydispersions, see Eq. (25), we express the volume fraction of loosely packed agglomerates as

$$\phi_s = (\lambda f_\phi(X_{63}, \tilde{m}_s, \sigma_g) / 6\pi E(\tilde{m}_s)) K_{ext} \quad (40)$$

where $E(\tilde{m}_s) = -\text{Im}[\tilde{G}(\tilde{m}_s)]$. The value of ϕ_c, which is normally of greater interest, is obtained by inserting

Eq. (40) into Eq. (34). In view of Eq. (39) we find

$$\phi_c = (\lambda f_\phi(X_{63}, \tilde{m}_s, \sigma_g)) / (6\pi\, E(\tilde{m}_c)) K_{ext} \quad (41)$$

where it is to be noted that f_ϕ, which accounts for the departure of the extinction of larger particles from the Rayleigh prediction, is based on $\tilde{m}_s$ and E is based on $\tilde{m}_c$. The values of f_ϕ (see Table 7) are bounded by the limits $0.59 < f_\phi < 1.0$ for $0.2 < X_{63} < 2.0$ and $0.10 < \sigma_g < 0.70$ when $\tilde{m}_s = \tilde{m}_c = 1.7 - 0.68i$. The nearness of f_ϕ to unity is fortuitous in that it applies only to absorbing particles. However, as $\tilde{m}_s$ is scaled downward the value of f_ϕ is even more narrowly confined near unity as is apparent from the values displayed in the last column of Table 8.

We conclude that the use of the Rayleigh expression for ϕ_c with $f_\phi = 1$ and E evaluated on the best available value of $\tilde{m}_c$ will often approach experimental accuracy even when the low density agglomerates are considerably larger than Rayleigh sizes. Finally, for small X_{63} when f_ϕ is precisely unity in the true Rayleigh limit, Eq. (41) shows that the value of ϕ_c is precisely determined by using $E(\tilde{m}_c)$. In this case a knowledge of $\tilde{m}_s$ is neither required nor available from the experimental observations.

In view of the definitions D_V and f_N, along with the use of Eqs. (27) and (39), we find the volume equivalent diameter is given by

$$D_V = \frac{\lambda}{\pi} \left[\frac{4\pi\, E(\tilde{m}_c)\, Q_{vv}(90\, \text{deg})}{f_N\, F(\tilde{m}_c)\, K_{ext}} \right]^{1/3} \quad (42)$$

and depends on the measured quantities as well as $\tilde{m}_c$ and f_N regardless of the value of $\tilde{m}_s$. Furthermore $[E(\tilde{m}_c)/F(\tilde{m}_c)]^{1/3} \sim 1.00$ and thus f_N is the principle uncertain factor in the determination of D_V.

Finally we see that N, which is determined from ϕ_c and D_V, requires a knowledge of both $\tilde{m}_c$ and f_N. Thus we conclude that careful observations made on low density agglomerates and the use of $\tilde{m}_c$, rather than $\tilde{m}_s$, will yield a reliable value of ϕ_c while D_V and N depend on the additional factors $f_N^{1/3}$ and f_N, respectively. Thus of the three quantities ϕ_c, D_V, and N, the first ϕ_c is measured with the least uncertainty, while the last N is subject to the largest uncertainty.

We summarize in brief the examination of the optical properties of low density agglomerate as follows. The properties ϕ_c, D_v, and N can be measured for small agglomerates when $X_{63} < 0.2$ by using Rayleigh theory by evaluating E and F based on $\tilde{m}_c$ provided f_N is known. However, no information on f_N or σ_g is available from the optical experiment if the particles are in the Rayleigh regime. Low density agglomerates for which $X_{63} > 0.2$ will display values of R_{12} and R_{13} larger than unity which together with Q_{vv}/K_{ext} may provide information on σ_g.

Data Reduction Using the Levenberg-Marquardt Algorithm

The various observed optical ratios are functions of the properties of the parameters of the particle dispersion which, in the data reduction process, constitute the independent variables whose values we seek. Thus, for example, R_{12} can be expressed as

$$R_{12} = R_{12}[\, X_{63}, \tilde{m}_s(\lambda), \sigma_g, ; \tilde{m}_c(\lambda), \lambda \,] \qquad (43)$$

where X_{63}, $\tilde{m}_s(\lambda)$, and σ_g are the unknown independent variables and $\tilde{m}_c(\lambda)$ and λ are specified parameters. We express the above relations in a general form as

$$y_j = y_j(x_1, x_2, \ldots x_i, \ldots x_n, ; B_1, B_2, \ldots B_k) \qquad (44)$$

where $j = 1$ to m with $n<m$ and the values of B represent the known parameters. At any one point in the flame the values of the ratios such as R_{12}, R_{13}, and $Q_{vv}(90\ \text{deg})/K_{ext}$ (designated as y_j) are measured. These measurements contain uncertainties of a wide variety of origins some of which are beyond control and may be difficult to estimate. The nonlinear functional form of Eqs. (43) and (44) is awkward in that it cannot, when Mie theory applies, be expressed in simple algebraic expressions. In the case at hand, the complex quantity $\tilde{m}_s$ possesses constraints on the manner in which its real and imaginary components are varied as has been discussed earlier. The problem is then to determine the "best" values x_i, which are appropriate for the observed dependent variables y_z^{obs}.

The Levenberg-Marquardt algorithm[17] (LMA) solves the problem of minimization of a prescribed error ε when the y_j are nonlinear functions of x_i. This approach has been implemented in a general form[18,25] in which the optical ratios expressed by Eq. (43) and scaling of $\tilde{m}_s$ are provided by approprite subroutines. We presently use

$y_1 = R_{12}$, $y_2 = R_{13}$, and $y_3 = Q_{vv}(90 \text{ deg})/K_{ext}$ to determine $x_1 = X_{63}$, $x_2 = \tilde{m}_s$, and $x_3 = \sigma_g$ for prescribed $\tilde{m}_c$. Improved values of x_i and more realistic values of ε would be yielded if additional optical measurements containing independent information were to become available. Further experience is required to determine if additional observations can yield a genuine improvement in accuracy at a realistic cost in effort and instrumentation.

When the values of $R_{12} \sim R_{13}^{1/2} \sim 1$ then a unique solution does not exist and it is necessary to reformulate the problem. Under these circumstances it may be possible to use Rayleigh theory but f_N or σ_g must be specified. For example, near the initial particle source a small value of σ_g has been observed by sampling the particles and the value of f_N is presumably narrowly bounded $1 < f_N < 2$.

Comparison with Experimental Data

Extensive laser scattering/extinction measurements have been made of the soot particle field in an ethene/air laminar diffusion flame. Measurements of extinction and scattering at three angles (45 deg, 90 deg, and 135 deg)

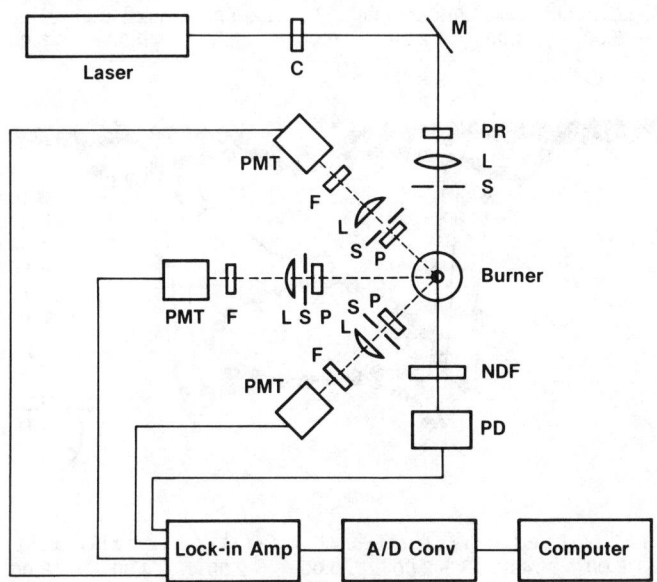

Fig. 1 Schematic of laser scattering-extinction system. C-chopper, M-mirror, PR-polarization rotator, L-lens, S-spatial filter, (circular aperture), NDF-neutral density filter, PD-photodiode, p-polarizer, F-narrowband filter, PMT-photomultiplier.

have been made in the flame as a function of position. The scattering measurements have also been obtained as a function of the laser light polarization orientation for both the incident and scattered light.

The experimental apparatus used in these experiments has been described in detail elsewhere[4] and only a brief description will be provided here. The apparatus used in these studies is shown in Fig. 1. The laser was an argon

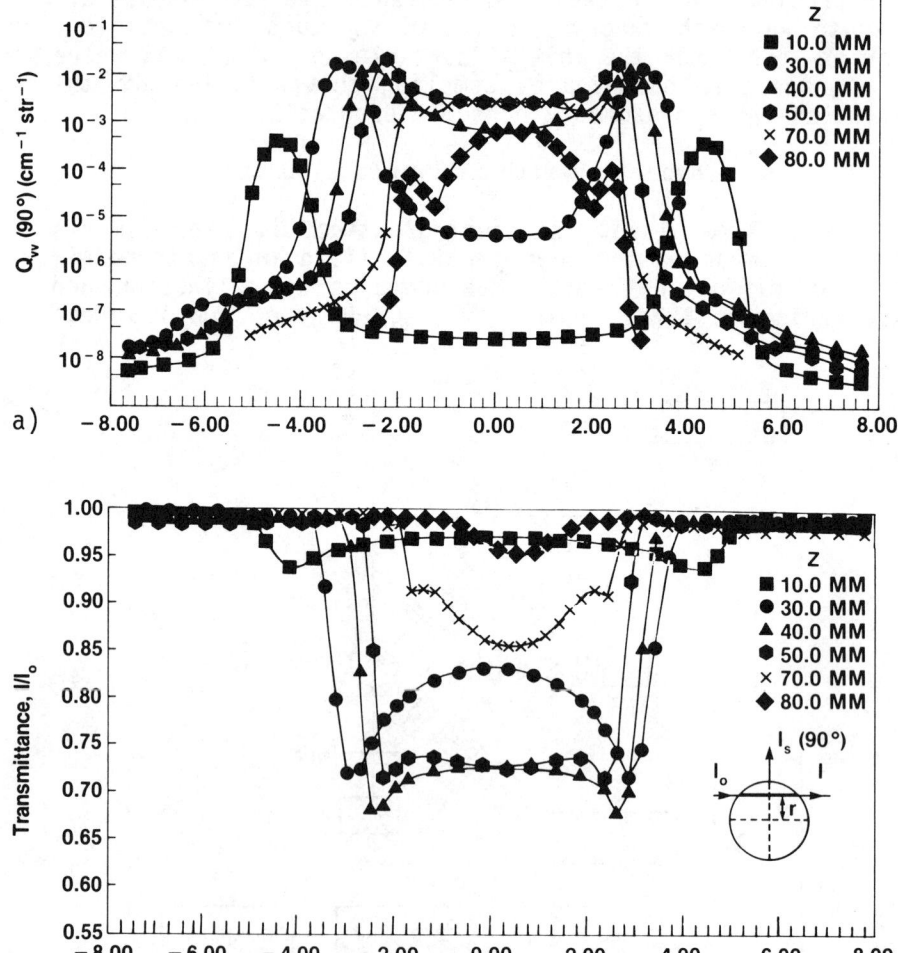

Fig. 2 (a) The volumetric scattering cross section, $Q_{VV}(90\ deg)$ as a function r and z for the non-sooting ethene diffusion flame. (b) Transmittance as a function of r and z.

ion laser operated at the 514.5-nm laser line at an output power of 0.5 W. The laser beam was modulated using a mechanical chopper operating at 1015 Hz. After passage through the co-annular diffusion flame burner, the transmitted and scattered light signals were detected by appropriate photodetectors and the resulting electronic signals were processed using a lock-in detection technique. Finally the signals were digitized and recorded using a minicomputer data acquisition system. A polarization rotator located in the incident beam path and polarizers placed in the scattered light path allow selection of the incident and scattered light polarization orientations.

The co-annular laminar flame burner consists of two concentric brass tubes of 11.1- and 101.6- mm inner diameter with the fuel flowing through the central tube and air through the outer passage. The ethene fuel and airflow rates for these experiments were 3.85 and 713.3 cm^3/s, respectively, resulting in a luminous yellow flame 88 mm in height.[4]

Figure 2 shows the radial profiles for Q_{vv} and the transmission measurements as a function of height in this flame. One important aspect of these profiles is the occurrence of an annular region low in the flame which is first observed to contain soot particles. In this paper we will present data taken at the position of maximum volumetric scattering cross section (i.e., maximum soot concentration) for the profiles at the 7-, 10-, 30-, and 50-mm heights. Table 9 shows the values obtained at each height for the various scattering cross sections Q_{ij} taken for the 90 deg angle as well as the angular dissymmetry and the value of Q_{vv}/K_{ext}. Also shown are the polarization and cross polarization ratios for the 90 deg angle.

The data shown in Table 9 provide a clear example of the conflict in the light scattering data for soot particles in flames. The first point of interest is the finite values observed for the cross polarization volumetric scattering cross section (Q_{hv} and Q_{vh}). Similar observations have been previously reported by others[5,7,26] and has been explained as anisotropy resulting from particle shape and/or internal particle structure. Spheres of equivalent volume are expected to show no cross polarization,i.e., $Q_{hv}(\theta) = Q_{vh}(\theta) = 0$ for all θ. Since the values of $Q_{hh}(90$ deg) that we observe are only perceptably larger than $Q_{hv}(90$ deg) and $Q_{vh}(90$ deg), we conclude that the magnitude of $Q_{hh}(90$ deg) results mainly from anisotropy and not from size effects. For this reason ρ_2 equal to $Q_{hh}(90$ deg)/$Q_{vv}(90$ deg) cannot be used as a measure of σ_g. This is unfortunate because ρ_2 resulting from polydispersity is a sensitive measure of σ_g for X_{63} ranging from 0.2-1.0.

From the results given in Table 9, we obtain an upper bound estimate of ρ_2 attributable purely to polydispersity and sphere size (as opposed to nonspherical shape) to be of the order of 1×10^{-2}. However, the results for polydisperse spheres show that, for the observed values of R_{12}, $\rho_2 = 0.05$ to 0.15 which considerably exceeds the upper bound estimate that has been observed.

Finally we use the values of $Q_{vv}(90 \text{ deg})/K_{ext}$ for the 30-mm height shown in Table 9 to estimate particle size. We find $X_{63} = 0.8$ to 1.8 as σ_g ranges from 0.1-0.7. On the other hand, from $R_{12} = 2.2$, we find $X_{63} = 1.4$-2.0 over the same range of σ_g. These results illustrate the conflict reported previously.[5]

There exist, then, two conflicts in the data we report. These are (1) the conflict in size deduced from R_{12} (or R_{13}) and $Q_{vv}(90 \text{ deg})/K_{ext}$, and (2) the fact that the observed upper bound value of ρ_2 is too small for the sizes that follow from the observed R_{12}. In addition to these conflicts we also note the failure of θ_m to shift from 90 deg for the larger sizes as has been reported by others.[7] All these conflicts are eliminated if the refractive index of the soot agglomerate were to be reduced below the value considered applicable to the particulate

Table 9 Experimental light scattering/extinction measurements

Height(mm)	7	10	30	50
$Q_{vv}(90 \text{ deg})(cm^{-1}sr^{-1})$	6.27(-5)	2.2(-4)	1.6(-2)	1.5(-2)
$Q_{vh}(90 \text{ deg})(cm^{-1}sr^{-1})$	-	2.0(-6)	3.5(-4)	4.1(-4)
$Q_{hh}(90 \text{ deg})(cm^{-1}sr^{-1})$	-	2.2(-6)	3.9(-4)	4.4(-4)
$Q_{hv}(90 \text{ deg})(cm^{-1}sr^{-1})$	-	2.0(-6)	3.8(-4)	3.9(-4)
R_{12}	1.06	1.1	2.2	2.3
R_{13}	1.28	1.4	4.0	4.2
$Q_{vv}/K_{ext}(sr^{-1})$	1.13(-3)	3.3(-3)	1.64(-2)	1.60(-2)
ρ_2	-	1.0(-2)	2.5(-2)	3.0(-2)
ρ_v	-	0.9(-2)	2.4(-2)	2.6(-2)

INTERPRETATION OF OPTICAL MEASUREMENTS OF SOOT 231

material. We conclude that the above conflicts constitute three reasons to seek an adjusted value for the refractive index of the loosely packed soot agglomerates that is based on their low effective density.

We illustrate the application of the LMA to the data for R_{12}, R_{13}, $Q_{vv}(90$ deg$)$ and K_{ext} given in Table 9 at the radial position of maximum soot concentration. These data have been analyzed for the following four cases:

(1) Rayleigh theory for a monodispersion: This is the widely used method that is based on Eqs. (25), (27), and (28) with $f_\phi = 1.0$, $D_{63} = D$, $f_N = 1.0$, and where it is assumed $\sigma_g = 0$. Values of R_{12} and R_{13} are not used and, $\tilde{m}_s = \tilde{m}_c$.

Table 10 Particle properties determined from experimental measurements[a]

Height	Method[b]	ϕ	D_v (nm)	N (cm^{-3})	σ_g	ϵ
7.0	1	5.12(-7)	38.85	1.68(10)	0	9.5%
	2	5.12(-7)	30.84	3.36(10)	0.2775	9.5%
	3	5.24(-7)	11.57	6.46(10)	0.647	2.15%
	4[c]	4.84(-7)	13.90	3.44(11)	0.6	0
10.0	1	1.14(-6)	56.16	2.23(10)	0	15.6%
	2	1.14(-6)	44.58	2.46(10)	0.2775	15.6%
	3	1.11(-6)	20.40	2.50(11)	0.609	2.30%
	4[d]	1.05(-6)	27.33	9.82(10)	0.5	0
30.0	1	8.04(-6)	94.72	1.80(10)	0	108%
	2	8.04(-6)	75.17	3.61(10)	0.2775	108%
	3	6.78(-6)	29.36	5.11(11)	0.995	2.46%
	4[e]	6.10(-6)	90.10	1.59(10)	0.45	0
50.0	1	1.16(-5)	94.01	2.66(10)	0	114%
	2	1.16(-5)	74.60	5.33(10)	0.2775	114%
	3	1.01(-5)	23.61	1.47(12)	1.077	2.46%
	4[f]	8.83(-6)	92.80	2.11(10)	0	0

[a] $\tilde{m}_c = 1.70 - 0.68i$ for all cases.
[b] See text for identification of method.
[c] $\tilde{m}_s = 1.70 - 0.68i$, $n_v = 1.0$.
[d] $\tilde{m}_s = 1.50 - 0.38i$, $n_v = 0.69$.
[e] $\tilde{m}_s = 1.33 - 0.23i$, $n_v = 0.46$.
[f] $\tilde{m}_s = 1.31 - 0.21i$, $n_v = 0.43$.

(2) Rayleigh theory for a preselected polydispersion: The values of D_{63} and D_{30} are distinguished and are related by assuming $f_N = 2.00$ which corresponds essentially to the self-preserving size distribution function and is approximated by the logarithmic normal distribution when $\sigma_g = 0.2775$. Equations (25), (27), and (28) are used, but values of R_{12} and R_{13} are ignored.

(3) Mie theory for an unspecified polydispersion with $\tilde{m}_s = \tilde{m}_c$: The LMA is used to search through the $X_{63} - \sigma_g$ space for the minimum value of ε using all optical data $\tilde{m}_s = \tilde{m}_c$ and all y_j are calculated from this theory for polydispersions.

(4) Mie theory for an unspecified polydispersion with $\tilde{m}_s$ scaled: The LMA is used to search through $X_{63} - \tilde{m}_s - \sigma_g$ space for the minimum value of ε. The value of $\tilde{m}_s$ is scaled on the prescribed value of $\tilde{m}_c$ and all optical data are used. All y_j are calculated using Mie theory for polydisperse spheres.[25]

Examination of the results of applying these methods, see Table 10, reveals interesting trends. All four methods provide essentially the same value of ϕ_c which depends strongly on K_{ext}. The value of D_v and N given by methods (1) and (2) differ as a result of the value of σ_g that was assumed. Thus methods (1) and (2) must be described as theoretically invalid since they require R_{12} to equal R_{13}, although at the higher two heights they give better results than might be expected. Method (3) satisfies the values of the optical observations by finding a very large σ_g, particularly at the higher heights which drastically reduces D_v while increasing N. The values of σ_g at these heights as given by method (3) are considered unrealistic. Thus the results of method (4) are considered to be more plausible at this time.

The results obtained with method (4) indicate that the refractive index for the soot particles is decreasing with increasing height. This effect is attributed to the formation of low density agglomerates as previously described. A second interesting feature of this calculation is the large value of σ_g obtained even low in the flame. Previous studies[10-12] have shown that the particles, when first formed, are characterized by a narrow Gaussian distribution. Recent theoretical studies,[13] of the time-dependent behavior of the combined effects of free molecular agglomeration and sustained particle nucleation, do predict the rapid evolution of a particle distribution of similar σ_g on the time scale of these measurements low in the flame.

Table 11 Comparison of the observed and predicted polarization ratios

Height (mm)	ρ_2 (observed)	ρ_2 (method 4)
10	1.0 (-2)	0.3 (-2)
30	2.5 (-2)	2.9 (-2)
50	3.0 (-2)	2.6 (-2)

Also shown in Table 10 are the values for the aggregate relative error ε which gives a measure of the deviation of the calculated optical observations (i.e., R_{12}, R_{13}, etc.) from the experimentally measured values. In methods (1) and (2), the values for ε are large since these methods imply $R_{12} = R_{13} = 1.0$. Method (4) gives an unrealistically low value of ε because the number of dependent and independent variables are equal (in this case three variables are used).

As an additional comparison of the results of method (4) for calculating the optical properties of the soot particles, Table 11 compares the experimentally measured polarization ratio ρ_2 with that calculated for a polydispersion arrived at with method (4). Given the uncertainty discussed earlier in measuring the polarization ratio these results are encouraging in that they do show the proper trend for the observed polarization ratio reported by other workers.

The above results are encouraging to the extent that they do show that Mie theory for polydispersed spheres is capable of yielding a self-consistent interpretation of the observed optical properties of soot particles in flames. However it should be pointed out that the present results are for a limited set of data and more work is required to further confirm the above approach. Of particular interest will be the extension of the present approach to randomly oriented spheroids. Spheroids will exhibit cross polarization effects similar to those observed in these flames.[7] Since homogenous spherical particles do not exhibit cross polarization effects, the consideration of shape effects is a logical extension of the present work. In addition, the above described trends need to be verified over a wider range of flame conditions to more fully test the consistency of our interpretation.

Summary

The conflicts between the observations of several investigators and the predictions of Mie theory for polydispersion soot agglomerates are now well documented. These difficulties cannot be attributed to the well-known discrepancies in the refractive index of soot material that have been reported by various investigators. However, these conflicts are resolved if the refractive index of the loosely packed agglomerates is scaled downward using a method suggested in the past for macroscopic aggregates of soot.[24] The application of this method, which is approximate and requires further examination for irregular particles of size equal to the wavelength of light, yields values of the properties of the polydisperse agglomerates that are internally consistent with the Mie theory for polydisperse spheres. In the Mie size range, information on the width of the polydispersion (σ_g) and the fraction of the optical volume-mean diameter that is occupied by the particulate material $(n_v)^{1/3}$ is also obtained. The method fails when the angular dissymmetry is weak $(R_{12} \simeq R_{13}^{1/2} \sim 1)$. In this case the Rayleigh theory for polydisperse spheres is applicable but no information on σ_g, and hence no information on f_N, is provided by the optical experiments. These quantities are needed to find D_v and N. If independent information on f_N is available, then the Rayleigh theory for polydisperse spheres may be applied to the low density agglomerates to yield ϕ_c, D_v, and N provided R_{12} and R_{13} are near unity.

The value of ϕ_c is given with fortuitously good accuracy by the Rayleigh expression when the sizes of the agglomerates are well into the Mie regime provided the particles are absorbing. The value of D_{63} is also easily measured in the Rayleigh regime but this quantity cannot be converted to D_{30} from which N is derived without a knowledge of f_N. Thus, in general, ϕ_c is more readily measured over a broad size range and N is more difficult to measure because of the need to evaluate f_N. A greater stress should be placed on the design of the optical experiment in such a way as to yield information on the width of the distribution function. The simultaneous observation of angular dissymmetry and local extinction has the potential to yield the required information in certain size regimes.

Acknowledgments

The authors are indebted to Dr. R. F. Boivert of the Scientific Computing Division of the National Bureau of Standards for recommending the Hiebert program for the Levenberg-Marquardt algorithm and for assistance with the implementation of this program. One of us (R.A.D.) expresses gratitude to the National Bureau of Standards for providing a stimulating environment and partial financial support during his sabbatical year. The partial support of this work by the Department of Energy, Division of Energy Conversion and Utilization Technology, is gratefully acknowledged.

References

[1] D'Alessio, A., DiLorenzo, A., Borghese, A., Berreta, F., and Masi, S., "Study of the Soot Nucleation Zone of Rich Methane-Oxygen Flames," The Sixteenth (International) Symposium on Combustion, The Combustion Institute, Pittsburgh, Pa, 1977, pp. 695-708.

[2] Haynes, B. S. and Wagner, H. Gg., "Sooting Structure in a Laminar Flame," Berichte Der Bunsen-Gesellschaft Fur Physikalische, Vol. 84, 1980, pp. 499-506.

[3] Kent, J. H., Jander, H., and Wagner, H. Gg., "Soot Formation in a Laminar Diffusion Flame," The Eighteenth (International) Symposium on Combustion, The Combustion Institute, Pittsburgh, Pa, 1981, pp. 1117-1136.

[4] Santoro, R. J., Semerjian, H. G., and Dobbins, R. A., "Soot Particle Measurements in Diffusion Flames," Combustion and Flame, Vol. 51, 1983, pp. 203-218.

[5] D'Alessio, A., DiLorenzo, A., Sarafim, A. F., Beretta, F., Masi, S., and Venitozzi, C., "Soot Formation in Methane Oxygen Flames," The Fifteenth (International) Symposium on Combustion, The Combustion Institute, Pittsburgh, Pa, 1975, pp. 1427-1438

[6] Bonczyk, P. A., "Measurement of Particle Size by in Situ Laser-optical Methods: a Critical Evaluation Applied to Fuel Pyrolyzed Carbon," Combustion and Flame, 1979, Vol. 35, pp. 191-206.

[7] D'Alessio, A. "Laser Light Scattering and Fluorescence Diagnostics of Rich Flames," Particulate Carbon Formation During Combustion, edited by D. C. Siegla and G. W. Smith, Plenum Press, New York, 1981, pp. 207-256.

[8] Lai, F. S., Friedlander, S. K., Pich, J., and Hidy, J., "The Self-Preserving Particle Size Distribution for Brownian Coagulation in the Free Molecular Regime," Journal of Colloid and Interface Science, 1972, Vol. 39, pp. 395-405.

[9] Graham, S. C., and Robinson, A., "A Comparison of Numerical Solutions to the Self-Preserving Size Distribution for Aerosol Coagulation in the Free-molecular Regime," Journal of Aerosol Science, 1976, Vol. 7, pp. 261-273.

[10] Wersborg, B. L., Howard, J. B., and Williams, G. C., "Physical Mechanisms in Carbon Formation in Flames," The Fourteenth (International) Symposium on Combustion, The Combustion Institute, Pittsburgh, Pa, 1973, pp. 929-940.

[11] Chippett, S. and Gray, W. A., "The Size and Optical Properties of Soot Particles," Combustion and Flame, 1978, Vol. 31, pp. 149-159.

[12] Lahaye, J., Prado, G., and Donnet, J. B., "Nucleation and Growth of Carbon Black Materials During Thermal Decomposition of Benzene," Carbon, 1974, pp. 27-35.

[13] Dobbins, R. A., and Mulholland, G. W., "The Evolution of Size Distribution During Particle Formation Accompanied by Coagulation in Flames," to be published, Combustion Science and Technology, 1984.

[14] van de Hulst, H. C., Light Scattering by Small Particles, J. Wiley and Sons, New York, 1957.

[15] Kerker, M., The Scattering of Light and other Electromagnetic Radiation, Academic Press, New York, 1969.

[16] Bevington, P. R., Data Reduction and Error Analysis for the Physical Sciences, McGraw-Hill, New York, 1969.

[17] More, J. J., "The Levenberg-Marquardt Algorithm: Implementation and Theory," Numerical Analysis, Lecture Notes in Mathematics Vol. 630, edited by G. A. Wilson, Springer-Verlag, New York, 1977, pp. 105-116.

[18] Hiebert, K. L. "Subroutine SNLSE," in "The SLATEC Common Mathematical Subprogram Library: SNLA Implementation," Haskell, K. H., Vanderender, W. H., and Walton, E. L., Sandia Technical Report SAND 80-2792, Sandia National Laboratories, Albuquerque, NM, 1980.

[19] Dalzell, W. H., and Sarofim, J. "Optical Constants of Soot and Their Application to Heat Flux Calculations," Journal of Heat Transfer, Transactions of the American Society of Mechanical Engineers, Vol. 91, 1969, pp. 100-104.

[20] Lee, S. C., and Tien, C. L., "Optical Constants of Soot in Hydrocarbon Flames," The Eighteenth (International) Symposium on Combustion, The Combustion Institute, Pittsburgh, Pa, 1981, pp. 1159-1166.

[21] Pluchino, A. B., Goldberg, S. S., Dowling, J. M., and Randall, J. M., "Refractive-Index Measurements of Micron-Sized Carbon Particles," Applied Optics, Vol. 19, 1980, pp. 3370-3372.

[22] Menna, P., and D'Alessio, A., "Light Scattering and Extinction Coefficients for Soot Forming Flames in the Wavelength Range from 200 nm to 600 nm," The Nineteenth (International) Symposium on Combustion, The Combustion Institute, Pittsburgh, Pa, 1982, pp. 1421-1428.

[23] Medalia, A. I., and Richard, L. W., "Tinting Strength of Carbon Black," Journal of Colloid and Interface Science, Vol. 40, 1972, pp. 233-252.

[24] Graham, S.C., "The Refractive Indices of Isolated and Aggregated Soot Particles," Combustion Science and Technology, Vol. 9, 1974, 159-163.

[25] Dave, J.V., "Subroutines for Computing the Parameters of the Electromagnetic Radiation Scattered by a Sphere," Report 320-3237, IBM Palo Alto Scientific Center, Palo Alto, Ca, May 1978.

[26] Muller-Dethlefs, K., "Optical Studies of Soot Formation and the Addition of Organic Peroxides to Flames," Ph.D. Thesis, Imperial College, London, 1979.

In Situ Measurement of the Complex Refractive Index of Combustion Generated Particulates

E. A. Powell* and B. T. Zinn†
Georgia Institute of Technology, Atlanta, Georgia

Abstract

An in situ optical technique for measurement of the complex refractive index ($m = n - ik$) of combustion generated particulates is discussed in this paper. This technique involves measurement of light scattered by the smoke particles in the plane perpendicular to the polarized incident beam of an argon-ion laser operating at 488-nm wavelength. In particular, 90-deg scattering intensities are measured in the plane parallel to ($I_\parallel$) and perpendicular to ($I_\perp$) the plane of polarization of the incident light beam; for micron and submicron sized particles the ratio $I_\parallel / I_\perp$ is shown to be sensitive to the imaginary part k of the complex refractive index for $0 < k < 0.4$. In addition the forward scattering ratio method is used to obtain the mean particle diameter (D_{32}), and transmitted light measurements at two widely separated wavelengths (488 and 633 nm) are used to obtain the corresponding ratio of optical densities OD_R / OD_B. Using the Mie scattering theory, these three measurements are combined to yield the refractive index n and the absorption index k of the smoke particles. Typical results for flaming and nonflaming combustion are used to illustrate the application of the method in the cases of nonabsorbing, moderately absorbing, and highly absorbing smoke particulates.

Introduction

Particulate matter or smoke is an undesirable by product of most practical combustion processes and is a major hazard of accidental fires in buildings, aircraft, ships, and other confined

---

Presented as Paper 83-1518 at the AIAA 18th Thermophysics Conference, Montreal, Canada, June 1-3, 1983. Copyright © American Institute of Aeronautics and Astronautics, Inc., 1983. All rights reserved.
*Senior Research Engineer, School of Aerospace Engineering.
†Regents' Professor, School of Aerospace Engineering.

spaces. One of the principal hazards of smoke produced during accidental fires in enclosed spaces is the loss of visibility caused by the scattering and absorption of light by the smoke particles. The degree of light attenuation due to smoke is specified by the optical density per unit optical path length, which is proportional to the volume fraction and extinction efficiency of the smoke particulates. The extinction efficiency depends upon the size and shape of the smoke particles and their optical properties as well as the wavelength of incident light. For spherical particles the extinction efficiency is readily calculated using the Mie scattering theory if the ratio of particle diameter to wavelength and the particle complex refractive index are known. The optical density, and hence loss of visibility produced by burning a known quantity of a given material in an enclosed space can therefore be estimated if the mean particle diameter, particle volume fraction and particle complex refractive index are known. Thus the smoke hazards of various building and furnishing materials may be assessed.

The physical properties of the smoke such as mean particle size, refractive index, and particulate volume fraction can be obtained by means of small scale tests. In situ optical techniques are especially suited to this task, since they do not disturb the particulate suspension as do conventional sampling techniques. For several years such a technique has been in use by the authors for the simultaneous measurement of mean particle diameter, refractive index, and particle volume fraction of smokes produced by the combustion of polymeric materials in a steady ventilation airflow.[1-3] The original technique[4] used the forward scattering ratio method[5,6] to obtain the average particle size (which varies with time during the test) coupled with light transmission measurements at two wavelengths[7,8] for refractive index determination. A curve of particle volume fraction vs time was then determined from the transmission measurements at one of the wavelengths using the previously determined values of average size and refractive index. Integrating this curve with respect to time and dividing by the initial unburned sample weight yielded the total volume of particulates produced per unit mass of material burned. This quantity, called the specific total particulate volume (STPV), can then be used along with the particle size and refractive index to determine the optical density produced by the material in actual fire situations.

The determination of particle volume fraction using the combined light-scattering/attenuation method as described above is based on the assumption that the smoke particles are spherical and do not absorb light; that is, they are perfectly transparent with a real refractive index. This may be the case for smoldering or nonflaming combustion (oxidative pyrolysis) of some materials where the smoke particles are spherical liquid droplets composed of mixtures of organic compounds, many of which are nonabsorbing with refractive indices between 1.35 and 1.50. For flaming combustion of organic materials, the smoke particulates consist of irregularly shaped aggregates of spherical primary soot particles, which strongly absorb light over a wide range of wavelengths. For these particles the refractive index is complex; for soot a value m = 1.57-0.56i[9] is often used. Even for nonflaming combustion, it is suspected that the smoke particles produced by many materials may be moderately absorbing, since the refractive indices based on the two-color transmission measurements are often too low (m < 1.35) for the types of organic compounds expected. Furthermore, collected samples of such smoke particulates often exhibit a yellow color, indicating absorption of blue and red light.

Even if the smoke particles are nonabsorbing, an additional assumption was made in the two-color transmission method. The refractive index was assumed to be the same at both of the wavelengths at which the transmission measurements were made. Since the refractive index of most substances varies significantly with wavelength, it is desirable to measure the refractive index using a technique involving only one wavelength.

If one assumes that the refractive index is complex (i.e., m = n-ik), but does not vary with wavelength, a new unknown parameter is introduced into the analysis. This is the absorption index k, which must be determined by means of additional measurements. On the basis of investigations being carried out at the National Bureau of Standards,[10] scattering measurements made in the plane normal to the polarized laser beam appeared especially promising for determination of the absorption index. These additional 90-deg scattering measurements have been incorporated into the forward-scattering, two-color transmission technique. A description of this new technique illustrated with typical results is given in the remainder of this paper.

Experimental

In order to determine the imaginary part of the complex refractive index (i.e., the absorption index), additional scattered light measurements are made at an angle of 90 deg to the polarized incident light beam (originally a He-Ne laser, λ = 0.633 μm; later an argon-ion laser, λ = 0.488 μm). Thus, two additional detectors have been incorporated into the present system (see Fig. 1). These detectors measure the 90-deg scattering intensities parallel to ($I_\parallel$) and perpendicular to ($I_\perp$) the plane of polarization of the incident light beam. The ratio of $I_\parallel / I_\perp$ is used along with the mean particle size D_{32} (from the forward scattering measurements) and the ratio of optical densities at the two laser wavelengths (λ_1 = 0.458 or 0.488 μm and λ_2 = 0.633 m) to obtain the complex refractive index m of the smoke particles.

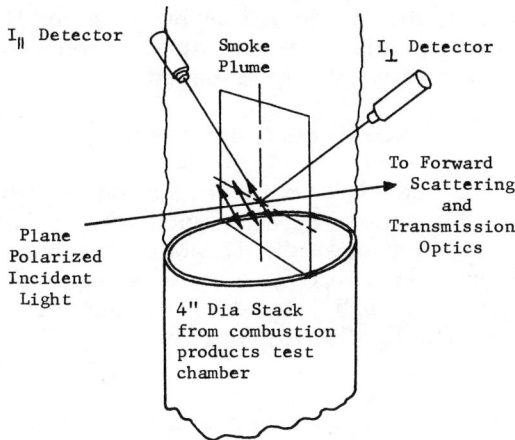

Fig. 1 Optical system for 90-deg scattering measurements.

The 90-deg scattering detectors each consist of an achromatic focusing lens, a 600-μm pinhole aperture, a collimating lens, a laser-line interference filter, and a photodetector. In the original system, the photosensitive element was a United Detector PIN 6DP photodiode used with a 0.633 μm He-Ne laser line filter. However the power of the He-Ne laser and the sensitivity of the photodiode detector proved to be inadequate for these measurements, and they were replaced by RCA 931 A photomultiplier tubes and a 0.488-μm argon-ion laser (25-mW power) with corresponding interference filter. The optical axes of the detectors are at 45-deg angles to the axis of the smoke plume, and they intersect the incident light beam at the center of the smoke plume emerging from the 11.4-cm-diam stack leading from the test chamber. The laser is provided with a broadband polarization rotator to rotate the initially vertical polarization plane to the required 45-deg angle. With this system, 90-deg scattered light is detected over a cylindrical scattering volume about 1 mm in diameter and about 1 cm in length.

Since the sensitivities of the $I_\parallel$ and $I_\perp$ detectors are different, the 90-deg scattering optical system must be calibrated. This is done by passing a dioctyl phthalate (DOP) aerosol through the laser beam and measuring the ratio of detector outputs ($S_\parallel / S_\perp$) first with the beam vertically polarized (VP) and then with the beam horizontally polarized (HP). From symmetry the intensities of the light scattered to the two detectors are equal for both the VP and HP beam configurations. Thus a calibration factor is obtained by taking the average of the VP and HP detector output ratios. This calibration factor is then checked by measuring $I_\parallel / I_\perp$ for the same DOP aerosol with the beam polarization at 45-deg and comparing with the theoretical value obtained from the known refractive index of DOP and the DOP mean particle diameter obtained from simultaneous measurements of the forward scattering ratio.

Theoretical Calculations

In order to develop data reduction techniques, the dependence of the ratio $I_{\parallel}/I_{\perp}$ upon the smoke particle characteristics was investigated[11]. The Mie scattering intensities for single particles were integrated over an assumed upper limit distribution function (ULDF)[12] to obtain calculated values of $I_{\parallel}$ and $I_{\perp}$. Thus the ratio $I_{\parallel}/I_{\perp}$ is given by:

$$I_{\parallel}/I_{\perp} = \int_0^{D_\infty} i_{\parallel}(m,\alpha,90^\circ)N(D)dD \bigg/ \int_0^{D_\infty} i_{\perp}(m,\alpha,90^\circ)N(D)dD \quad (1)$$

where $i_{\parallel}(m,\alpha,\theta)$ and $i_{\perp}(m,\alpha,\theta)$, respectively, represent the Mie scattering functions for the polarization component parallel to the scattering plane and the polarization component perpendicular to the scattering plane, D is the particle diameter, and N(D) is the particle number size distribution. The Mie scattering functions depend on the size parameter $\alpha = \pi D/\lambda$ (where λ is the wavelength of the incident light), the complex refractive index m, and the scattering angle θ, which is 90 deg in this case. The upper limit distribution function (ULDF) is given by:

$$N(D) = C \left\{ \frac{\exp\left[-\{\delta \ln[aD/(D_\infty - D)]\}^2\right]}{D^4(D_\infty - D)} \right\} \quad (2)$$

where a and δ are two adjustable parameters, C is a constant, and D_∞ is the maximum particle diameter. Since only the ratio of scattering intensities is desired, the constant C is not needed. The value of D_∞ which appears in Eqs. (1) and (2) is related to the volume-surface mean diameter D_{32} by

$$D_{32} = D_\infty/[1 + a \exp(1/4\delta^2)] \quad (3)$$

where D_{32} is defined by

$$D_{32} = \int_0^{D_\infty} N(D)D^3 \, dD \bigg/ \int_0^{D_\infty} N(D)D^2 \, dD \quad (4)$$

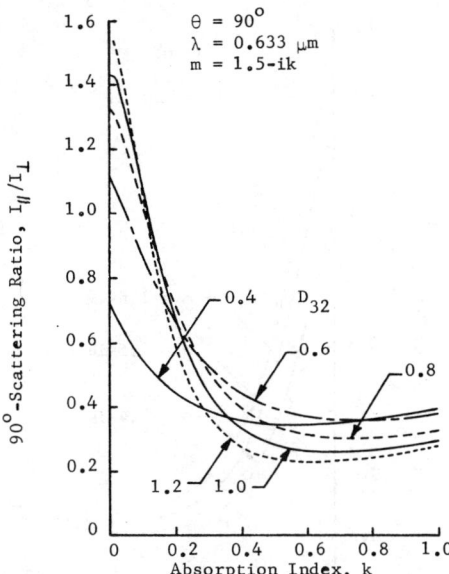

Fig. 2. Dependence of $I_\parallel/I_\perp$ upon absorption index and D_{32}.

Typical values of the parameters in Eqs. (2) and (3) are $a = 1.13$ and $\delta = 1.26$ [used in Ref. (7)] which gives $D_{32}/D_\infty = 0.431$.

As in the case of the forward scattering intensities and the ratio of optical densities at two wavelengths,[4] moderate variations in the ULDF parameters a and δ had little effect on the calculated ratios $I_\parallel/I_\perp$. Thus the remaining calculations were performed for the "standard" distribution function given by $a = 1.13$ and $\delta = 1.26$. The dependence of $I_\parallel/I_\perp$ upon absorption index k for $m = 1.5 - ik$ is shown in Fig. 2 for mean particle diameters (D_{32}) typical of smoke. Similar curves were also obtained for $m = 1.4 - ik$ and $m = 1.6 - ik$ (see Fig. 3). These results show that measurements of $I_\parallel/I_\perp$ can be used to determine the absorption index k of typical smokes for absorption indices up to about 0.4.

To use measurements of $I_\parallel/I_\perp$ to determine k, the real part n of the refractive index must also be known. This was previously determined from the ratio of optical densities at two wavelengths by assuming that the smoke particles are nonabsorbing ($k = 0$). This ratio is obtained from the Mie theory as

$$\text{OD}(\lambda_2)/\text{OD}(\lambda_1) = \frac{\ln(I_o/I)_{\lambda_2}}{\ln(I_o/I)_{\lambda_1}} = \frac{\bar{Q}_{ext}(m,D_{32},\lambda_2)}{\bar{Q}_{ext}(m,D_{32},\lambda_1)} \quad (5)$$

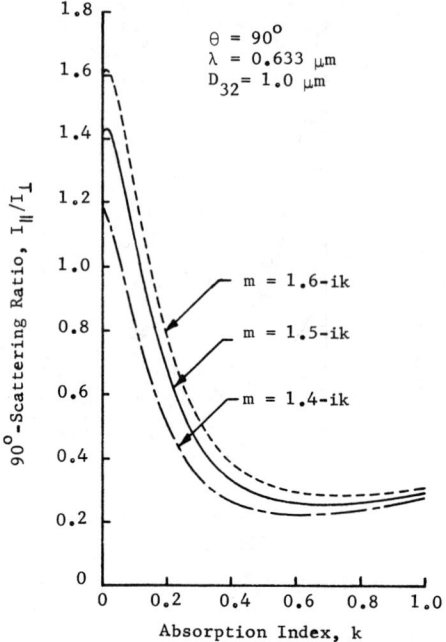

Fig. 3 Effect of refractive index upon $I_\parallel / I_\perp$.

where $\bar{Q}_{ext}$ is the mean extinction efficiency defined by

$$\bar{Q}_{ext} = \int_0^{D_\infty} Q_{ext}(m,D,\lambda) N(D) D^2 dD \bigg/ \int_0^{D_\infty} N(D) D^2 dD \qquad (6)$$

Calculations of the ratio of optical densities at two widely spaced wavelengths ($\lambda_1 = 0.458$ and $\lambda_2 = 0.633$ μm) using Eqs. (5) and (6) indicate that $OD(\lambda_2)/OD(\lambda_1)$ depends upon k as well as n (see Fig. 4). Thus both 90-deg scattering data ($I_\parallel / I_\perp$) and optical density data must be considered together with the previously measured value of D_{32} (the forward scattering technique is insensitive to refractive index) in order to determine the complex refractive index m = n-ik. A computer program has been developed to perform these additional calculations.

For measured values of D_{32}, $I_\parallel / I_\perp$ and $OD(\lambda_2)/OD(\lambda_1)$, the computer program obtains the refractive index n and absorption index k by the following algorithm. First the absorption index k is assumed to be zero. Then the refractive index is calculated by two independent methods: 1) from the measured optical density ratio $OD(\lambda_2)/OD(\lambda_1)$ to yield n_{tr} (transmission) and 2) from the measured 90-deg scattering ratio

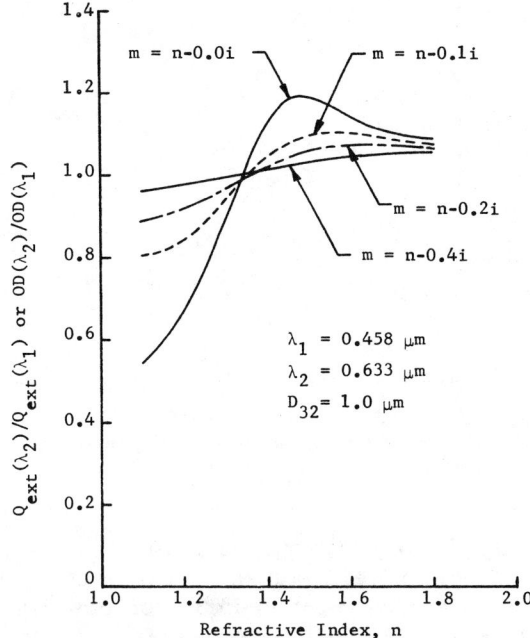

Fig. 4 Effect of absorption index upon ratio of extinction efficiencies at two wavelengths.

$I_\parallel / I_\perp$ to yield n_{sc} (scattering). If the solutions are equal, then $k = 0$ and $n = n_{tr} = n_{sc}$. If $n_{tr} \neq n_{sc}$, then the difference $n_{tr} - n_{sc}$ is calculated and new values of n_{tr} and n_{sc} are computed assuming a small value of k, say $k = 0.01$. Again $n_{tr} - n_{sc}$ is computed and compared with the previous value to determine a new trial value of k using linear interpolation. This process is repeated until $n_{tr} - n_{sc}$ is less than about 2×10^{-4}. It is assumed in making these calculations that both n and k are independent of wavelength.

Typical Smoke Refractive Index Measurements

As a test of the 90-deg scattering optical system and calibration procedure, a number of DOP aerosol tests were conducted in which the mean particle size D_{32} was obtained from the forward scattering ratio and the refractive index n_B ($\lambda = 0.488$ μm) was calculated from the 90-deg scattering ratio assuming that $k = 0$. For most of these tests, D_{32} ranged between 0.70 and 0.86 μm and measured values of n_B were between 1.467 and 1.521 with an average value of 1.489. This is in close agreement with the published value for DOP, $n_D = 1.4853$ (sodium D-line).

Results are now presented for smokes produced by oxidative pyrolysis of selected polymeric materials, which was induced by

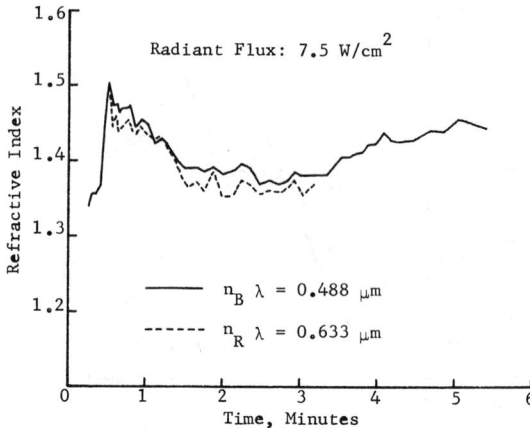

Fig. 5 Variation of smoke refractive index during nonflaming combustion of cellular polyphosphazene insulation material.

exposing small samples of these materials to a radiant heat flux of 5-7.5 W/cm^2 in room temperature air. Two cases are considered to illustrate the measurement technique for both absorbing and nonabsorbing smokes. For a polyphosphazene foam wall insulation material, measured values of $OD(\lambda_2)/OD(\lambda_1)$ or OD_R/OD_B and $I_{\parallel}/I_{\perp}$ are consistent with nonabsorbing particles. For this material, it was initially assumed that the particles were nonabsorbing (k = 0), and the measured values of $I_{\parallel}/I_{\perp}$ in blue light (λ_1 = 0.488 μm) along with the simultaneously determined mean particle diameters (D_{32}) were used to calculate the corresponding refractive index n_B. Measured OD_R/OD_B values (λ_2 = 0.633 μm) were also used to obtain the refractive index, assuming again that k = 0 and furthermore that n does not vary significantly with wavelength (i.e., $n_R = n_B$). Near the time of peak smoke optical density, average values of n_B and n_R of 1.41 were obtained from the optical density ratio, while the corresponding value of n_B from the 90-deg scattering ratio was 1.45. In order to reconcile these two values, the assumption of k = 0 was relaxed and the Mie theory was used in an attempt to determine n and k to simultaneously satisfy the optical density and 90-deg scattering data. This proved to be impossible for most of the data, requiring a negative value of k for a solution. A different approach was taken by requiring that k = 0, but allowing n to vary with wavelength. In this method, n_R was calculated from the optical density ratios using the values of n_B obtained previously from the $I_{\parallel}/I_{\perp}$ data. This method was successful in fitting both the OD_R/OD_B and $I_{\parallel}/I_{\perp}$ data. As shown in Fig. 5, the refractive index in red light is slightly smaller than the refractive index in blue light, and it exhibits similar variations with time during the test. Such variations in refractive index

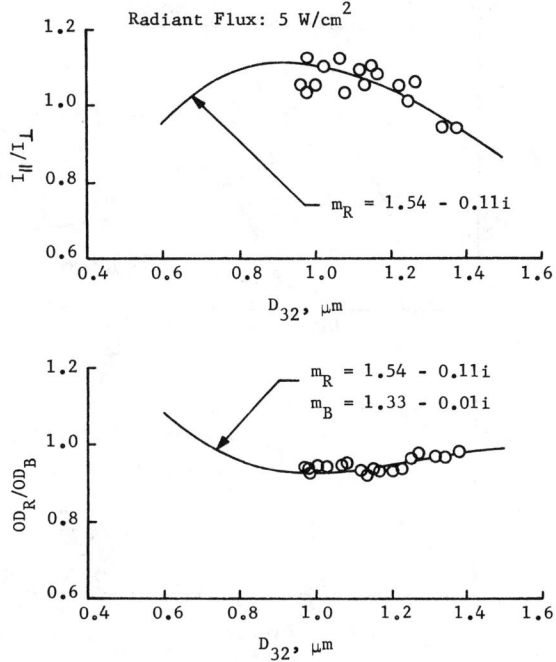

Fig. 6 Illustration of curve fitting technique for nonflaming combustion of PVC cable jacket.

indicate corresponding variations in the chemical composition of the smoke particles during the test. The averaged values of the refractive indices at peak optical density are $n_B = 1.452$ and $n_R = 1.439$.

The second case of oxidative pyrolysis is presented to illustrate the measurement technique for moderately absorbing particles. Here a polyvinyl chloride (PVC) cable jacket material was exposed to 5-W/cm^2 radiant flux in room-temperature ventilation air. Under the assumption of $k = 0$, values of the refractive index n obtained from both the optical density ratio and the 90-deg scattering ratio were generally too low for the organic compounds expected from pyrolysis of this material. Furthermore, the Mie theory was unable to determine values of n and k to simultaneously satisfy the optical density and 90-deg scattering data when the complex refractive index $m = n-ik$ was assumed to be independent of wavelength. Finally, a plot of OD_R/OD_B and $I_\parallel /I_\perp$ as a function of mean particle diameter D_{32} was constructed. Various values of the complex refractive index at the two wavelengths (m_R and m_B) were assumed until a reasonably good curve fit was obtained for the data points near the time of maximum optical density. This resulted in

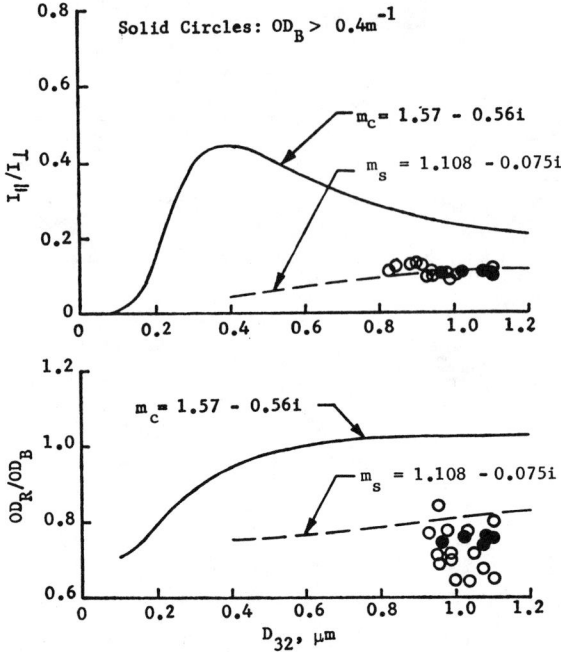

Fig. 7 Optical density ratios and 90-degree scattering ratios for flaming combustion of cellular polyphosphazene insulation material.

m_R = 1.54 - 0.11i (from the $I_\parallel / I_\perp$ data at λ_2 = .633 μm) and m_B = 1.33 - 0.01i; the experimental data and the corresponding theoretical curves of optical density ratio and 90-deg scattering ratio are shown in Fig. 6. It should be noted that these values of m_R and m_B are not unique, as there are more unknown quantities ($n_R, k_R, n_B,$ and k_B) than there are constraints (OD_R^R/OD_B^B and $I_\parallel^R/I_\perp^R$). Furthermore, it is necessary to assume that m_R and m_B do not change with time during the test. Nevertheless, these results indicate some general optical characteristics of the smoke particles produced by this PVC material under nonflaming conditions.

For the case of <u>flaming</u> combustion of organic polymers the soot particles produced are highly absorbing, and the determination of the complex refractive index (n and k) directly from the measured values of OD_R/OD_B and $I_\parallel / I_\perp$ is difficult and unreliable. Fig. 7 shows measured values of $I_\parallel / I_\perp$ and OD_R/OD_B plotted vs D_{32} for a typical flaming test of polyphosphazene insulation conducted in room temperature air. Also plotted in Fig. 7 are curves of $I_\parallel / I_\perp$ and OD_R/OD_B vs D_{32} which were calculated using the Mie scattering theory with m = 1.57 - 0.56i. The measured values of $I_\parallel / I_\perp$ are seen to cluster about 0.1 which is about half as large as the theoretical values. In addition, the measured optical density ratios lie about 25% below the theoretical curve for these smoke particles.

In a recent paper by Santoro, et al.[13] it was shown that similar discrepancies between experimental observations and the Mie theory for polydispersions of absorbing spheres can be resolved. There it was concluded that the loosely packed, low-density soot agglomerates have an effective refractive index m_s that is significantly reduced below the refractive index m_c of the particulate material of which they are composed. This downward scaling of the refractive index was applied to the interpretation of the data shown in Fig. 7 using the method of Graham.[14] Here the Lorentz-Lorenz formula was used to relate m_s and m_c as a function of η_v, the fraction of the optical mean volume that is occupied by the particulate material. Using $m_c = 1.57 - 0.56i$, measured values of $I_\parallel/I_\perp$ and D_{32} yield a unique value of $m_s = n_s - ik_s$ from which η_v can also be determined. The best fit to the $I_\parallel/I_\perp$ data shown in Fig. 7 is given by $m_s = 1.108 - 0.0749i$ for which $\eta_v = 0.184$. Using this same value of m_s, the Mie theory gives significantly lower values of OD_R/OD_B which are only about 8% above the experimental data (Fig. 7). Although this small remaining discrepancy may be due in part to variations in the effective refractive index with wavelength, it is within the expected experimental error. Thus the present data confirms the results of Ref. 13 and indicates that the 90-deg scattering technique provides a means of measuring the <u>effective</u> refractive index of loosely packed soot agglomerates typical of flaming combustion.

Summary and Conclusions

A new laser technique has been developed for in situ measurement of the complex refractive index of smoke particles, which is needed for calculation of particle concentration (i.e., volume fraction) from light transmission and mean particle size measurements. This technique uses measurements of light scattered in the plane perpendicular to the polarized incident laser beam at azimuthal angles parallel to and perpendicular to the beam polarization. This method is an improvement over an earlier technique based on light transmission measurements at two wavelengths which was applicable only for nonabsorbing particles.

The following conclusions regarding the applicability of this technique to the determination of the refractive index of combustion generated particulates have been obtained:
1) For <u>nonabsorbing</u> smokes, the 90-deg scattering measurements coupled with the particle size measurements will yield the refractive index. In this case the 90-deg scattering method offers two advantages over the earlier transmission method: a) the scattering method is more sensitive and can be

used at lower smoke concentrations, and b) it is not necessary to assume that the refractive index is independent of wavelength since only one wavelength is used.

2) For moderately absorbing smokes, both the 90-deg scattering measurements and the two-color transmission measurements are needed to determine the complex refractive index. It is also necessary to assume that both the real and imaginary parts of the refractive index are independent of wavelength in order to obtain a unique solution. In some cases, where particle size varies considerably during the experiment, a curve fitting technique may be used to obtain the complex refractive index at both wavelengths. In this latter case the solution may not be unique and the refractive index must not vary with time during the experiment.

3) For the highly absorbing soot agglomerates characteristic of flaming combustion the new technique is able to measure the <u>effective</u> complex refractive index, which is considerably lower than the refractive index of the particulate material of which the agglomerates are composed. In addition the method yields the fraction of the optical mean volume that is occupied by the particulate material.

Acknowledgments

The development of this technique was supported by the National Bureau of Standards under Grant No. G8-9003. The measurements presented in this paper were obtained as part of a project funded by the Naval Research Laboratory under Contract No. N00014-78-C-0771.

References

[1] Zinn, B. T., Powell, E. A., Cassanova, R. A., and Bankston, C. P., "Investigation of Smoke Particulates Generated During the Thermal Degradation of Natural and Synthetic Materials," <u>Fire Research</u>, Vol. 1, March 1977, pp. 23-36.

[2] Bankston, C. P., Powell, E. A., Cassanova, R. A., and Zinn, B. T., "Detailed Measurements of the Physical Characteristics of Smoke Particulates Generated by Flaming Materials," <u>Journal of Fire and Flammability</u>, Vol. 8, Oct. 1977, pp. 395-411.

[3] Powell, E. A., Bankston, C. P., Cassanova, R. A., and Zinn, B. T., "The Effect of Environmental Temperature upon the Physical Characteristics of the Smoke Produced by Burning Wood and PVC Samples," <u>Fire and Materials</u>, Vol. 3, March 1979, pp. 15-22.

[4] Powell, E. A., Cassanova, R. A., Bankston, C. P., and Zinn, B. T., "Combustion-Generated Smoke Diagnostics by Means of Optical Measurement Techniques," <u>Experimental Diagnostics in Gas Phase</u>

Combustion Systems: Progress in Astronautics and Aeronautics, Vol. 53, edited by Ben T. Zinn, AIAA, New York, 1977, pp. 449-463.

[5] Hodkinson, J. R., "Particle Sizing by Means of the Forward Scattering Lobe," Applied Optics, Vol. 5, May 1966, pp. 839-844.

[6] Gravatt, C. C., "Real Time Measurement of the Size Distribution of Particulate Matter by a Light Scattering Method," Journal of the Air Pollution Control Association, Vol. 23, Dec. 1973, pp. 1035-1038.

[7] Dobbins, R. A. and Jizmagian, G. S., "Optical Scattering Cross Sections for Polydispersions of Dielectric Spheres," Journal of the Optical Society of America, Vol. 56, Oct. 1966, pp. 1345-1350.

[8] Walters, P. T., "Optical Measurement of Water Droplets in Wet Steam Flows," Heat and Fluid Flow in Steam and Gas Turbine Plant, Conference Publ. 3, Institution of Mechanical Engineers, London, 1973, pp. 66-74.

[9] Dalzell, W. H. and Sarofim, A. F., "Optical Constants of Soot and Their Application to Heat-Flux Calculations", Journal of Heat Transfer, Trans. ASME, Vol. 91, Feb. 1969, pp. 100-104.

[10] Mulholland, G., Private Communication, Jan. 1976.

[11] Kerker, M., The Scattering of Light and Other Electromagnetic Radiation, Academic Press, New York and London, 1969.

[12] Mugele, R. A. and Evans, H. D., "Droplet Size Distribution in Sprays," Industrial Engineering Chemistry, Vol. 43, June 1951, pp. 1317-1324.

[13] Santoro, R. J., Semerjian, H. G., and Dobbins, R. A., "Interpretation of Optical Measurements of Soot in Flames," AIAA paper 83-1516, AIAA 18th Thermophysics Conference, June 1983.

[14] Graham, S. C., "The Refractive Indices of Isolated and Aggregated Soot Particles," Combustion Science and Technology, Vol. 9, 1974, pp. 159-163.

Chapter IV. Combustion Diagnostics Applications

Temperature and Concentration Measurements in an Internal Combustion Engine using Laser Raman Spectroscopy

A. zur Loye* and D. A. Santavicca†
Princeton University, Princeton, New Jersey

Abstract

Gas temperature and carbon monoxide concentration were measured in an internal combustion engine using laser Raman spectroscopy. The measurements were taken in the center of the clearance volume in the postflame gases at 20, 40, and 80 deg after top dead center and at equivalence ratios of 1.15, 1.26, and 1.36. The measured CO concentration was found to increase with equivalence ratio, to decrease during the expansion stroke and to remain at or very close to equilibrium until the gas temperature falls below 1300 K, after which it appears to freeze.

Introduction

The postflame reactions in an internal combustion engine are important in determining the exhaust pollutants[1] and in terms of their contribution to the overall heat release[2]. It is generally assumed that the CO concentration is in equilibrium immediately after the flame, due to the high temperature, but that during the expansion stroke, due to the rapid cooling, the CO does not stay in equilibrium. Theoretical studies of the chemical kinetics of the expansion process have predicted such behavior[3] but experimental studies have been inconclusive[4,5]. The

---

Presented as Paper 83-1551 at the AIAA 18th Thermophysics Conference, Montreal, Canada, June 1-3, 1983. Copyright © American Institute of Aeronautics and Astronautics, Inc., 1983. All rights reserved.

*Graduate Student, Department of Mechanical and Aerospace Engineering.

†Research Engineer, Department of Mechanical and Aerospace Engineering.

experimental studies have consisted of in-cylinder measurements of the postflame gases using fast acting gas sampling valves. They have shown that the CO concentration decreases during the expansion stroke but an accurate comparison with the corresponding equilibrium concentration has not been possible because the gas temperature has only been known approximately through thermodynamic calculations based on the measured pressure. In addition, sampling valve measurements are subject to uncertainties due to 1) possible reactions within the sampling probe, 2) the relatively long sampling time, e.g., typically 1 ms or 7 crankangle deg at 1200 RPM, and 3) the fact that the probe is either mounted flush with the wall and thereby extracting an unrepresentative sample, or it protrudes into the combustion chamber and perturbs the local chemistry and fluid dynamics[6].

The measurements in this study were made using laser Raman spectroscopy. The feasibility of obtaining Raman spectra from the pre-ignition and postflame gases in an internal combustion engine was first demonstrated by Setchell[7] in 1978. The following year Johnston[8] reported the first quantitative fuel-to-air ratio measurements in a motored internal combustion engine using Raman scattering, and in 1980 Smith[9,10] made the first Raman temperature measurements in a firing internal combustion engine using the Stokes/anti-Stokes method. In this study quantitative CO concentration measurements in the postflame gases of a firing internal combustion engine are reported for the first time.

Gas temperature and carbon monoxide concentration were measured at several crankangles and for several equivalence ratios in the postflame gases in an internal combustion engine. The measurements were taken in the center of the clearance volume and are spatially and temporally resolved. Equilibrium CO concentrations were calculated based on the known equivalence ratio and the measured temperature and pressure. Comparisons between the measured CO concentration and the equilibrium CO concentration are presented.

Experiment

The Raman and engine experimental setup is shown in schematic form in Fig. 1. The Raman measurements were made in one of the transparent-piston, transparent-head engines in the Engine Combustion Laboratory at Princeton University. A schematic drawing of the piston, cylinder and head assembly is shown in Fig. 2. The piston-cylinder assembly was mounted on a high speed Waukesha CFR-48 crank-

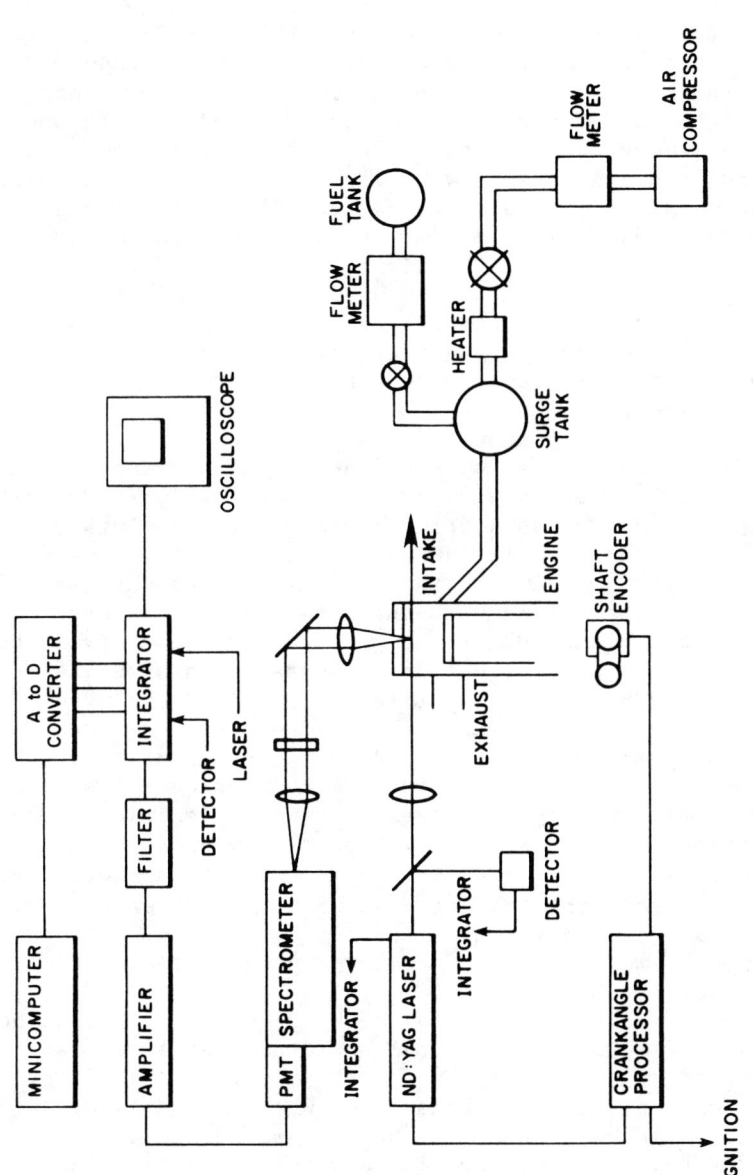

Fig. 1 Raman and engine experimental system.

case and had an 8.26-cm dia bore and an 11.43-cm stroke. Air intake and exhaust gases passed through ports in the cylinder wall as in a two cycle engine. Extensive optical access was possible in this engine, i.e., through windows mounted in the head, the piston, and the cylinder; however, in this study the only windows required were a 2.54-cm dia window in the head for the collection of the Raman scattered light, and two windows in the cylinder wall through which the laser beam passed. The combustion chamber was open and "pancake" shaped with a compression ratio of 8.5. The intake ports were directed 30 deg from the cylinder radius to produce swirl and upward at 30 deg to improve the scavenging efficiency. The timing of the exhaust and intake ports was ±113 and ±126 deg, respectively, where 0 deg is top dead center (TDC). Teflon impregnated bronze rings were used, enabling the engine to be run without cylinder lubrication and thereby reducing the problem of window fouling.

For the work presented in this paper the engine was run at 1200 RPM with a TDC swirl number of 4. The intake air was supplied by an air compressor at 16.5 gm/s which corresponds to 100% volumetric efficiency. The intake manifold pressure was 125 kPa, the intake manifold temperature was 25°C and the exhaust manifold pressure was 99 kPa. Propane was mixed with the intake air in a surge tank upstream of the engine. The engine was run at equivalence ratios of 1.15, 1.26, and 1.36 with spark timing at 24 deg before TDC. The engine was fired every sixth cycle to insure complete scavenging. The engine crankangle was monitored by an optical shaft encoder the output of which was input to a crankangle trigger module which is used to provide variable trigger pulses to control ignition and laser pulse timing. Note that the laser was pulsed every third engine cycle whereas the engine was fired every sixth cycle, therefore the measurements were made alternately in a firing and a non-firing engine cycle. In-cylinder pressure was measured with a water-cooled piezoelectric transducer.

The laser used in this work was a Quanta-Ray frequency doubled Nd:YAG with a 10 ns pulse width. The laser was operated at approximately 50 mj per pulse. Higher laser powers could not be used because of what are believed to be "hot spots" in the laser beam which damaged the windows. The laser beam was focused with a 2-m focal length lens with the beam waist positioned 20-cm before the engine. This gave a slightly diverging 0.7-mm wide by 3-mm high beam diameter in the engine. The spatial resolution along the beam was 3.3-mm which was determined by the spectrometer slit height (10-mm) and the magnification of the collection optics. The Raman signal was collected at 90 deg with an F3.0 lens. The collected light was passed through a high pass filter (Schott KV550) and a 3/4-m single monochromator and detected by a photomultiplier tube (RCA C31034). The photomultiplier tube output was amplified, filtered, integrated, digitized, and then recorded by a minicomputer. The purpose of the filter was to remove low frequency noise due to combustion generated light. The measurements, which were recorded pulse by pulse, were then sorted according to whether they occurred in a firing or a nonfiring cycle. To account for pulse-to-pulse fluctuations in the laser power, the Raman signal was normalized by simultaneous laser intensity measurements.

The gas temperature was determined by comparing measured and theoretical nitrogen vibrational Q-branch

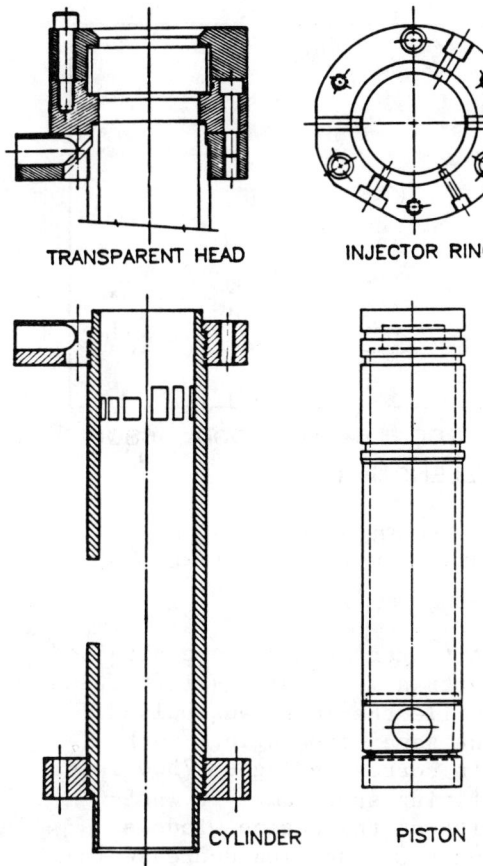

Fig. 2 Schematic drawing of piston, cylinder, and head assembly.

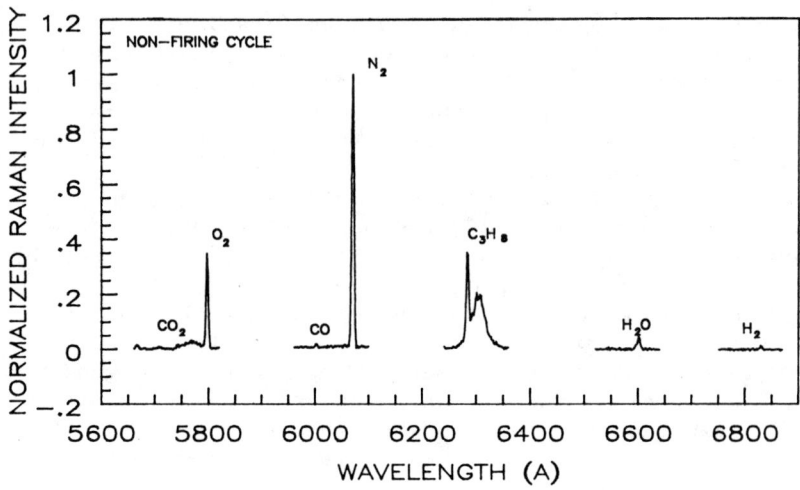

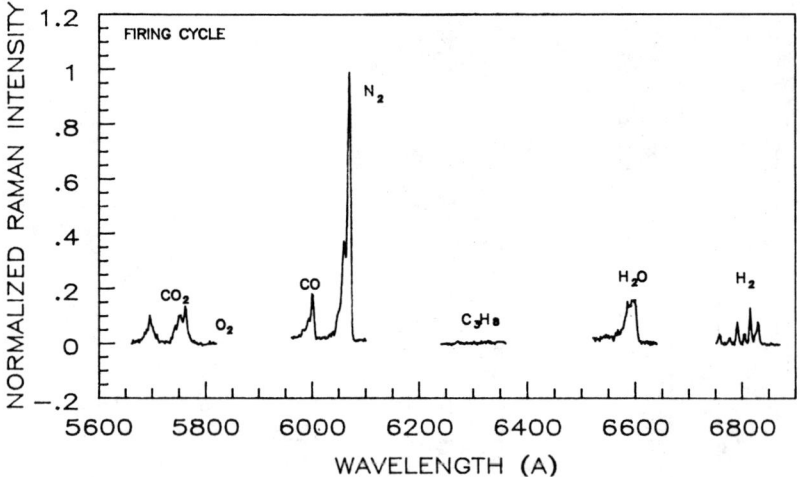

Fig. 3 Raman scattering spectra, 1200 RPM, equivalence ratio = 1.5, propane fuel, ignition at 24 deg BTDC, laser pulse at 50 deg ATDC, spectrometer slit width = 4.4 Å.

Raman spectra. The measured nitrogen spectra were obtained by scanning the spectrometer with a 4.4 Å wide slitfunction (full width at half height), while the laser was pulsed every third engine cycle at the same crankangle. Such scans took approximately 40 min corresponding to 7500 laser pulses in each firing and nonfiring spectrum. It was observed that the transmissivity of the engine windows decreased during the scans, and the resulting decrease in collection efficiency was nearly linear in time. To account for this the nitrogen peak intensity was measured before and after each run and the spectrum adjusted accordingly.

The CO concentration was determined by first measuring the intensity of the CO Raman signal at 6000 Å using a 21 Å wide slitfunction. Measurements of the nitrogen Raman signal at 6068 Å and the background noise at 5900 Å and 6150 Å were also made with the same slit width, before and after each CO measurement. The background noise was subtracted from the CO and nitrogen measurements and then the contribution of the nitrogen Raman spectrum at the CO wavelength was subtracted from the CO measurement. This latter correction was less than 1%. Next the CO measurement was normalized by the nitrogen measurement. The reason for normalizing by the nitrogen measurement was to account for changes in the collection efficiency both during and between runs, e.g., due to changes in the alignment of the collection optics or changes in the transmissivity of the windows in the engine. Using the measured temperature, the ratio of the CO to nitrogen Raman intensities was adjusted to give an equivalent intensity ratio at 300 K. For the 21 Å wide slitfunction, the 6000 Å CO measurement, and the 6068 Å nitrogen measurement the temperature correction factor was between 0.985 and 1.0, and therefore the CO to nitrogen Raman intensity ratio was

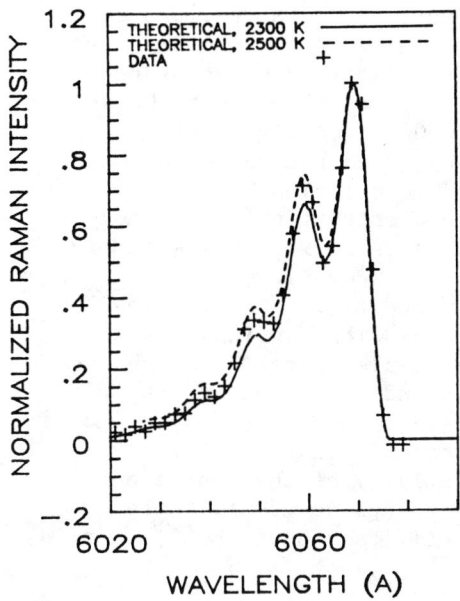

Fig. 4 Vibrational Raman nitrogen spectrum in postflame gases, 1200 RPM, equivalence ratio = 1.36, propane fuel, ingition at 24 deg BTDC, laser pulse at 20 deg ATDC, spectrometer slit width = 4.4 Å, 250 readings per data point.

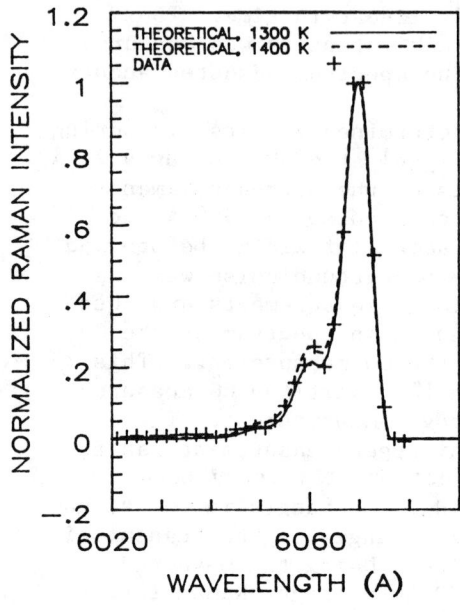

Fig. 5. Vibrational Raman nitrogen spectrum in postflame gases, 1200 rpm, equivalence ratio = 1.36, propane fuel, ignition at 24 deg BTDC, laser pulse at 80 deg ATDC, spectrometer slit width = 4.4 A, 250 readings per data point.

relatively insensitive to errors in the temperature measurement. The Raman system was calibrated in terms of the CO to nitrogen intensity ratio at room temperature using a calibration gas. The equilibrium nitrogen concentration, which was calculated from the known equivalence ratio and the measured temperature and pressure, was then used together with the calibration results to convert the temperature corrected intensity ratio to a CO molar concentration. The CO measurements were all made at the same crankangle and averaged over 2000 consecutive firing cycles, and similarly the nitrogen and background noise measurements were averaged over 1000 cycles.

Since the equilibrium CO concentration is extremely sensitive to the initial equivalence ratio, it was important that the equivalence ratio be known accurately and be repeatable from run to run. In addition to monitoring the individual fuel and air flow rates with Hastings linear mass flow meters, the actual equivalence ratio was checked by gas chromatographic analysis and the repeatability of the equivalence ratio was checked by measuring the ratio of the propane to nitrogen Raman signals in the nonfiring cycle in each run. The repeatability of the equivalence ratio for the same flowmeter settings was found to be within the 2% repeatability specification of the flowmeters.

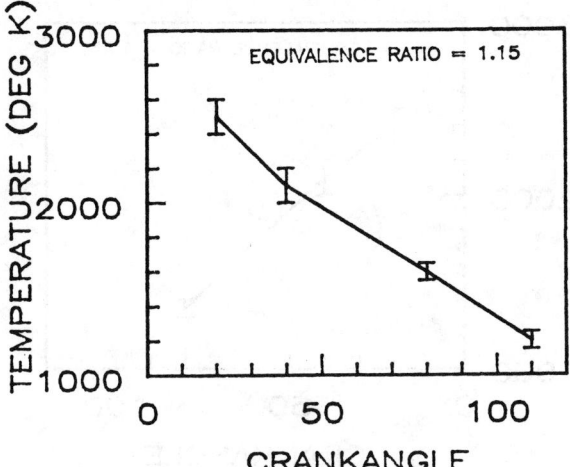

Fig. 6 Measured temperature in postflame gases vs crankangle, 1200 RPM, propane fuel, ignition at 24 deg BTDC, equivalence ratio = 1.15.

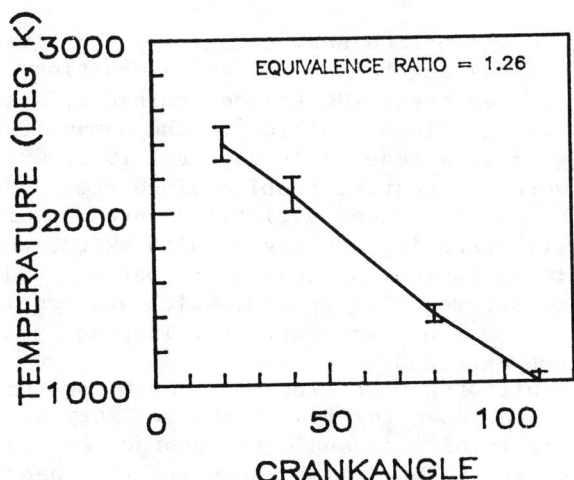

Fig. 7 Measured temperature in postflame gases vs crankangle, 1200 RPM, propane fuel ignition at 24 deg BTDC, equivalence ratio = 1.26.

Results

Typical Raman spectra from the firing and non-firing cycles are shown in Fig. 3. The spectra are normalized by the peak nitrogen intensity. In the nonfiring cycle, which occurred three cycles after the firing cycle, there is evidence of residual hydrogen, carbon monoxide, and carbon

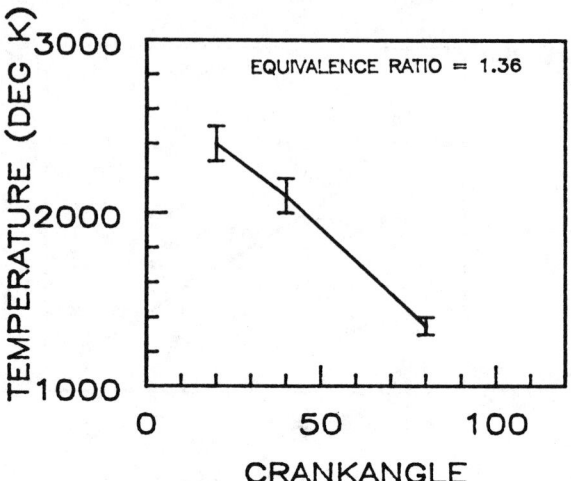

Fig. 8 Measured temperature in postflame gases vs crankangle, 1200 RPM, propane fuel ignition at 24 deg BTDC, equivalence ratio = 1.36.

dioxide due to incomplete scavenging. In the firing cycle, ignition was at 24 deg before TDC and combustion was complete by 20 deg after TDC (as determined by shadowgraph photographs of the flame position). The Raman measurements shown in Fig. 3 were made at 50 deg past TDC. Cyclic variations were a potential problem since the measurements were cycle averaged. These variations were minimized, however, by 1) operating the engine with swirl, which gave peak pressure variations of only a few percent and by 2) taking the measurements after combustion was complete, thus staying away from the flame where the largest cyclic variations are expected.

Typical nitrogen vibrational Q-branch spectra from the firing cycle are shown in Figs. 4 and 5. They are for an equivalence ratio of 1.36 and were recorded 20 and 80 deg, respectively, after TDC. Also shown are the theoretical spectra which best fit the measured spectra. Based on these comparisons the measured gas temperatures are estimated to be 2400 ±100 K at 20 deg after TDC and 1350 ±50 K at 80 deg after TDC. By comparing the measured and theoretical spectra in this fashion, the gas temperature was determined from measurements made at 20, 40, and 80 deg after TDC in the firing cycle. The measured gas temperature is plotted vs crankangle for the equivalence ratios 1.15, 1.26, and 1.36 in Figs. 6-8, respectively.

The equilibrium concentrations were calculated with the NASA Chemical Equilibrium Code[11] using the measured

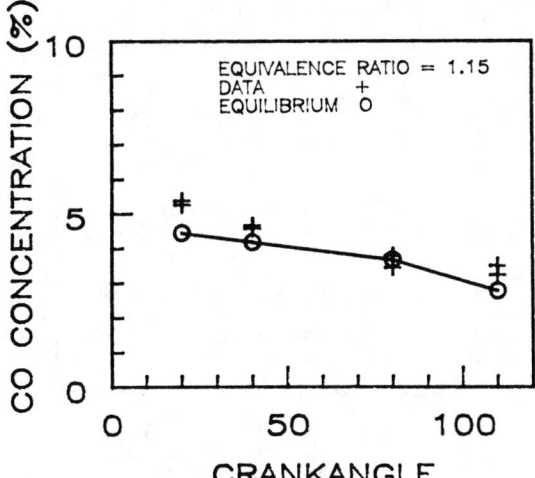

Fig. 9 Measured CO concentration in postflame gases and calculated equilibrium CO concentration vs crankangle, 1200 RPM, propane fuel, ignition at 24 deg BTDC, equivalence ratio = 1.15.

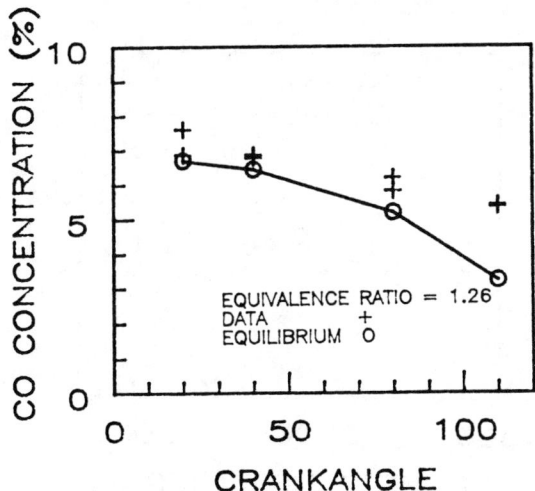

Fig. 10 Measured CO concentration in postflame gases and calculated equilibrium CO, concentration vs crankangle, 1200 RPM, propane fuel, ignition at 24 deg BTDC, equivalence ratio = 1.26.

equivalence ratio, temperature, and pressure. There is an uncertainty in the calculated equilibrium CO concentration of less than 0.5 mole percent which is due to the uncertainty or error in the measured equivalence ratio, to which it is most sensitive, and in the measured temperature. The equilibrium calculation is relatively insensitive to the pressure.

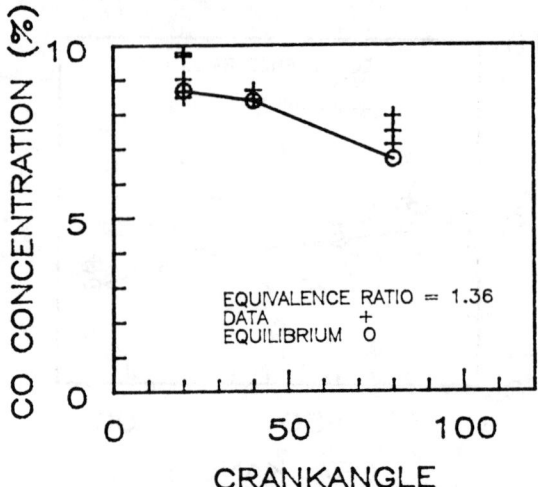

Fig. 11 Measured CO concentration in postflame gases and calculated equilibrium CO concentration vs crankangle, 1200 RPM, propane fuel, ignition at 24 deg BTDC, equivalence ratio = 1.36.

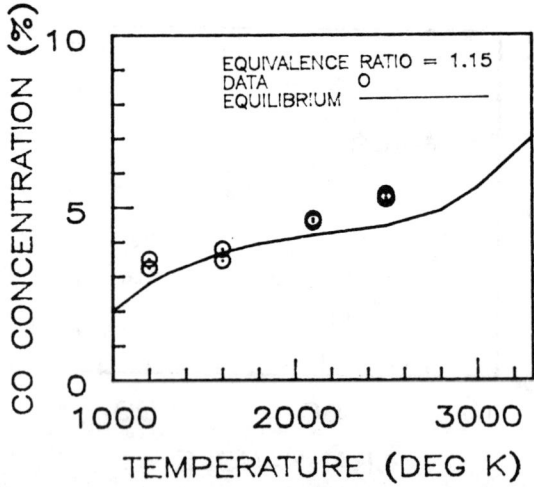

Fig. 12 Measured CO concentration in postflame gases and calculated equilibrium CO concentration vs temperature, 1200 RPM, propane fuel, ignition at 24 deg BTDC, equivalence ratio = 1.15.

The measured CO concentrations and the corresponding equilibrium CO concentrations are plotted vs crankangle in Figs. 9-11 for the equivalence ratios 1.15, 1.26, and 1.36, respectively. The experimental data points have an uncertainty of between 0.25 and 0.65 mole percent, due primarily to statistical error. The measured CO con-

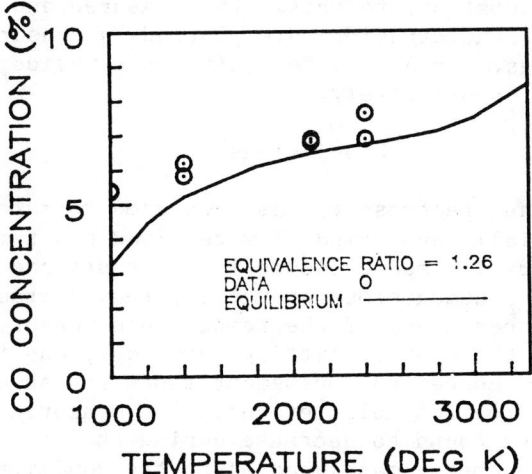

Fig. 13 Measured CO concentration in postflame gases and calculated equilibrium CO concentration vs temperature, 1200 RPM, propane fuel, ignition at 24 deg BTDC, equivalence ratio = 1.26.

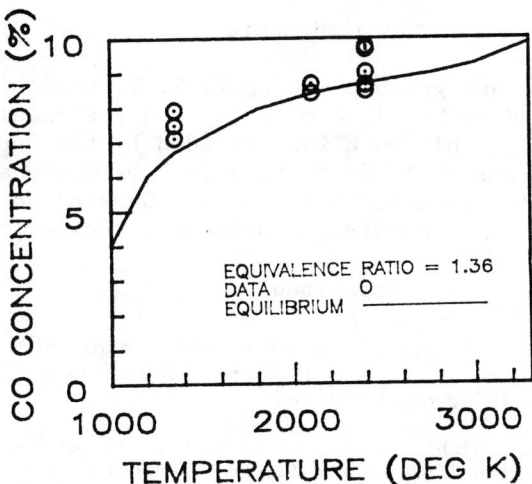

Fig. 14 Measured CO concentration in postflame gases and calculated equilibrium CO concentration vs temperature, 1200 RPM, propane fuel, ignition at 24 deg BTDC, equivalence ratio = 1.36.

centrations, as indicated by the least-squares fit straight line, indicate that the CO concentration is decreasing during the expansion stroke and that it is at or very close to the equilibrium concentration at temperatures above 1300 K. The CO concentration seems to begin to freeze at lower temperatures, but there is not sufficient data available at the moment to substantiate this. The measured and equilibrium CO concentrations are plotted vs measured temperature in Figs. 12-14 for the equivalence ratios 1.15, 1.26, and 1.36, respectively.

Conclusions

Laser Raman spectroscopy has been used to make cycle averaged, spatially and temporally resolved gas temperature and CO concentration measurements in the postflame gases of a spark ignited, homogeneous charge internal combustion engine. The uncertainty of the temperature measurements, obtained using the contour fitting technique, was ±50 to ±100 K. The CO concentration measurements had an uncertainty of ±.25 to ±.65 mole percent. The measured CO concentrations were found to decrease during the expansion stroke and were approximately equal to the equilibrium CO concentration calculated using the measured equivalence ratio, temperature and pressure until the gas temperature falls below 1300 K, after which it appears to freeze.

Acknowledgments

The authors are grateful to Prof. F. V. Bracco for his encouragement and to Mr. J. Semler for his technical support. Support for this work was provided by the Department of Energy (Contract DE-AC-04-81AL16338), the National Science Foundation (Grant CPE 80-03483), General Motors, Volkswagenwerk, FIAT, Komatsu, and Cummins Engines.

References

1. Heywood, J. B., "Pollutant Formation and Control in Spark-Ignited Engines," *Progress in Energy Combustion and Science*, Vol. 1, 1976, pp. 135-164.

2. Beckel, S. A., Shizhi, M., Matthews, R. D., and Peters, J. E., "Combustion, the Efficiency Rule, and Thermodynamic Modeling of Piston Engines," Paper No. WSS/CI-82-75, Fall Meeting, Western States Section, The Combustion Institute, Livermore, CA, Oct. 1982.

3. Newhall, H. K., "Kinetics of Engine-Generated Nitrogen Oxides and Carbon Monoxide," Twelfth Symposium (International) on Combustion, The Combustion Institute, 1968, pp. 603-613.

4. Samaga, B. S., Murthy, B. S., and Mahadevan, K., "A Combustion Gas Sampling System for a Study of CO and NO History During Engine Expansion Process," *Israel Journal of Technology*, Vol. 14, 1976, pp. 270-274.

5. Turk, S. L., "Time Resolved Measurements of Carbon Monoxide During the Expansion Stroke of an Internal Combustion Spark-Ignited Engine," S.M. Thesis, Department of Mechanical Engineering, M.I.T., Cambridge, Mass., 1972.

6. Johnson, P. J., "A Comparison of In-Cylinder Species Concentration Measurements Taken with Two Gas Sampling Valves," Paper No. CSS/CI-79-17, Spring Meeting, Central States Section, The Combustion Institute, Columbus, Indiana, April 1979.

7. Setchell, R. E., "Initial Measurements Within an Internal Combustion Engine Using Raman Spectroscopy," Report No. SAND78-1220, August 1978.

8. Johnston, S. C., "Precombustion Fuel/Air Distribution in a Stratified Charge Engine Using Laser Raman Spectroscopy," SAE Paper No. 790433, February 1979.

9. Smith, J. R., "Instantaneous Temperature and Density by Spontaneous Raman Scattering in a Piston Engine," AIAA Paper No. AIAA-80-1359, July 1980.

10. Smith, J. R., "Temperature and Density Measurements in an Engine by Pulsed Raman Spectroscopy," SAE Paper No. 800137, 1980.

11. Gordon, S. and McBride, B. J., "Computer Program for Calculation of Complex Chemical Equilibrium Compositions, Rocket Performance, Incident and Reflected Shocks, and Chapman-Jouget Detonations," NASA SP-273, NTIS, No. N71, 37775, 1971.

Rayleigh Thermometry with Low-Power Laser Sources

D. Benhachmi,* N. Younes,* H. Yakout,* P. E. Emmerman,† and R. Goulard‡
The George Washington University, Washington, D.C.

Abstract

Rayleigh thermometry has been demonstrated as a practical diagnostic technique in unsteady or turbulent flames with large CW lasers (1 W or more). The aim of the present experiment is to test this technique on a steady, premixed laminar air-methane flame, with a 15 mW He-Ne laser. Because of the drastic loss of signal, the noise sources in the experiment were analyzed and some of them reduced. The main thrust of this work is the reduction of the electronic dark current by the use of an adjustable offset amplifier and by cooling the photomultiplier. The paper presents our results at three different sections of the flame. Temperature profiles are reproduced with decreasing accuracy as the temperature increases. Photomultiplier cooling is indispensable: when used, signal-to-noise ratios of the order of 10 were obtained. Further areas of work are discussed, including other filtering techniques, increased observation bandwidth and better controlled experimental conditions. Signal-to-noise ratios of the order of 50 are expected.

Presented as Paper 83-1554 at the AIAA 18th Thermophysics Conference, Montreal, Canada, June 1-3, 1983. Copyright © American Institute of Aeronautics and Astronautics, Inc., 1983. All rights reserved.
*Graduate Student, School of Engineering and Applied Science.
†Research Associate, School of Engineering and Applied Science.
‡Professor, School of Engineering and Applied Science.

Introduction

The need for an accurate determination of the temperature of flames has become highly important in combustion research. The importance of an accurate temperature measurement in a flame is obvious since the fundamental properties such as density, concentration, dissociation, and others are dependent on temperature. A considerable effort has gone into the development of adequate instrumentation for measuring temperature, which includes intrusive probes and optical systems.

Usually the simplest and least costly diagnostic system utilizes intrusive probes such as thermocouples and they are used if they meet the needed requirements. However, the temporal response of intrusive probes is poor; their size limits the spatial resolution, they introduce disturbances in the medium, and they survive with difficulty in the hostile environment of combustion systems. In addition, reactions may be induced in the flowfield and the properties of the medium can be changed.

Optical techniques have the capability to overcome many of the probe limitations for a large number of cases. The principal advantages of these techniques are that they do not disturb the system being studied and they have no high-temperature limit. More recently, with the advent of lasers and the rapid improvements in detectors, techniques which use some form of optical scattering are being developed to measure properties (density, concentration, temperature, etc.) in combustion systems. These optical techniques provide excellent time and space resolution and have excellent time response from essentially a point measurement.

The application of laser-based diagnostics is an area of intensive research and several papers have been published on the subject.[1,2] These references and others lead to the conclusion that well-conceived applications of light scattering techniques are important to the progress of combustion technology. The best known techniques for density and temperature measurement are Raman and Rayleigh scattering.

From a practical point of view, Rayleigh diagnostics suffer from a lack of molecular specificity and the spectral overlap from Mie scattering. For these reasons, the technique has had very limited applications and has been employed in only very clean flames. However, its major advantage resides in its relatively high signal strength (the Rayleigh cross section is about 1000 times larger than the vibrational Raman cross section). Interpretation of density and temperature from the Rayleigh signal is relatively straightforward.

Rayleigh thermometry has been performed by a number of investigators for temperature measurements in premixed and diffusion flames.[3-7] Robben[3] obtained satisfactory turbulent density and temperature measurements with Rayleigh scattering and less satisfactory ones with Raman scattering. Smith[4] determined temperature profiles in a hydrogen diffusion flame, using the numerical model developed by Miller and Kee.[5] More recently, Dibble and Hollenbach[6,7] reported that Rayleigh scattering can be applied to turbulent premixed and diffusion flames.

In all studies mentioned above, relatively high-power argon lasers were used as a light source. A 7.5-W laser was used in the first one, 5 W in the second one, and 1 W in the third, all at 488-nm wavelength. Also, sophisticated instrumentation was used to collect and analyze the scattered signal.

The references just described illustrate the usefulness of Rayleigh scattering for combustion studies. One of its limitations, at least in its simplest form, is to be limited to reactions which insure a constant average Rayleigh cross section throughout reactions, products, and the ambient atmosphere. Although some practical fuels come close to this requirement (such as methane-air mixtures), a truly accurate procedure depends on the choice of a "Rayleigh-designed" flame.

Another aspect of these studies is that the background sources (luminescence, PM dark current) are kept small in comparison with the useful scattered signal by the use of powerful lasers. This option is an expensive one. In the present paper, we explore a hopefully less expensive approach, that of the more available low-power lasers (e.g., 15 mW) in conjunction with noise reducing techniques. The early phase of this work was reported in detail in Ref. 1.

Theory

Rayleigh thermometry, as proposed by Dibble and Hollenbach[6], relies on the linear relationship between Rayleigh scattered signals and the density of the gases being observed. The perfect gas law gives an equally simple relationship between density and temperature. If all other properties involved stay constant during the scanning of the flame, <u>in time and space</u>, a remarkably simple flame temperature diagnostic is available. This section considers the relevant relationships and properties in the case of a premixed methane-air flame.

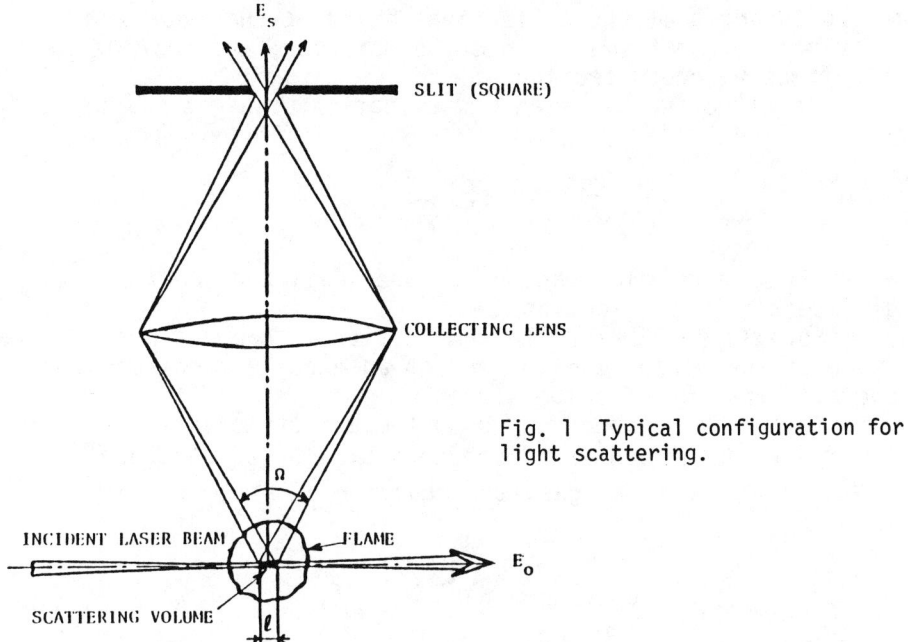

Fig. 1 Typical configuration for light scattering.

Rayleigh Scattering Theory

Rayleigh scattering is an elastic photon-molecule interaction whereby light is scattered at the same wavelength as the incident radiation. Scattering from particles is known as Mie scattering.

A general schematic for scattering measurement at right angles is shown in Figure 1. An incident laser beam, which causes scattering, is a high-powered pencil of light. A portion of the beam is viewed by the collection optics, thus defining a sample volume of the order of 1 mm^3. Light scattered from gas molecules in the sample volume is collected and analyzed to obtain the desired information.

The energy E_S Rayleigh-scattered into the collection optics by a gas mixture at 90 deg from the incident laser beam, is given by[4,6]

$$E_s = C\, E_o\, N\, \Omega\, \ell\, \left(\frac{d\sigma}{d\Omega}\right)_{eff} \tag{1}$$

where C is the calibration constant for the optics, E_o the incident laser energy, N the molecular number density, Ω

the solid angle of the collection optics, ℓ the length of laser beam segment imaged into the detector, and $(d\sigma/d\Omega)_{eff}$ the effective cross section.

The effective Rayleigh cross section of a gas mixture is

$$\left(\frac{d\sigma}{d\Omega}\right)_{eff} = \sum_i x_i \left(\frac{d\sigma_i}{d\Omega}\right) \quad (2)$$

where x_i is the mole fraction of species i and $(d\sigma_i/d\Omega)$ its differential cross section.

For gases in which the composition is known a priori, Rayleigh scattering can be used as an excellent measure for overall density and temperature.

An implicit dependence upon temperature exists for Equation (1) through the dependence of N on temperature. If we assume as ideal gas, the equation of state is then

$$P = \frac{N}{A_0} RT \quad (3)$$

where A_0 is the Avogadro number and R the universal gas constant. Substituting N from Equation (3) into Equation (1) gives

$$E_s = C \, E_0 \, \Omega \, \ell \, \frac{PA_0}{RT} \left(\frac{d\sigma}{d\Omega}\right)_{eff} \quad (4)$$

Since the laser intensity is assumed to be constant, and the variation of the pressure through the flame is negligible, variations in Rayleigh scattering intensity result from temperature variation or species variation or both. Equation (4) becomes

$$E_s = \frac{K}{T} \left(\frac{d\sigma}{d\Omega}\right)_{eff} \quad (5)$$

where K is a constant.

If the scattered signal E_s is measured first at a point at room temperature ($E_{s,air}$), and then at a point (r,z) inside the flame $E_s(r,z)$, it is possible to write from Equation (5):

$$\frac{E_s(r,z)}{E_{s,air}} = \frac{\left[\left(\frac{d\sigma}{d\Omega}\right)_{eff,flame(r,z)}\right]}{\left[\frac{T(r,z)}{T_{air}} \left(\frac{d\sigma}{d\Omega}\right)_{air}\right]} \quad (6)$$

Table 1 Rayleigh scattering cross section for combustion gases at $\lambda = 6328$ Å

Gases	$(d\sigma/d\Omega)$, $10^{28} cm^2$	Ref.
CH_4	6.507	8
H_2O (vapor)	2.221	9
Air	2.948	10
CO_2	7.105	10
O_2	2.560	10
N_2	3.052	11

This equation (and Eq. 7) offers a calibration procedure for $T(r,z)$ in terms of T_{air}, as long as the dependence of the cross section on temperature is known.

For premixed flames, the variation of Rayleigh intensity is primarily due to the variation in temperature; a factor of 2100 K/300 K = 7 is typical.[6] The variation in the effective cross section from reactants to products is often less than 10% for simple hydrocarbon-air systems. This condition is largely due to the fact that the major constituent is nitrogen in reactants, intermediates, and products.

For a stoichiometric mixture of methane and air, the variation in the cross section is less than 1% from reactants to products (see Ref. 1). This variation is negligible and the cross section can be assumed constant throughout the whole flame. Thus, the procedure of determining the temperature in the flame is reduced to measuring the Rayleigh scattered intensity for each point of interest relative to that of the reference, i.e., air at room temperature. In this case the temperature can be easily computed from Equation (6) by:

$$T(r,z) = T_{air} \frac{\left[E_{s,air} \left(\frac{d\sigma}{d\Omega} \right)_{eff,flame} \right]}{\left[E_{s,flame} \left(\frac{d\sigma}{d\Omega} \right)_{air} \right]} \quad (7)$$

The values of some typical cross sections used here to compute the effective cross sections of the reactants and products are given in the next section.

Cross Section of Reactants and Products

In the case of a stoichiometric mixture, methane burns with air according to the overall reaction: $CH_4 + 2(O_2 +$

3.76 N_2) → Intermediates → CO_2 + 2 H_2O + 7.52 N_2. Here we assume that the reaction is complete and only CO_2 and H_2O are present in the products and need to be included in the summation. The scattering cross sections of typical combustion molecules were reviewed in Ref. 1 and Table 1 summarizes the values of interest in our experiment.

In the methane-air reaction considered here, the resulting effective (average) cross sections are computed in Table 2.

Thus the variation of the effective cross section is 0.23% which is very small, However, it should be kept in mind that STP air has a cross section of 2.95×10^{-28} cm^2 (Table 1). Hence, using a constant mixture cross section (3.29×10^{-28} cm^2) in Eq. (7) overpredicts T in the outside layers, up to a maximum of 11% outside the flame. However, since T_{air} is known independently, we made no attempt to match this part to the profile and we used the mixture cross section throughout.

Experimental Procedure

The flame used in the experiment was a steady premixed laminar methane-air flame diffusing in atmospheric air. The accuracy and resolution of temperature measurements were tested at several sections. Figure 2 shows qualitatively the expected temperature variations induced by the flame front.

Figure 3 illustrates how the concept shown in Figure 1 was demonstrated in a practical experiment. In order to scan the flame across different sections, the burner was moved horizontally so that the flame moved by known increments across the laser beam, at different elevations, while

Table 2 Effective cross sections for reactants and products

Element	x_i	$x_i (d\sigma_i/d\Omega)$, $10^{28} cm^2$
Reactants		
CH_4	0.0951	0.6185
O_2	0.1901	0.4867
N_2	0.7148	2.1817
mixture	1.0000	3.2869
Products		
CO_2	0.0951	0.6754
H_2O	0.1901	0.4222
N_2	0.7148	2.1817
mixture	1.0000	3.2793

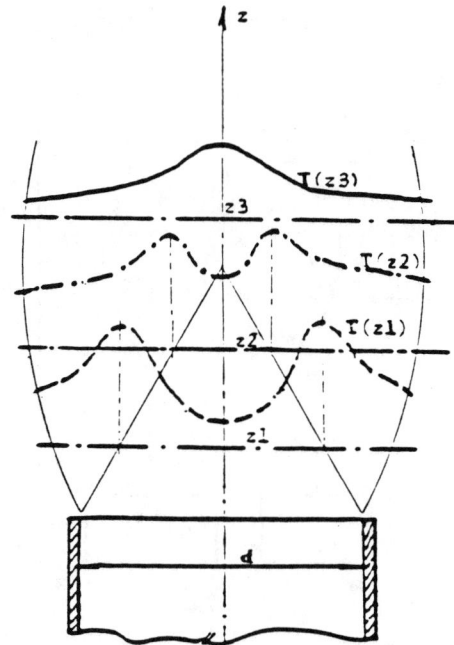

Fig. 2 Typical temperature profile for a premixed flame.

the optics remained fixed. All the components shown on Figure 3 are standard scattering diagnostics equipment. A detailed discussion of their characteristics and performance is available in Chap. 3 of Ref. 1, as well as the description of the burner and flow control systems.

In a practical experiment, the Rayleigh scattering light of the flame must be separated from other unwanted contributors: (1) the Mie-scattered light from particles in suspension (including soot); (2) the luminescence of soot (if any is present); amd (3) the contribution from molecular-excited states in the front region. Generally speaking, the methods used to eliminate or reduce these background contributions are based on the idea of separating them from the Rayleigh contributions. This can be done:

1) By using their spectroscopic characteristics: here we used both an interference filter (frequency) and a polarizer lens. This approach eliminates a great deal of contributions 2 and 3, which are broadly spread over the spectrum. Also, since 2 and 3 are not laser generated as scattering is, the remainder of their contribution (around the laser frequency) can be measured separately by turning off the laser, and then subtracting this signal from the signal received when the laser is on. All these procedures were

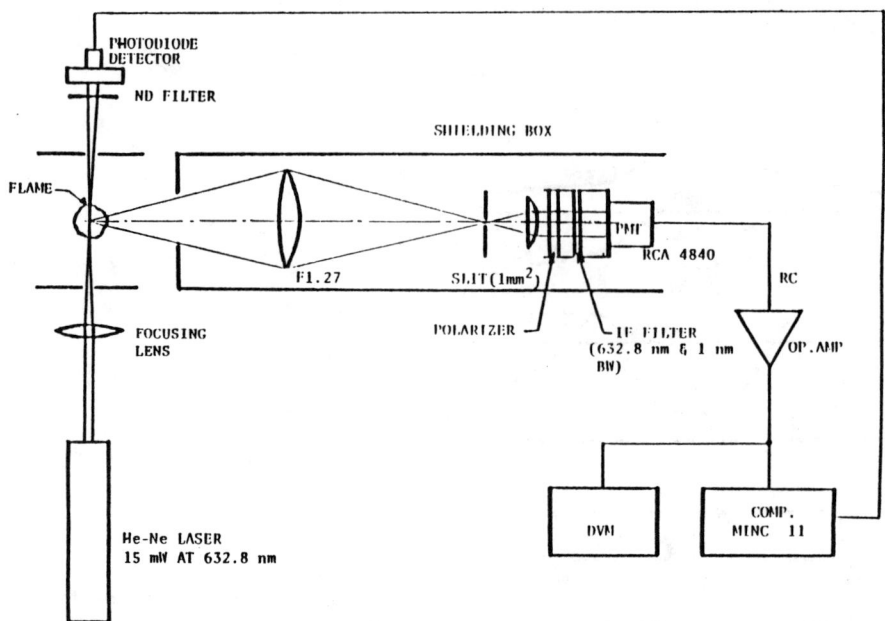

Fig. 3   Experimental layout.

used in these experiments (see section entitled The Interference Filter and the Polarizer).

2) By using their time-dependent characteristics: typically a modulation can be introduced in the laser beam (e.g., by a chopper), and the proper frequency treatment of the signal should eliminate most of the broadband "natural" contributions of the 1 and 2 type.

3) By making additional measurements, which bring out the difference in frequency dependence between Rayleigh and Mie scatterings (contribution 1). This procedures entails a substantial experimental complication, since several frequencies must be emitted and measured. Furthermore, the particles properties, shape and distribution in size are usually unknown: one would like to leave it that way.

In general the problem is avoided, as we have tried to do, by eliminating as many particulates as possible, either by installing filters on the atmospheric air supply, or by using high purity bottled air (Fig. 4). Soot, being generated inside the flame, is much harder to control.

Another pesky problem is the flame instabilities inherent to burners when they are not surrounded by a large annular sheath of stable, quite air. In our case, fluid dynamic instabilities were not entirely eliminated by our time-averaging procedures. They obscured at times our evaluation of signal noise control procedures.

4) By reducing the considerable <u>electronic noise</u> generated by the thermionic emission of the photomultiplier cathode. Because the background contribution, even frequency-filtered, is most often larger than the Rayleigh one, the desired measurement is a small difference between two large signals: under these circumstances it is critical to eliminate as much noise as possible in these two signals. This was accomplished successfully by the use of a biased operational amplifier, in conjunction with a photomultiplier cooling unit.

In the following paragraphs, we shall discuss those instruments most essential to our attempts at noise control. Finally, a short description of the experimental procedure will be given.

The Interference Filter and the Polarizer

An interference filter and a polarizer (Oriel Corporation, Models 5272 and 2734, respectively) were used to reduce the background luminescence from the flame.

The spectral response of the filter indicates that it has a central wavelength of 632.9 nm, 45% peak transmis-

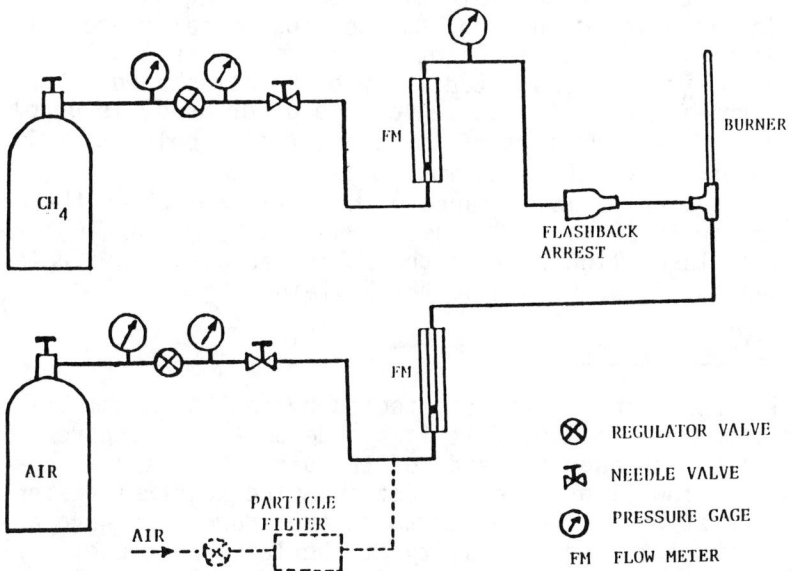

Fig. 4 Schematic of flow and pressure control system.

sion and, 1-nm bandwidth at half-height. The filter has a highly reflective (mirror-like) side and a darker absorptive side (colored side). The reflective side has to be placed toward the scattering volume; in this manner the majority of the unwanted radiation from the flame is reflected rather than absorbed and converted to heat in the filter itself. The specifications of this filter correspond to its use with normal incident rays. In the case of a converging beam, light rays fall on the filter within a range of angles off the normal; this causes the bandwidth to broaden and the peak wavelength to shift to shorter ones[12], sometimes outside of the bandwidth, causing a drastic reduction of transmitted light.[13] This can be cured by mounting the filter holder on a vertical hinge and by adjusting the angle between the normal to the filter and the optical axis, so as to catch the maximum amount of rays. In our case, we simply added a 15-mm focal length convergent lens behind the slit; in this fashion we obtained a parallel beam falling perpendicularly on the interference filter for maximum transmission.

The polarizer used was a nearly uv-visible linear polarizer which has a wavelength range from 300-700 nm. The polarizer's task is to divide the incident beam into two orthogonal components, transmit one and absorb or divert the other.[14] This is accomplished, in this type of polarizer, by dichroism (the electric field component parallel to the direction of the molecules constituting the polarizer material is allowed through, but not the normal component).

Since the scattered light conserves its original polarization, it is transmitted after being attenuated. The background flame luminescence, on the other hand, is unpolarized, and a large part of it is absorbed: only a small part is viewed by the photomultiplier.

The typical transmittance of nonpolarized light through a polarizer is 20% at 632.8 nm,[14] whereas it reaches 50% for polarized laser light. Thus the polarizer helps in a relative way but it weakens also the Rayleigh signal.

The Photomultiplier

The scattered light is detected by an RCA photomultiplier tube (Model 4840). It is a side-on type which receives light through the side of the glass bulb and is designed for low light level detection. The spectral response characteristics of the photocathode are shown[15] to have a 14-mA/W absolute responsivity at the laser wavelength. A complete housing (Pacific Precision Instruments, Model 3150), containing a voltage divider circuit, encases the photomultiplier tube, and shields it from magnetic fields. The housing is connected to a dc high-power supply (P.P.I., Model 205) which provides negative high voltage for photomultiplier exitation and plus-minus 15 V at 250 mA for amplifier operation. The voltage can be regulated as a constant value from 0-2010 V with an output current from 0 to 10 mA. The maximum voltage to be applied between the anode and cathode of the tube is 1250 V, but in the experiment the power supply was operated at 1000 V. Extrapolated values at the wavelength of interest here are listed in Ref. 1. Increasing the applied voltage increases the responsitivity of the detector; but, it also increases the dark current and the noise associated with it.

Even though the photomultiplier is in complete darkness, it generates a dark current that may originate from the following:[15]

1) Thermionic emission of electrons from the photocathode or dynodes.

2) Leakage current (ohmic leakage) resulting form imperfect insulation of the glass stem and the socket.

3) Ionization of residual gases inside the tube by electrons.

The thermionic component is the principal contributor to the dark current; it varies with the temperature T in accordance with the Richardson equation (Ref. 15):

$$j = [4\pi emK^2T^2/h^3] \exp(-\phi/KT) \qquad (8)$$

where j is the thermionic current density, e the electron charge, m the electron mass, K Boltzman's constant, h Planck's constant, and ϕ the work function (1-1.5 eV).

The dark current was increasing during the experiment and 2-3 h were required to reach an equilibrium. An average is 3×10^{-8} A after 1-h warm-up (this value was obtained without the operational amplifier) which is slightly less than the typical value (4.5×10^{-8} A) specified by RCA.

Photomultiplier Cooling

The equivalent noise input (ENI) is the radiant power which would generate at the anode a current equal to the photomultiplier dark current. In the case of the RCA 4840, it is the order of 1.6×10^{-15} W for a 1-Hz bandwidth and at ambient temperature (23°C). This is much less than 3×10^{-13} W, which is the Rayleigh energy predicted for our system by Eq. (1). It means a satisfactory signal to noise ratio.

In practice, however, one is very far from the perfect optics and fine co-focusing assumed by Eq. (1). A realistic estimate of the Rayleigh signal, as verified by our measurements in the Experimental Results section, shows that the dark current is an appreciable fraction of the expected Rayleigh signal. Hence a poor signal-to-noise situation.

An improvement is to reduce the dark current by using a photomultiplier cooling unit. In our case, Figure 5 shows that for our multialkali photocathode ($Na_2KSb: Cs$), a decrease from ambient temperature to $-10°C$ would reduce the dark current by a factor of 10, in accordance with the Richardson equation. Accordingly, we acquired a thermoelectric cooling unit from Pacific Precision Instruments (Model 3461), with a $-30°C$ capability. Since no further noise reduction is expected from cooling below $-10°C$, we

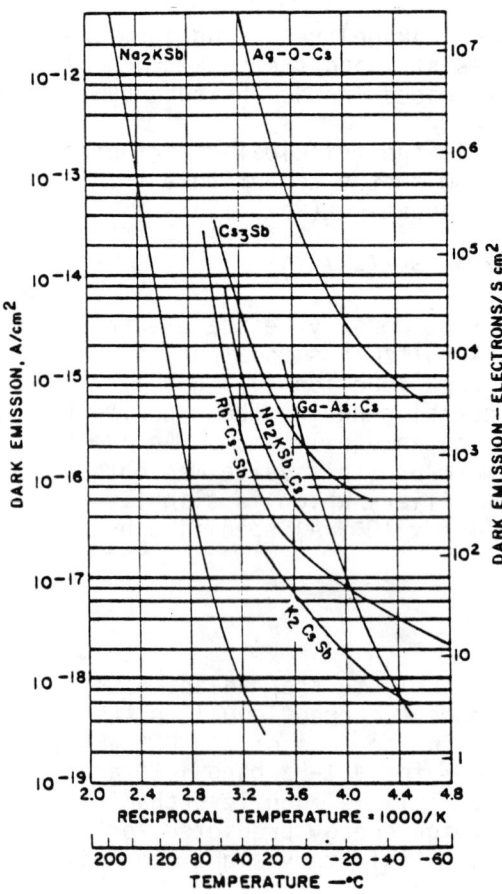

Fig. 5 Variation of thermionic-emission current density from various photo cathodes (Fig. 16 from Ref. 15, courtesy of RCA New Products Division, Lancaster Pa.).

used this setting in our experiments. The results were spectacular (Fig. 6).

Signal Handling System

As the pure scattering signals from the hot region of the flame are small, they generate anode currents of the order of 10^{-9} A, and the voltage readings sometimes disappear in the background noise. It was necessary then to amplify the output of the photomultiplier. An operational amplifier was built for this purpose; its circuit is shown in Figure 7. The $\pm$ 15 V were supplied by a low voltage power supply (Lambda, Model LPD 422 A-FM) which can be set at either constant current or voltage.

A voltage offset adjustment was integrated to the circuit, in order to remove as much of the unwanted voltage of the dark current as possible. In this manner, the voltage which is amplified to fill the dynamic range of the Minc digitizer ($\pm$ 5.12 V) will be the quantity of interest to us. In the noncooled case, it was done by setting the offset voltage at 400 mV: this fixed value was close and

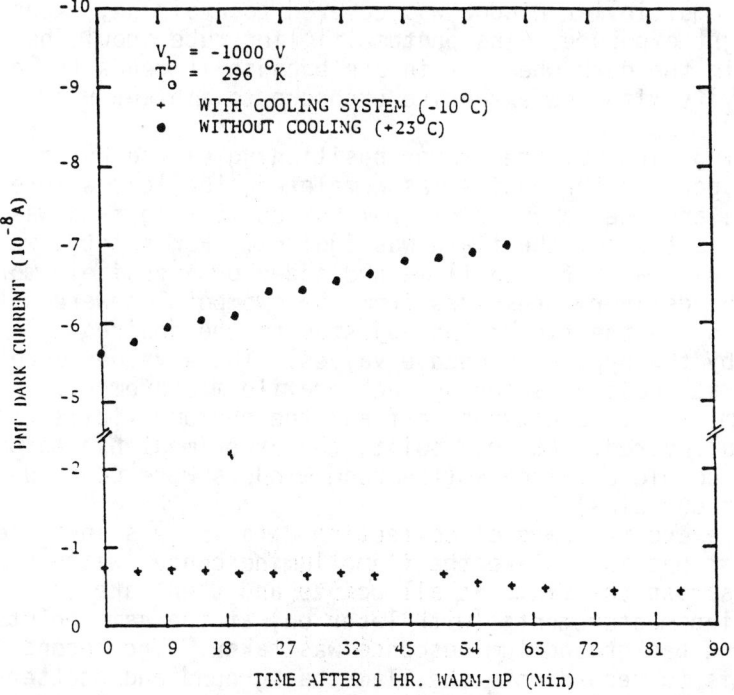

Fig. 6 Photomultiplier dark current with and without cooling.

always less than the dark voltage. A gain of the order of 10 was achieved.

The RC time constant of the circuit is 2.24s, thus filtering effectively the frequency components of the noise above 10 Hz [$f \geqslant 1/2\pi RC$, see Ref. 16]. The impedance value of the amplifier circuit ($R = 10^7 \Omega$) delivers to the amplifier, as an example, a voltage of the order of 10 mV for the smallest anode current discussed above, 10^{-9} A ($V=RI$). After amplification, 100 mV values were obtained (see later, Table 4).

The offset feature was used only in the first part of the experiment, when the photomultiplier was not cooled and the dark current was sizable. Once cooling practically eliminated the dark current, only the amplification and frequency filtering features of the amplifier were maintained.

Operation of the System

After the optics alignment is complete, see Ref. 1, the laser and the photomultiplier have to be started and left to warm up for about an hour. The supply voltage to the photomultiplier has to be increased slowly by steps of 100 V in order to avoid photomultiplier saturation. Also, the photomultiplier window was covered to avoid any excessive light exposure. The photomultiplier tube should be stored in the dark when not in use because it tends to lose sensitivity with the extensive exposure to ambient room light.[15]

During warm-up, the proper positioning of the burner with respect to the optics was completed, the lenses were cleaned, and the computer program for collecting data was called. Later on, the flame was ignited. For safety, we started with a diffusion flame and added progressively more air. The delivery pressures from the two bottles were set at 5 psig and the flow rates adjusted to the desired values by the precision needle valves. These values were checked at least once during each profile measurement. The room lights were turned off and the photomultiplier window uncovered. At this point, the experiment had to be done in complete darkness (the room windows were covered by black curtains).

There are two ways of collecting data in this instance. The first one is to take the flame luminescence (without laser) across the flame at all points and then take the scattering measurements (with laser on) at the same points where the background luminescence was taken. The second method is to record both the flame background and scattering at each point before moving to the next one. The former method could not be achieved in our case since the dark

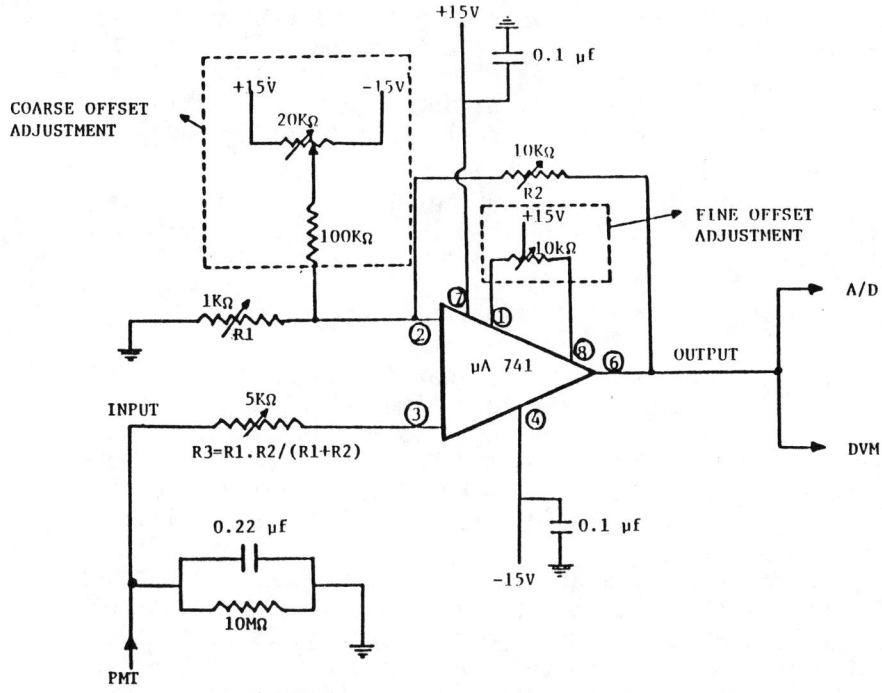

Fig. 7 Operational amplifier circuit with offset adjustment.

current was slowly increasing during the test. In the latter one, the dark current increase between the two measurements (background and scattering) was negligible and did not affect their difference (Rayleigh scattering signal); this method was consequently adopted.

Because of the photomultiplier response time, we had to wait for several seconds for the anode current to settle to a steady value before taking data. These considerations dictated the pace at which a measurement traverse could proceed across the flame.

The computer displayed the average values and their rms fluctuations for the background and the scattered signals for each point measurement. The difference between the two values corresponds to the Rayleigh scattering signal E_s.

After background and scattering were measured at one point, the burner was moved to the next one. The spacing between points was 0.5 mm, of the order of the beam radius. At the end of each test the flame was extinguished and the scattering from air at room temperature was recorded; this was done by flowing clean air (from the bottle) in the burner. This value was considered as the reference for all scattering measurements from the flame (Eq. 7). The values of E_s, from the flame and air, were corrected for the effect of laser drift by multiplying them by the ratio L_1/L_i, where L_1 and L_i are the laser intensities during readings 1 and i, respectively.

To map another section of the flame, the burner was moved up or down to the desired location and the same procedure was followed.

Experimental Results

The characteristics of laminar flames have been established and discussed for a long time, especially in the steady state.[17,18] Current work is mostly concerned with turbulent and unsteady flames. Because of the low powers used in this experiment, we used time-averaging extensively and strived to correlate our results with classical steady flame results.

The temperature measurements reported here were taken at three axial locations, z = 15, 25, and 30 mm downstream of the burner exit. These point measurements were made by moving the burner along the scattering axis, i.e., across the laser beam. At each point, one hundred readings were made at a rate of 10 Hz. The spatial interval between point measurements was 1/50 in. ($\simeq$ 0.5 mm). The first series of measurements (A) were made without PM cooling and were reported in detail in Ref. 1. The second (B) were done with cooling. The corresponding flow conditions and the operational parameters are summarized in Table 3.

Our principal aim was to assess the effect of photomultiplier cooling; thus we tried to reproduce in the second series (B) combustion conditions identical to the first (A). However, the substitution of an improved burner of a slightly different design created minor differences in the flow conditions. Columns A and B illustrate these variations;

Table 3 Summary of experimental conditions

Parameter	A(No cooling)	B(Cooling)
Burner internal diam.(mm)	5.0	5.0
Methane flow-rate(cm^3/min)	295	320
Airflow rate(cm^3/min)	2625	2605
Average velocity(cm/s)	247.9	248.3
Equivalence ratio	1.07	1.17
Pipe Reynolds number	786	787

the differences were thought to be small enough to consider flames A and B as identical.

Results Obtained at Section z/d = 5 With and Without Photomultiplier Cooling

Figure 8 corresponds to two half-profiles, with and without cooling, in the same combustion configuration. It shows a potential (volts) which is proportional to the density at the points across two half-sections of the axisymmetrical flame at z/d = 5, one with cooling, the other without. No effort was made to calibrate the voltage V in terms of the beam power E_S, since E_S is linear in V and only the ratio of E_S is used in Eq. (7).

The potential voltage averaged over 100 measurements at each location is shown as a point, whereas the vertical

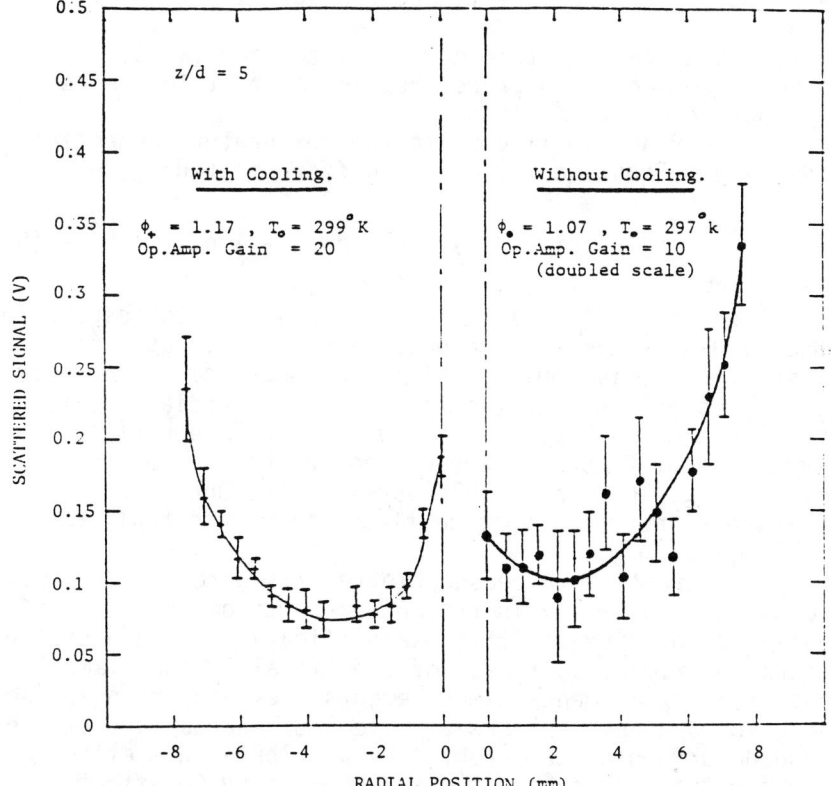

Fig. 8 Rayleigh scattering signal at flame section z/d=5, with and without cooling (scale doubled on the right, to reproduce the same gain as on the left).

error bar corresponds to the r.m.s. of the signal fluctuations at that location (see Error Analysis section). To curve fit these points into a profile across the flame, a smoothing algorithm was used (see Ref. 1, Appendix D). It minimizes the r.m.s. (of the points with respect to the profile) with a constraint on the curvature. A parameter S indicates the relative role of the constraint in this algorithm, in a scale from 0 to 1. Here, as elsewhere in this work, it was chosen to be 0.7.

Observe the large noise reduction effected by cooling the photomultiplier. It shows in two ways:

1) The r.m.s. without cooling is much larger than with cooling. In both cases, the noise (i.e., the r.m.s.) stays fairly constant across the flame, whereas the signal itself is the smallest at the location of lowest density. Thus, as expected, the minimum signal-to-noise ratio is observed at the flame front. One observes also a fluctuation increase at the edge of the flame, probably due to mixing instabilities.

2) The averages obtained without cooling (dots) do not fit well the smoothing procedure, in contrast to the results obtained with cooling.

Figures 9 and 10 illustrate the temperature dependence on density. Equation (6) can be differentiated:

$$dT = d\left(\frac{K}{N}\right) = -\frac{K}{N^2}dN = -\frac{T^2}{K}dN \qquad (9)$$

Clearly a small measured r.m.s. (dN) will correspond to large values of temperature uncertainties (dT) at large values of T. This amplification compounds the weakness of the results obtained at the flame front, especially with a noncooled photomultiplier (Fig. 9). Also, the profile aberrations of Figure 9, are not noticeable in the cooled PM case (Fig. 10): one could surmise that uncooled photomultipliers incur temperature fluctuations which affect their consistency.

In contrast, some reasonable results were obtained with PM cooling, especially in the cooler parts of the flame (center and outside). But the amplification of (dT) creates temperature r.m.s. up to 15% of the signal in the flame front area; experimental improvements are being planned (see Conclusion and Future Tasks). Note the fluctuations of the smoothed curve near the front. We attribute this effect to flame fluctuations during the time necessary to effect a flame cross-section profile ($\sim$ 30 min.).

Figure 11 contrasts the two sets of results. It shows that noncooled photomultipliers cannot bring out the most characteristic aspect of a laminar flame, i.e., the existence of two sharply separated flame fronts, at least at this section of the flame.

Results of Several Stations of the Flame

Figure 12 illustrates the results obtained for section $z/d = 3$. At this station, the two sides of the front are separated enough so that even the noncooled photomultiplier accounts for the low temperatures at the core of the flame. In this case, the better accuracy of the profile obtained by the cooled PM is mostly useful as a basis for a better estimation of the thermal diffusion process from the front region to the cool reactants area in the center.

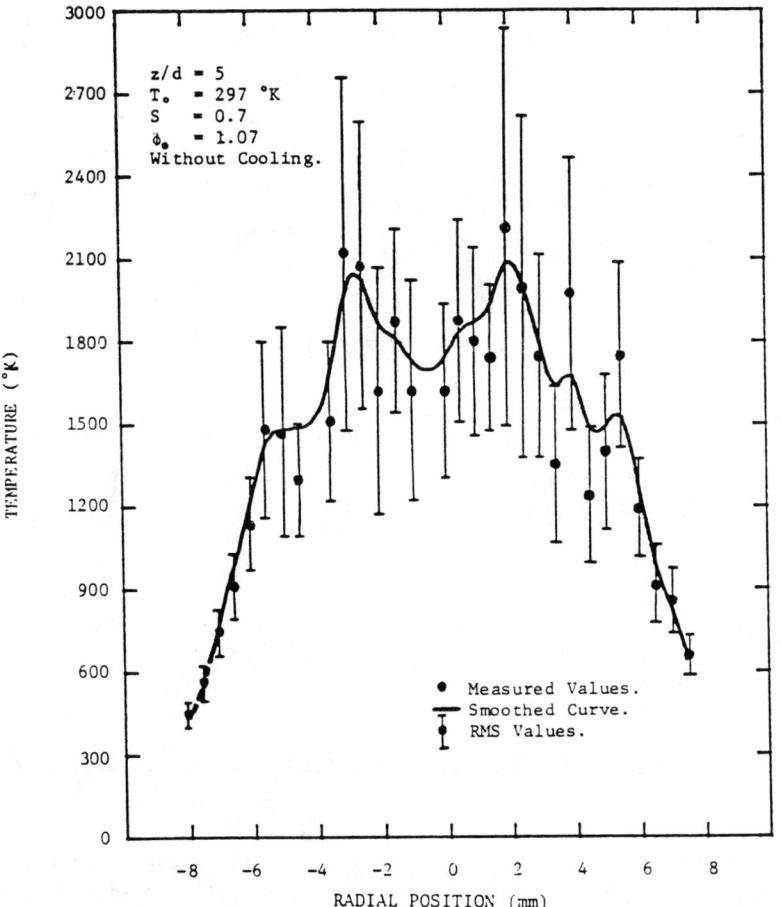

Fig. 9 Radial temperature profile at z=25 mm (no cooling).

Figure 13 shows the profiles obtained near the tip of the flame. The noncooled PM results are too scattered to be usable. The best fit to the cooled PM results displays appreciable irregularities, more than at the other sections. We believe it is due to the fact that this part of the flame front displays more developed fluid dynamic instabilities than the others.

Finally, Figure 14 shows the three profiles on the same graph. The convergent character of the flame front (Fig. 2) is reproduced. We also detect a slight temperature increase as the front reaches the axis, probably as a result of up-

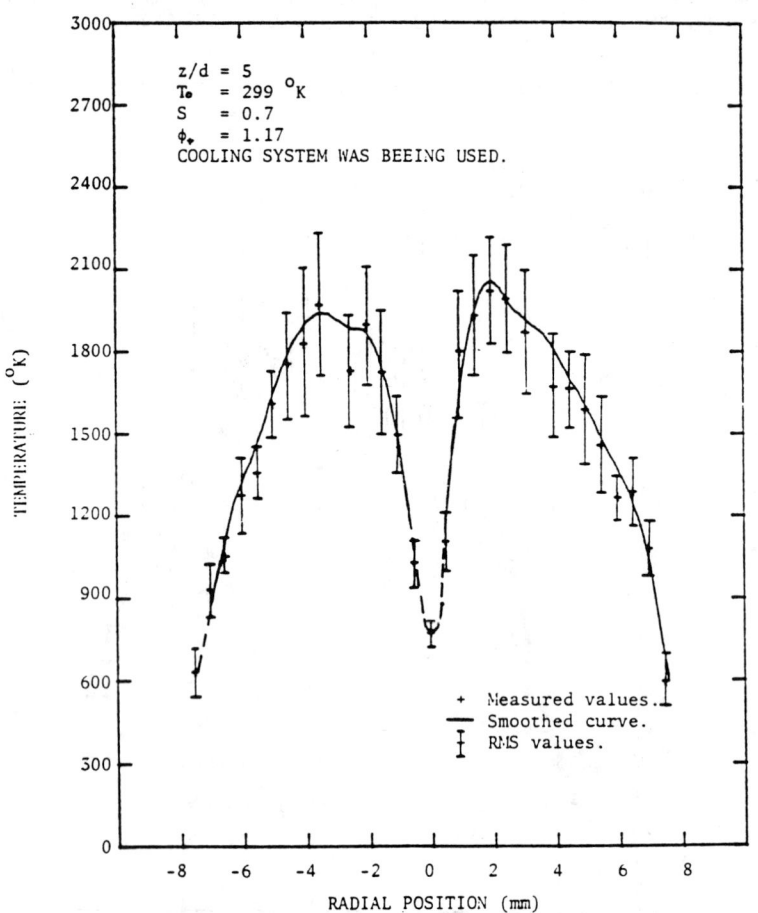

Fig. 10 Radial temperature at z = 25 mm (with cooling). The dashed line at the center was done by visual fit; the smoothing factor S = 0.7 is excessive in this part of the profile. Note that T_{ad} = 2210 K (Ref. 22).

stream thermal diffusion from the outside layers inward.

In the next section, we shall analyze the probable causes for error associated with our results.

Error Analysis

These preliminary measurements give little better than a tentative confidence in Rayleigh thermometry with low-power lasers. Thus, it is now important to identify and, if possible, eliminate all losses of signal strength or sources of noise. As discussed in classical books,[19] such a list is extensive, especially in weak signal techniques. Such are, for instance, the lens quality and proper

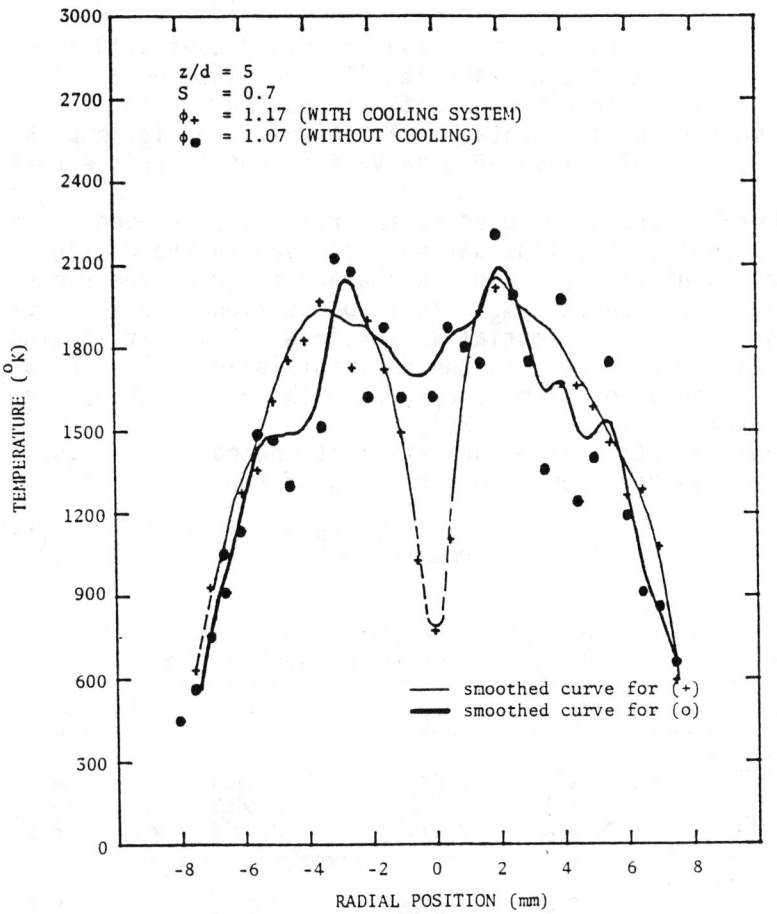

Fig. 11 Radial temperature profiles at z=25 mm.

alignment of the optics, an unquantified but essential aspect of optical measurement accuracy.

The Rayleigh Signal as a Difference of Two Large Signals

As outlined in the section entitled Operation of the System, the procedure we used to measure the scattering potential V_S, was to subtract the value V_B obtained with the laser off (i.e., background luminescence + dark current only) from the value V_T obtained with the laser on (i.e., Rayleigh scattering + background luminescence + dark current). The difference V_S is due entirely to scattering:

$$V_S \equiv V_T - V_B \tag{10}$$

Table 4 shows typical values averaged over a 10-s period on the axis of the flame at different sections. The data was collected with and without cooling. This table is shown mostly to illustrate, qualitatively, the large background-to-signal ratio BSR $\equiv V_B/V_S$ inherent to this experiment.

More interesting to us is the r.m.s. corresponding to these signals. The fluctuations observed in the PM signal at each point of the flame are characterized by the r.m.s. current at the anode I_{rms}. It is proportional to the corresponding r.m.s. potential V_{rms} (since V=RI). It is also equal to μi_{rms} where μ is the photomultiplier gain and i_{rms} is the photocathode r.m.s. current. The photocathode current itself is i, and $V = R\mu i$.

Poisson statistics indicate that the cathode r.m.s. current expected from a current i is of the form:

$$i_{rms} = (2eiB)^{\frac{1}{2}} \tag{11}$$

Table 4 Photomultiplier average potentials, in volts and background-to-signal ratios

z,mm	V_B	V_T	V_S	BSR
No cooling				
15	1.9934	2.0763	0.0829	24
25	0.7391	0.7852	0.0461	16
30	0.5424	0.6447	0.1023	5.3
With cooling				
15	0.7828	0.8729	0.0901	8.7
25	0.6920	0.7632	0.0712	9.7
30	o.5193	0.5973	0.0780	6.7

where B is the bandwidth of the measurement in Hz and e the electron charge. In our case: $B = 1/2\tau$, τ being the time duration of our observation (a square wave).

These statistics apply to thermally generated currents such as the luminescence signal i_ℓ and the thermionic signal i_d [Eq. (8)]. In addition, we know[15,19] that the variance σ^2 of independent additive currents is also additive:

$$\sigma^2 \equiv i^2_{rms} = i^2_{d,rms} + i^2_{\ell,rms}$$

Thus, in <u>the laser-off case:</u>

$$i_{rms} = [2e(i_d + i_\ell)B]^{\frac{1}{2}} = (2ei_b B)^{\frac{1}{2}}$$

where the "background" current $i_b \equiv i_d + i_\ell$.

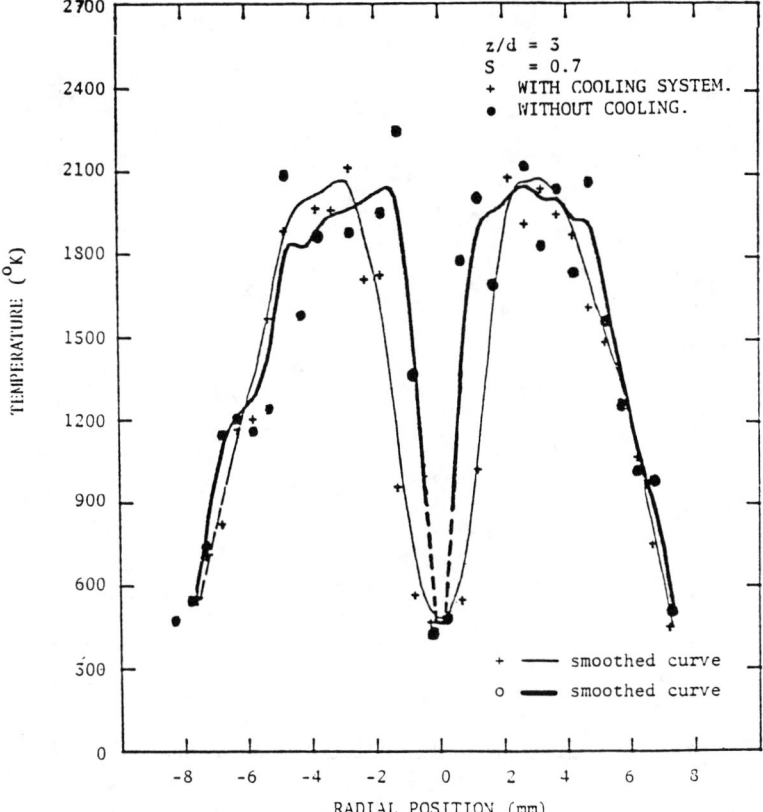

Fig. 12 Radial temperature profiles at z=15 mm, with and without cooling.

The signal-to-noise ratio is then:[20]

$$SNR_\ell \equiv \frac{i_\ell}{i_{rms}} = \frac{i_\ell}{[2e(i_d + i_\ell)B]^{\frac{1}{2}}} \qquad (12)$$

This formula shows the role of the time of exposure $\tau = (2B)^{-1}$ in increasing the accuracy of the measurement.

In the <u>laser-on case</u>, the signal of interest is a signal difference: $i_s \equiv i_t - i_b$, where the intensities are proportional to the potentials V_S, V_T and V_B defined above. Thus,

$$i_{rms} = [2e(i_t + i_B)B]^{\frac{1}{2}} \qquad (13)$$

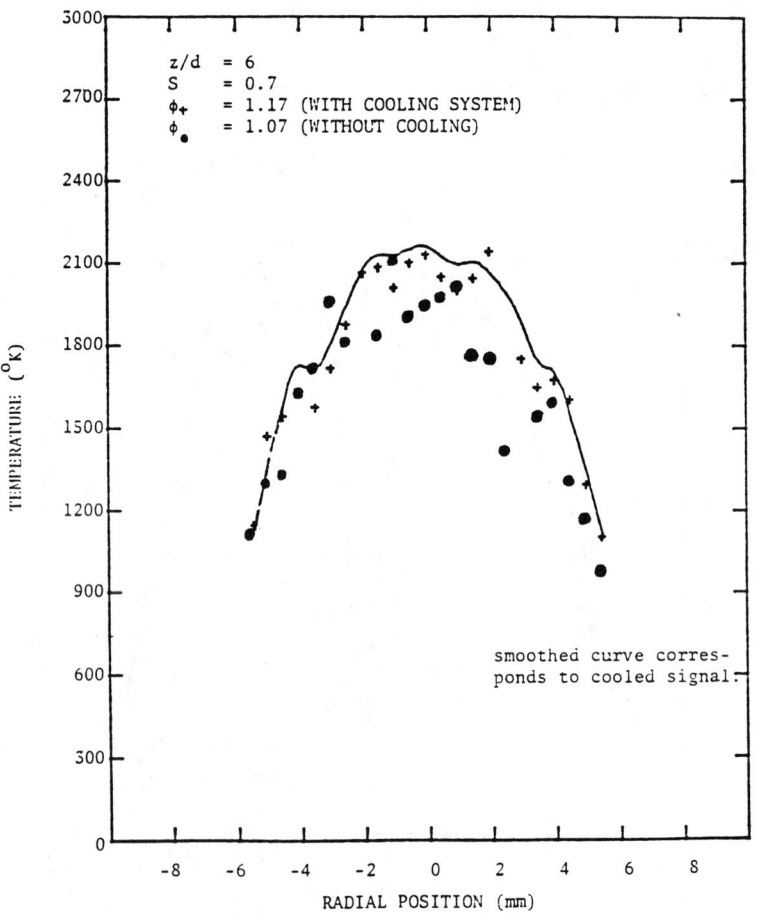

Fig. 13 Radial temperature profiles at z=30 mm, with and without cooling.

and:

$$SNR_S = \frac{i_S}{i_{rms}} = \frac{i_S}{[2e(i_t+i_b)B]^{\frac{1}{2}}} \quad (14)$$

If $i_s \ll i_b$, then

$$SNR_S = i_S/(4e\, i_b B)^{\frac{1}{2}} \quad (15)$$

a familiar result in measurements based on signal differences, such as absorption.[21]

In our measurements, both SNR_ℓ and SNR_S were available. In the case of interest (laser on), we found SNR_S values of about 5 with cooling and 3 without (see Fig. 8). The difference can be attributed to the larger dark current i_d in the non-cooled case, leading to a larger r.m.s.

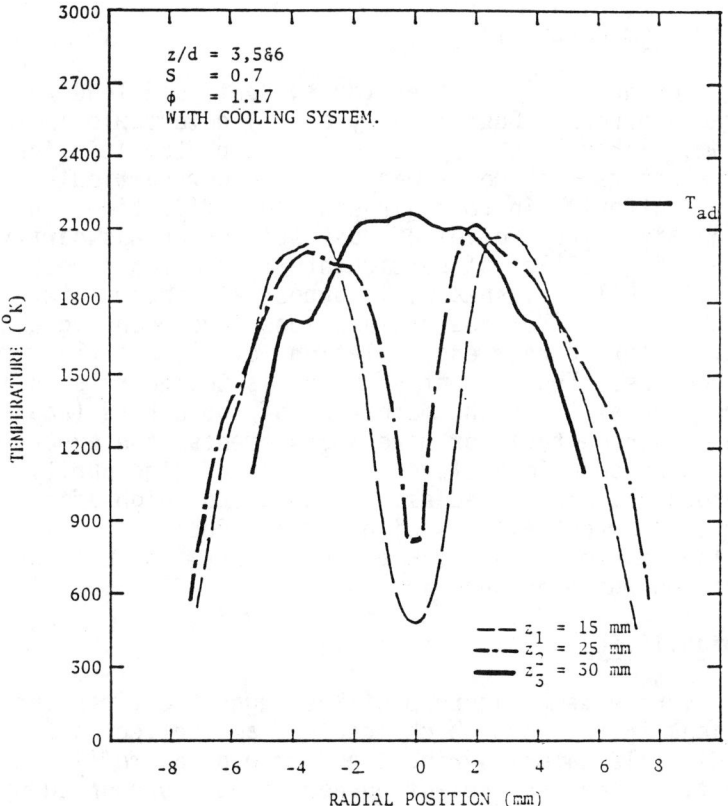

Fig. 14 Temperature profile variations at different flame sections: 15, 25 and 30 mm.

value. It can be shown theoretically, if we use the expression $V=R\mu i$ and the values: V_B = 700 mV (from Table 4) with cooling; V_B^1 = 1240 mV (adding a typical dark current (Fig. 6) to V_B); $V_S \simeq$ 60 mV (from Table 4); R = 10 MΩ; μ = 2.9 x 10^7 (RCA PM 4840 handbook); and B = 1 Hz.

On the basis of these numbers, Eq. (15) yields a SNR value of 5.26 for the cooled case, and 3.96 for the uncooled case, fairly close to the measured results. The point of this comparison is to show that as long as the background is more than 10 times larger than Rayleigh scattering, the standard deviation of the data will be too large: the 95% confidence interval ($\pm$ 2σ) will be 40% on both sides of the average.

One way to improve the SNR, would be to extend the effective aperture time of the data handling system. The actual sampling time of a digitizer is only a fraction of the total measurement time.

Scattering From Particulates

The presence of impurities (dust, soot) can lead to considerable errors. Dust is very easily detectable in the laser beam, either by the eye or with the photomultiplier, since it scatters much more than gas. In our particular case, Mie scattering in room air was about five times that from clean air. The problem of dust scattering was partly solved by using bottles of particulate-free air and methane. Still, yellow flashes would appear at random intervals in the flame. Various explanations have been suggested (welding flux, burner edge ablation, etc.), but the problem remains. The current solution is to ignore signals which are several times the magnitude of the others (and redo the measurement). For more improvements, the experiment should be done in a clean room gases of high purity levels procured, and if necessary, the use of high efficiency particle arrest filters with air and fuel. The flame should also be clean; this can be met by burning fuels or fuel mixtures which produce no soot.

Flow Instabilities

Our average temperature profiles suggest environmental fluctuations in the room which create flame deflections of an unpredictable nature during the course of a profile management. Since the present burner is not surrounded by an annulus of quiet gas flow, as is standard combustion practice, we are not too surprised by this limitation.

Laser Drift

Equation (7) states the way in which all power measurements $E_{s,flame}$ are calibrated against $E_{s,air}$. In fact, we have observed with the photodiode shown on Fig. 3 that the laser we use drifted up to 5% during the measurement period of one profile.

Since we measure $E_{s,air}$ only once before scanning each profile, the value E_s, as calculated by Eq. (4), is proportional to a decreasing value of E_0 as we proceed through the profile. Thus a correction factor $E_0(t=o)/E_0(t)$ should be added on the right of Eq. (7). An average value of $\bar{E}_0$ could also be established for each profile. With an allowance for a minor error, it is possible to use a single value of E_0 per profile.

Conclusion and Future Tasks

The first phase of our work has explored the possibility of reducing some noise components (especially from the photomultiplier) to the point where low but acceptable signal-to-noise ratios (about 10) were obtained. The problems discussed in the last section are all susceptible to amelioration with current resources: when these improvements are made, a better signal-to-noise ratio will be obtained. Specifically, we propose to:

1) improve the scattered isgnal by the use of an optical cavity,
2) upgrade the burner design so as to eliminate both fluctuations and particulate impurities,
3) develop mathematical models to deal appropriately with the laser drift, the variable scattering cross section, etc.,
4) to further explore filtering techniques, in particular laser modulation (choppers), and
5) to establish a theoretical foundation for noise measurements, especially with a view to increase the observation bandwidth, thus speeding up measurements and avoiding the effect of environmental changes.

We expect to accomplish these improvements without the need for expensive equipment, since the main point of this work is to demonstrate that Rayleigh thermometry at low cost is possible. A SNR of 50 seems to be a reasonable goal, especially if near steady experiments are considered.

Acknowledgment

We wish to acknowledge the help which we have received in designing, building and conducting this experiment from many colleagues, in particular J. Daily, R. Dibble, C. Garris, M. Noori, F. Robben and R. Santoro. This program was supported in part by the Air Force Office of Scientific Research, under Grant AF-77-3439.

References

[1] Benhachmi, D., "Rayleigh Thermometry in a Laminar Premixed Methane-Air Flame," M.S. Thesis, George Washington University, Washington, D.C., Dec. 1982.

[2] Hartley, D. L., "Laser Scattering Diagnostics for Temperature and Concentration Measurements," Experimental Diagnostics in Gas Phase Combustion Systems: AIAA Progress in Astronautics and Aeronautics, Vol. 53, edited by Ben T. Zinn, AIAA, New York, 1976, pp. 467-477.

[3] Robben, F., "Comparison of Density and Temperature Measurement Using Raman Scattering and Rayleigh Scattering," Combustion Measurement: Modern Techniques and Instrumentation, edited by Robert Goulard, Academic Press, New York, 1976, pp. 180-196.

[4] Smith, J. R., "Rayleigh Temperature Profiles in a Hydrogen Diffusion Flame," Laser Spectroscopy-Applications and Techniques: Proceedings of the SPIE, Vol. 158, edited by Howard Schlossberg, San Diego, Ca., Aug. 1978, pp. 84-90.

[5] Miller, J. A. and Kee, R. J., "Chemical Nonequilibrium Effects in Hydrogen-Air Laminar Jet Diffusion Flames," Journal of Physical Chemistry, Vol. 81, Dec. 1977, pp. 2534-2542.

[6] Dibble, R. W. and Hollenbach, R. E., "Laser Rayleigh Thermometry in Turbulent Flames," 18th Symposium (International) on Combustion, The Combustion Institute, University of Waterloo, Canada, Aug. 1980.

[7] Dibble, R. W. and Hollenbach, R. E., "Nonintrusive Temperature Measurement in a Turbulent Diffusion Flame," AIAA Paper 80-1362, 13th Fluid and Plasma Dynamics Conference, Snowmass, Colo., July 1980.

[8] Rudder, R. R. and Bach, D. R., "Rayleigh Scattering of Ruby-Laser Light by Neutral Gases," Journal of Optical Society of America, Vol. 58, Sept. 1968, pp. 1260-1266.

[9] Muller-Dethlefs, K. and Weinberg, F. J., "Burning Velocity Measurements Based on Laser Rayleigh Thermometry," 17th Symposium (International) on Combustion, The Combustion Institute, University of Leeds, England, Aug. 1978, pp. 985-992.

[10] Bridge, N. J. and Buckingham, A. D., "The Polarization of Laser Light Scattered by Gases," Proceedings of the Royal Society (London), Vol. 295, 1966, pp. 334-349.

[11] Born, M. and Wolf, E., Principles of Optics, Pergamon Press, Oxford, 1975, p. 87.

[12] Oriel Corporation, "Complete Catalog of Optical Systems and Components," 1978.

[13] Robben, F., Private Communication, Nov. 1982.

[14] Shurcliff, W. A., *Polarized Light*, Harvard University Press, Cambridge, Mass., 1962.

[15] R.C.A., "Photomultiplier Handbook, Theory-Design-Application," PMT-62, 1980.

[16] Holman, J. P., *Experimental Methods for Engineers*, 3rd ed., McGraw-Hill, New York, 1978.

[17] Gaydon, A. G. and Wolford, H. G., *Flames, Their Structure, Radiation and Temperature*, Chapman and Hall, London, 1978, pp. 10, 14.

[18] Fristrom, R. M. and Westenberg, A. A., *Flame Structure*, McGraw-Hill, New York, 1965.

[19] Wilson, E. B. Jr., *An Introduction to Scientific Research*, McGraw-Hill, New York, 1952.

[20] Wang, C. P., "Laser Applications to Turbulent Reactive Flows," *Combustion Science and Technology*, Vol. 13, 1976, pp. 211-227.

[21] Sulzmann, K. G. P., Lowder, J. E. L. and Penner, S. S. "Estimates of Possible Detection Limits for Combustion Intermediates and Products," *Combustion and Flame*, Vol. 20, Apr. 1973, pp. 177-191.

[22] Glassman, I., *Combustion*, Academic Press, New York, 1977.

Laser Tomography for Simultaneous Concentration and Temperature Measurement in Reacting Flows

S. R. Ray[*] and H. G. Semerjian[†]
National Bureau of Standards, Washington, D.C.

Abstract

Laser tomography, a new optical diagnostic technique based upon multiangular absorption spectroscopy has been developed. This technique allows the rapid measurement of both species concentration and temperature throughout a two- or three-dimensional nonuniform flowfield. Laser tomography involves making absorption measurements along M parallel rays at N equally spaced angles. These M x N measurements are used to reconstruct the spatially resolved two-dimensional property field. Experimental results using two optical geometries are reported for measurements of sodium concentration in a seeded premixed flat flame. Computer simulations of a complete tomography system are presented, including the effects of noise based upon the experimental measurements. Results indicate that within the flame zone both the concentration of absorbing species and temperature can be recovered to within ±3% without noise filtering and to ±1.7% with filtering.

Nomenclature

a   = ratio of collisional to Doppler line width
A   = spacing between parallel beams (spatial interval)
c   = speed of light

Presented as Paper 83-1553 at the AIAA 18th Thermophysics Conference, Montreal, Canada, June 1-3, 1983. This paper is declared a work of the U.S. Government and therefore is in the public domain.
    [*]Research Engineer, Chemical Process Metrology Division.
    [†]Group Leader, Chemical Process Metrology Division.

e	= electronic charge
E.R.	= equivalence ratio
E_J	= energy of Jth rotational level
$f_{J'J''}$	= oscillator strength
$f_{v'v''}$	= band oscillator strength
$F(x,y)$	= the property field $K_\nu = N_i Q_\nu$
$g_{J''}$	= degeneracy of rotational state J"
h	= Planck's constant
I_ν	= transmitted light intensity at frequency ν
I_ν^0	= incident light intensity at frequency ν
J'	= upper state rotational quantum number
J''	= lower state rotational quantum number
k	= Boltzmann's constant
K_0	= centerline absorption coefficient for a pure Doppler line
K_ν	= absorption coefficient at frequency ν
m_a	= molecular mass
m_e	= electronic mass
M	= number of parallel rays for each projection
N	= number of projections (viewing angles)
N_i	= molecular number density of species i
$N_{J''}$	= molecular number density at lower rotational state J"
N_L	= Loschmidt number
p	= partial pressure
p_t	= total pressure
p_o	= standard pressure
$P(r,\theta)$	= projection at angle θ
Q_i	= internal partition function
Q_r	= rotational partition function
Q_v	= vibrational partition function
Q_ν	= absorption cross section at frequency ν
r	= radial coordinate
R	= radius of measurement field
s	= absorption path length
S	= line strength
$S_{J'J''}$	= pure rotational transition probability
T	= temperature
T_0	= room temperature (300 K)
$T_{J'J''}$	= correction factor for vibration-rotation interactions
$V(x,a)$	= Voigt function
x	= relative separation from line center
x	= spatial coordinate
X_i	= mole fraction of species i
y	= spatial coordinate
α_C	= collisional semi-half-width
α_D	= Doppler semi-half-width

θ	= viewing angle of the projection
ν	= frequency
ν_0	= frequency of line center
ρ	= spatial frequency in the transform domain
ρ_{max}	= spatial frequency bandlimit of $P(\omega,\theta)$
ρ	= density
φ	= Shepp-Logan filter function
ω	= $2\pi\rho$
^	= Fourier transform operation

Introduction

As researchers continue to pursue a more detailed understanding of systems involving mixing, fluid dynamics and chemical reactions, there is an ever increasing need for sophisticated diagnostic techniques to provide high quality data. The requirements imposed on the measurement techniques are also becoming increasingly stringent. They must provide a high spatial resolution, preferably less than a millimeter; they must allow data acquisition at rapid rates, which for turbulent flows can be in the tens of kilohertz to resolve the fluctuations present; they must be species specific, to allow the concentration measurement of each participant in complex chemical reaction schemes; and they must be nonperturbing. For the above reasons, optical diagnostic methods, particularly those using lasers have proven particularly well suited as measurement approaches. Techniques which have had notable success include those based upon CARS,[1,2] Raman scattering,[3,4] fluorescence,[5,6] and Mie and Rayleigh scattering.[7] CARS, for example, has been shown to provide absolute measurements of temperature in a single laser pulse,[8] providing a high degree of both spatial and temporal resolution.

More recently, a further need is becoming apparent, which is to perform high-speed, spatially resolved measurements of concentration and temperature simultaneously over an extended region. This is particularly important in fluctuating systems where statistical methods must be employed with point measurement techniques. Recently laser tomography, an advanced diagnostic method based upon multiangular absorption spectroscopy, has been developed which has the potential of performing such measurements. Based upon the absorption of light through a field of interest, it is readily species specific simply by tuning the light to the transition frequency of any particular atom or molecule of interest. Since it is an optical technique, it is inherently nonperturbing. The spatial resolution is comparable to other optical techniques, and due to the basic principle of tomography it measures the local values of concentration and temperature over an extended region simultaneously. The repetition rate of such a measurement is limited only by the data acquisition and digitizing rates.

Principles of Laser Tomography

Absorption or emission measurements have been frequently used in the past for determination of concentration and temperature in a flowfield. The primary drawback has been that these measurements have always represented integrated values along the line of sight. Several techniques have been tried to extract local information from a line-of-sight measurement. One approach is Abel inversion,[9,10] or "onion peeling," which involves measurement along a number of parallel lines. The primary disadvantage of this approach is that it requires an axis of symmetry to exist perpendicular to the plane of measurements, severely reducing the usefulness the technique in practical systems.

In addition, the errors present in the reconstruction process are cumulative,[11,12] becoming greatest at the center of the field, which can often be the region of most interest. A more attractive alternative is to use the convolution technique, a method based upon Fourier transform theory. It has the advantages of being able to reconstruct nonuniform fields as well as being free of cumulative error effects. The original concept of using absorption measurements at a number of angles to retrieve local property values is attibutable to Cormack[13] who was considering medical imaging applications, and to Hounsfield[14] who worked independently on the same problem, leading to the CAT scanners in wide use today for brain imaging. Several improvements have since been made, notably in the mathematical reconstruction techniques,[15,16] which have greatly increased the speed of the computations. The principles of tomographic reconstruction methods have been presented earlier,[17,18] but for the sake of clarity a brief explanation for the case of optical absorption measurements will be given below. This particular case will be discussed because of its potential for use in practical chemically on reacting flows. In particular, attention will be focussed on the absorption of light by the OH molecule. This species is known to be an important participant in many high-temperature reaction systems, including all hydrocarbon oxidation reactions.

The multiangular absorption technique involves the measurement of absorption along M equally spaced parallel beams at N equally spaced angles, forming an MxN data set from which it is possible to retrieve or "reconstruct" the original property field. The absorption along each individual beam is governed by the Bouguer-Lambert-Beer law. This can be expressed as:

$$-\ln \frac{I_\nu(r,\theta)}{I_\nu^0} = P(r,\theta) = \int_{-\infty}^{+\infty} F(x,y)\,ds \qquad (1)$$

where $r = x\cos\theta + y\sin\theta$ and $s = -x\sin\theta + y\cos\theta$.

$P(r,\theta)$ is defined as the "projection" for the angle θ. Since the ratio I_ν/I_ν^0 can be measured experimentally, $P(r,\theta)$ is known and the solution of Eq. (1) for $F(x,y)$ will yield the desired property field, for example, the absorption coefficient, $K_\nu = N_i Q_\nu$, where N_i is the molecular number density and Q_ν is the absorption cross section. Thus the tomography technique is based upon the reconstruction of a property field, $F(x,y)$, from a set of its projections, $P(r,\theta)$. The reconstruction can be carried out using a number of different methods, including linear superposition, algebraic reconstruction, and two-dimensional Fourier transform techniques. The solution procedure used here utilizes the convolution technique developed by Ramachandran.[15] Although it is a Fourier transform approach, this technique does not actually require the evaluation of any transforms, and allows high-speed computation of the property field. Since details of the solution procedure have been presented before,[17,18] only the final result will be stated here.

By taking the Fourier transform of Eq.(1), one can rearrange the terms to be in the form of a two-dimensional Fourier transform of a product of two functions which are in turn Fourier transforms. Use of the convolution theorem then leads to a straightforward expression for the property field:

$$F(x,y) = \frac{1}{2\pi}\int_0^\pi d\theta \int_{-\infty}^{+\infty} P(\tau,\theta)\phi(r-\tau)\,d\tau \qquad (2)$$

where ϕ is defined by its Fourier transform as

$$\hat{\phi}(\omega) = |\omega|$$

and $\omega = 2\pi\rho$.

Since $P(r,\theta)$ is known only in the sampled domain, the integrals in Eq.(2) must be replaced by their discrete summation approximations. Thus,

$$F(x,y) = \frac{A}{2N} \sum_{j=1}^{N} \sum_{k=1}^{M} P(r_k, \theta_j) \phi (x\cos\theta_j + y\sin\theta_j - r_k) \quad (3)$$

where A is the spacing between uniform samples and

$$\theta_j = (j-1)\pi/N, \quad r_k = kA$$

If the summation of $P(r_k, \theta_j)$ alone were evaluated, without multiplying by ϕ, one is effectively just back-projecting the shadows cast by the absorbing field. For this reason, this reconstruction approach is often termed a back-projection method. Physically, the role played by the function ϕ is simply to compensate for the blurring of the back-projected image, which varies as 1/r where r is the lateral distance from a given line of sight. It is often called the deblurring function.

An accurate reconstruction will require that the Fourier transform of ϕ satisfies the condition $\hat{\phi}(\omega) = |\omega|$ for $|\omega| < 2\pi\rho_{max}$ where ρ_{max} is the radius of the transform space. The choice of the function $\phi(r_k)$ greatly affects the computational requirements as well as the accuracy of the results. In the present study the modified Shepp-Logan filter[16]

$$\bar{\phi}(r_k) = 0.4\phi(r_k) + 0.3\phi(r_{k+1}) + 0.3\phi(r_{k-1})$$

where

$$\phi(r_k) = -\frac{4}{\pi A^2 (4k^2 - 1)} \quad k = 0, \pm 1, \pm 2 \ldots$$

has been used as the weighting function.

Accurate reconstruction of the property field $N_i Q_\nu$ will be dependent on the proper choice of the sampling intervals, that is, the choice of M and N, as well as the choice of the appropriate form of the filter function ϕ.

From the sampling theorem,[19] a function can be uniquely recovered from its samples if it is sampled at a rate greater than twice the highest frequency component of the function (sampling rate > $2\rho_{max}$). The present function $F(x,y)$ is assumed to have a band limited Fourier transform, i.e., $F(x,y) \equiv F_\theta(\rho) = 0$ for $\rho > \rho_{max}$, while in the measurement or physical space $F(x,y) = 0$ if $x^2 + y^2 > R^2$ where

R is the radius of the measurement space. The proper choice of M and N under these conditions, which has been discussed previously,[17] requires that (4 $R\rho_{max}$) equally spaced rays are measured at ($2\pi R\rho_{max}$) equally spaced angles. If this criterion is not met, the reconstructed field is likely to show effects due to "aliasing."[20] Aliasing is a term used to describe the effects observed when a function is under-sampled. In such a case the reconstructed function contains contributions from the under-sampled high frequency components which appear as low-frequency contributions in the transform domain.

Spectroscopic Background

Implementation of the tomographic reconstruction technique for concentration and temperature measurements requires an understanding of how these two properties to the absorption coefficient, since it is the latter which is directly measured. Therefore, a brief presentation of the spectroscopic background is given here.

The value of the absorption coefficient at any point within a spectral line can be computed using the Voigt function, which accounts for both Doppler and collisional broadening processes. The Voigt function is given as[1]:

$$V(x,a) = \frac{K_\nu}{K_0} = \frac{a}{\pi} \int_{-\infty}^{+\infty} \frac{\exp(-t^2)}{a^2+(x-t)^2} \, dt \qquad (4)$$

where

$$K_0 = (S/\alpha_D)(\ln 2/\pi)^{1/2}$$

$$a = (\alpha_C/\alpha_D)(\ln 2)^{1/2}$$

$$x = [(\nu-\nu_0)/\alpha_D](\ln 2)^{1/2}$$

α_D and α_C are the Doppler and collisional semi-half-widths (HWHM) and S is the line strength. The spectral absorption coefficient K_ν can then be expressed in terms of the line strength:

$$K_\nu = \frac{S}{\alpha_D}\left(\frac{\ln 2}{\pi}\right)^{1/2} V(x,a) \qquad (5)$$

The Doppler half-width of a spectral line at ν_0 is given as:

$$\alpha_D = \nu_0 (2kT \ln 2 / m_a c^2)^{1/2} \qquad (6)$$

where m_a is the mass of the molecule and c the speed of light.

The collisional half-width is classically proportional to the collision frequency, i.e.,

$$\alpha_c(\rho,T) \propto \rho T^{1/2} \propto pT^{-1/2}$$

Here, the pressure refers to the partial pressure of the molecules colliding with the absorbing species. In a flame where the pressure of the important collision partners is constant, the temperature dependence of the collisional half-width can be expressed as:

$$\alpha_c(T) = \alpha_c(T_0)(T_0/T)^{1/2} \qquad (7)$$

It should also be noted that for x=0, i.e., at the line center, the Voigt function reduces to[21]:

$$V(0,a) = \exp(a^2) \, \text{erfc}(a) \qquad (8)$$

so that the absorption coefficient at the line center can be expressed as:

$$K_\nu = \left(\frac{S}{\alpha_D}\right) \left(\frac{\ln 2}{\pi}\right)^{1/2} \exp(a^2) \, \text{erfc}(a) \qquad (9)$$

The concentration of the absorbing species and temperature can be related to the line strength of a spectral line as[21]:

$$S \equiv \int K_\nu d\nu$$
$$= \frac{\pi e^2}{m_e c} f_{J'J''} N_{J''} \left[1 - \exp\left(-\frac{h\nu_0}{kT}\right)\right] \qquad (10)$$

where $N_{J''}$ is the concentration of the lower level (J'') of the transition, $f_{J'J''}$ the oscillator strength, e the electronic charge, and m_e the mass of the electron. The term in the parenthesis is the contribution due to induced emission, which can be neglected for most cases of interest here. $N_{J''}$ can be related to the total mole fraction of the absorbing species X_i and the temperature T through the Boltzmann distribution:

$$N_{J''} = \left(\frac{p_t}{p_o}\right)\left(\frac{T_o}{T}\right) N_L X_i \frac{g_{J''}}{Q_i} \exp\left(-\frac{E_{J''}}{kT}\right) \qquad (11)$$

where p_t is the total pressure, p_o and T_o are standard pressure and temperature, N_L the Loschmidt number, $g_{J''}$ and $E_{J''}$ the degeneracy and the energy of the J" state, respectively, and Q_i is the internal partition function.

The f number of an individual line in a band can be related to the pure rotational transition probability $S_{J'J''}$ as[22]:

$$f_{J'J''} = f_{v'v''} S_{J'J''} T_{J'J''}/4(2J''+1) \qquad (12)$$

where $f_{v'v''}$ is the band oscillator strength of the rotationless molecule, and is assumed to be constant over a given vibration-rotation band. $T_{J'J''}$ is a correction factor for vibration-rotation interactions. For the OH molecule, $S_{J'J''}$ values have been calculated by Dieke and Crosswhite[23] and are normalized such that:

$$\sum_{J'} S_{J'J''} = 4(2J''+1) \qquad (13)$$

The values of $T_{J'J''}$ have been computed by Learner,[24] and the $f_{v'v''}$ values have been determined for the (0,0) and (1,0) bands by Rouse and Engleman.[25]

Since the absorption coefficient of a medium is a function of both the absorbing species concentration and its temperature, one of the simplest implementations of the theory is in a case where one of these quantities is uniform. Two component isothermal mixing processes are thus quite amenable to such a measurement. The value of the absorption cross-section Q_ν can be evaluated at the known temperature and divided into the reconstructed absorption coefficient field K_ν to lead directly to an absolute measurement of the number density N_i. An extension of this approach is to measure a chemically reacting flow where the temperature field has been determined by an alternative method,[26] such as a thermocouple. Again, once the temperature field is known, the number density of absorbing species can be calculated from the measured absorption coefficient field.

A still more complex scheme which is perhaps of more practical interest in chemically reacting systems, is the case where both the concentration and temperature are unknown.[27] In this case, there are several approaches which can be used to deduce both quantities, involving at least two measurements of the test field. One approach, which is referred to as the line ratio approach, involves measurements at the center of two different absorption lines. The other approach requires measurements along a single line, termed the line profile approach. An earlier study[27] showed these two approaches to be comparable in terms of sensitivity and noise immunity. Therefore, in this work only the line ratio technique will be discussed.

The peak absorption coefficients of two spectral lines can be related to each other as:

$$\frac{K_{\nu 1}}{K_{\nu 2}} = \frac{S_1}{S_2} \frac{\alpha_{D2}}{\alpha_{D1}} \frac{V_1(0,a)}{V_2(0,a)} \qquad (14)$$

At the same temperature, this expression reduces to:

$$\frac{K_{\nu 1}}{K_{\nu 2}} = \frac{\nu_2}{\nu_1} \frac{g_1}{g_2} \frac{f_1}{f_2} \exp\left(-\frac{E_1 - E_2}{kT}\right) \qquad (15)$$

$$= \frac{\nu_2}{\nu_1} \frac{(S_{J'J''} T_{J'J''})_1}{(S_{J'J''} T_{J'J''})_2} \exp\left(-\frac{E_1 - E_2}{kT}\right) \qquad (16)$$

if the two lines are within the same vibrational rotational band and the collisional broadening parameters are the same for both lines. If absorption measurements are made over the same optical path length, and if the temperature and species concentration are constant over that path, then the temperature can be determined as:

$$T = \frac{E_2 - E_1}{k} \left\{ \ln\left[\frac{K_{\nu 1}}{K_{\nu 2}} \frac{\nu_1}{\nu_2} \frac{(S_{J'J''} T_{J'J''})_2}{(S_{J'J''} T_{J'J''})_1}\right] \right\}^{-1} \qquad (17)$$

Absorption can also be measured for a series of n spectral lines to improve the accuracy of such measurements. In this case, a more general form of Eq. (16) can be used:

$$K_n \propto S_n \propto (S_{J'J''} T_{J'J''})_n \exp(-E_{J''n}/kT) \qquad (18)$$

and the temperature can be determined from the slope of the $\ln(S_n/(S_{J'J''}T_{J'J''}))$ vs $E_{J''n}$ plot.[28,29] Once the temperature is determined, then the concentration can be measured directly from the absorption observed at any one of the spectral lines.

Results

The feasibility of making tomographic measurements was investigated by studying a flat premixed flame of methane, seeded with sodium atoms. This particular configuration has several advantages. First, the geometry of a flat flame allows long absorption paths, resulting in increased absorption signals. Working with sodium seeding means that the incident light is visible (~590 nm) which facilitates alignment. Eventually, work will be done in the uv region of the spectrum to study the OH molecule, but it was considered preferable to start work with visible light.

It was also considered worthwhile to assess the ability of the technique to reconstruct a field such as would be found in a flat flame, by means of a computer simulation. To be of maximum utility, this simulation was performed for the OH molecule in anticipation of the subsequent experiments.

Computer Simulations

To simulate the conditions in a flat premixed flame, a two-dimensional region of space was prescribed, containing a circular region of constant stoichiometry ($\phi = 0.9$) of 2.5-cm diam. Outside of this region, the equivalence ratio drops off in a Gaussian manner, rapidly dropping to an equivalence ratio well below the flammability limit. With this prescribed field, the equilibrium concentration of OH can be calculated, as well as the flame temperature, assuming a methane/air mixture at atmospheric pressure. The resulting OH concentration distribution is shown in Fig. 1. This information is then sufficient to determine the local value of the absorption coefficient throughout the flowfield, using Eqs. (9-12). To complete the simulation, the absorption coefficient is integrated along each parallel ray. The procedure is then repeated for M number of rays and N angles, generating N projections. In this simulation, M=101 parallel rays were used. Since the field was axisymmetric, the computations were only performed for a single angle, and N=20 identical projections were then

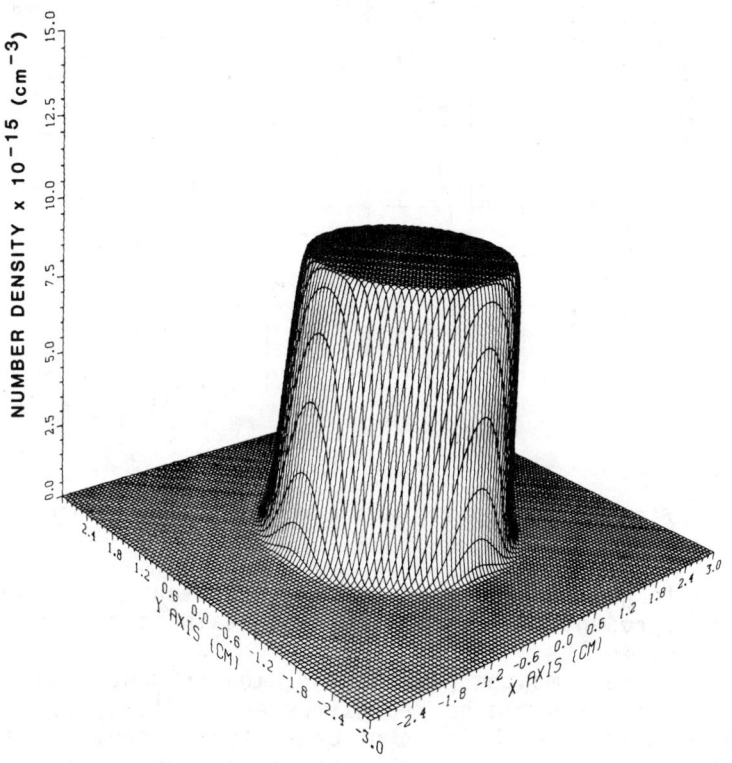

Fig. 1 Prescribed OH concentration field.

Table 1 Summary of spectroscopic data for the selected transitions in (0,0) $A^2\Sigma - X^2\Pi$ band of OH

Line	λ_0 (nm)	$S_{J'J''}$	$T_{J'J''}$	J''	$E_{J''}$ (cm^{-1})
$P_1(5)$	310.213	24.5	0.986	5.5	543.57
$Q_1(13)$	311.112	108.8	0.881	13.5	3319.35

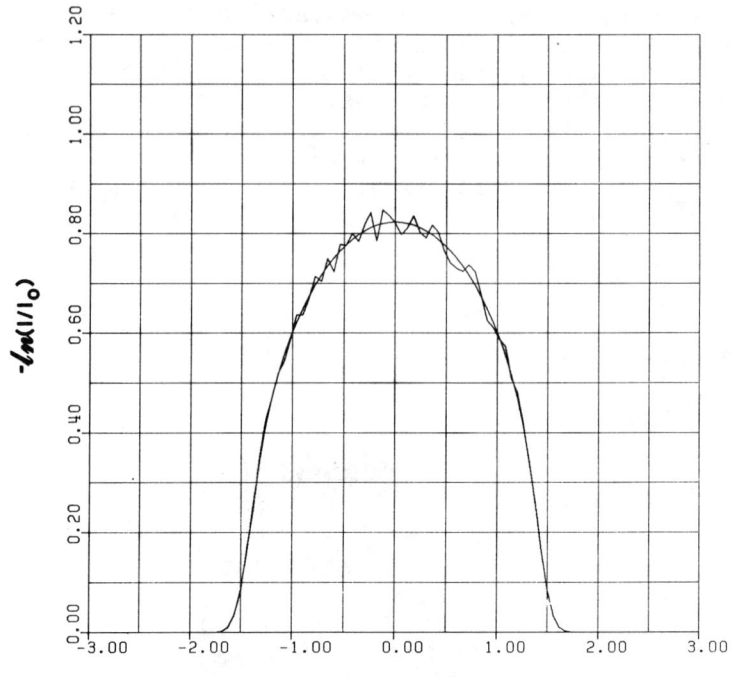

Fig. 2 Simulated projection with and without noise.

assumed to have been obtained. The resultant projections $P(r,\theta)$ represent the data one would obtain in an actual experiment.

From the computed projections, the original absorption coefficient field can be reconstructed as discussed in the section entitled Principles of Laser Tomography. To apply the line ratio technique, the simulation was repeated for two frequencies corresponding to the line center of the $P_1(5)$ and $Q_1(13)$ transitions of OH. These two lines are well isolated but near each other (in case rapid scanning of both lines is desired). In addition, the absorption coefficient for the $P_1(5)$ line decreases with temperature, since it originates in a low lying energy state, whereas the absorption coefficient for $Q_1(13)$ increases with temperature. This behavior provides increased sensitivity. The relevant spectroscopic data used in the computations are given in Table 1.

Twenty projections containing noise were independently generated from the same noise-free projection. These would most closely approximate a true set of measurements. The magnitude of the noise was based upon actual measurements which will be discussed later in this paper, determined to be 4% of the measured signal. A sample noisy projection used in the simulation is shown in Fig. 2, along with a noise-free projection. The reconstruction of the absorption coefficient field for the $P_1(5)$ line of OH is shown in Fig. 3. This was obtained by simply reconstructing the twenty noisy projections. As can be seen, the reconstruction is also quite noisy. A comparison of a cross section through this field with the originally prescribed properties is shown in Fig. 4, indicating a standard deviation of +3.6% from the true value. Using a similar reconstruction of the $Q_1(13)$ line, one can apply the line ratio technique to deduce the temperature field, and hence the concentration field. Cross sections of both these fields are shown in Fig. 5, with the original values plotted for comparison. One should note that the temperature field reconstruction was truncated and set to ambient values outside of a critical radius. This is because the ratio of the two absorption coefficients is quite sensitive to the

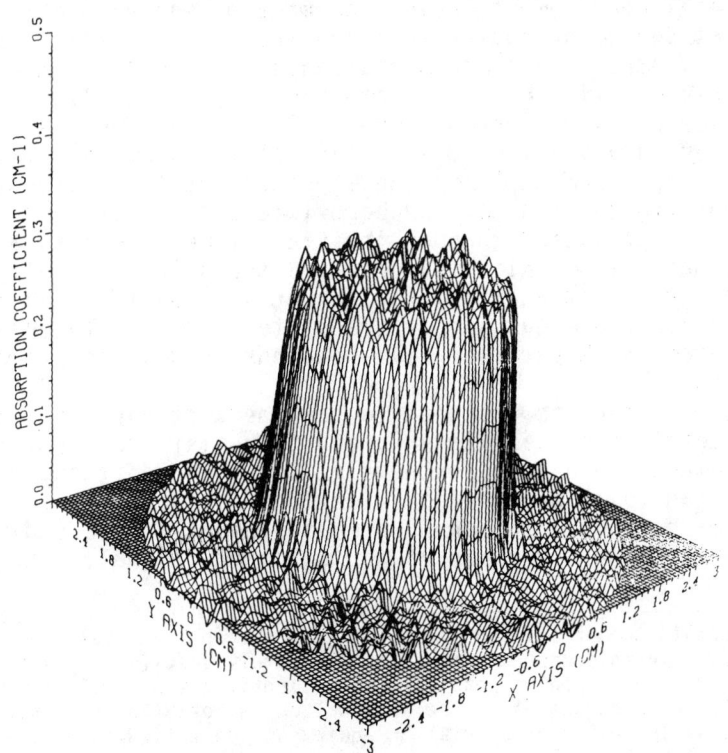

Fig. 3 Unfiltered reconstruction of absorption coefficient field.

aliasing "ripples" observable in regions of low signal. As can be seen from Fig. 5, this truncation only becomes necessary where the OH concentration has dropped to less than 1% of its peak value.

A significant improvement can be made in the reconstruction if some a priori knowledge about the field can be applied to the problem. For example, if the maximum gradients in the field are known (which can be estimated from known diffusivities) one can filter the projections in spatial frequency before reconstructing the fields. Such an approach was used, with the resulting cross sections shown in Fig. 6. Note that the gradients at the edge of the flame have still been passed essentially unfiltered, but spatial frequencies significantly higher than this have been attenuated.

Experimental Measurements

As a parallel effort to the computer simulations, experimental measurements have been made of the absorption of light at 589 nm by sodium injected into the premixed flat methane flame. The experimental apparatus used for much of the early work is shown diagrammatically in Fig. 7. The light source is a Coherent╪ model 699-03 ring dye laser pumped by an argon ion laser. Part of the beam is split off to a spectrum analyzer and a photodiode to monitor the incident light frequency and beam intensity. The rest of the beam is focussed through the flat flame region to a second photodiode which measures the transmitted light. In this early implementation, the burner was mounted on a motorized stage and was simply translated past the laser beam to provide the M parallel beam readings used in the reconstruction.

The burner was designed to provide a steady, laminar, axisymmetric premixed flame which, to satisfy the symmetry requirements, had to be extremely flat. In addition, the design had to allow for the injection of a salt solution to perform the initial absorption measurements at the sodium doublet. Therefore a honeycomb matrix of inconel, with

---

╪Certain commercial equipment, instruments, or materials are identified in this paper in order to adequately specify the experimental procedure. In no case does such identification imply recommendation or endorsement by the National Bureau of Standards, nor does it imply that the material or equipment identified is necessarily the best available for the purpose.

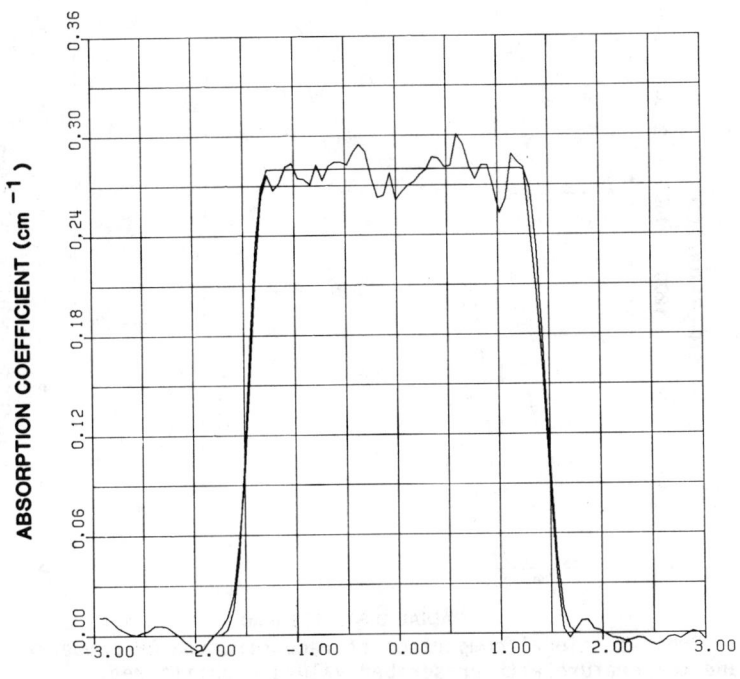

Fig. 4 Cross-sectional comparison of reconstructed absorption coefficient with prescribed value.

0.8-mm diam. holes, was used as the burner surface. Beneath the honeycomb is a layer of compressed metallic fibers which acts as a pressure barrier to provide flow uniformity, as well as to prevent flashback. Beneath the metal fiber layer is a simple inlet chamber into which the premixed gases are introduced.

Normally, the use of honeycombs to stabilize premixed flames leads to several problems. The major problem is that the flame heats the honeycomb to a temperature sufficient to melt it. Even if the honeycomb does not melt, there is an increased danger of flashback through the openings of the honeycomb since the honeycomb will preheat the gases. To avoid both of these problems, the flame temperature of our lean premixed methane flame was reduced by using a mixture of methane, oxygen and carbon dioxide. The larger specific heat of CO_2 as compared to N_2 brought the temperature down to a value at which the honeycomb remained reasonably cool. This type of gas mixture produced a stable, but cellular flame. Since the cellular structure clearly destroys the symmetry of the flame, helium was also

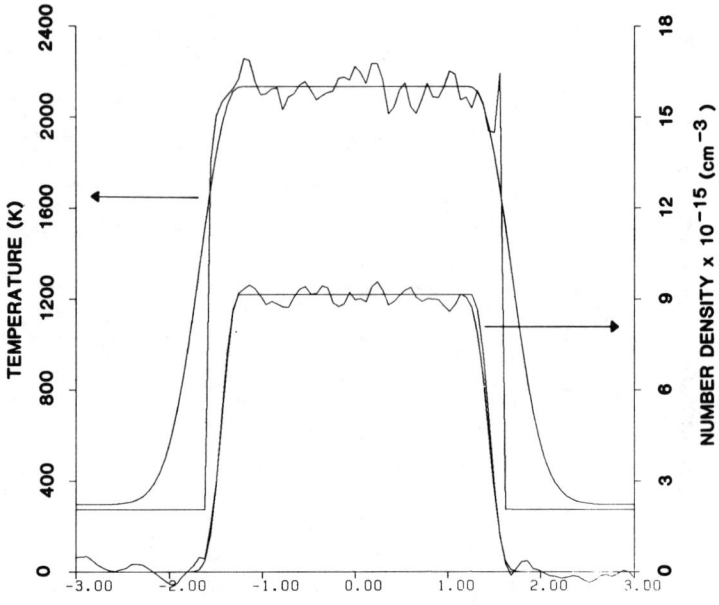

Fig. 5 Cross-sectional comparison of reconstructed OH concentration and temperature with prescribed values - unfiltered.

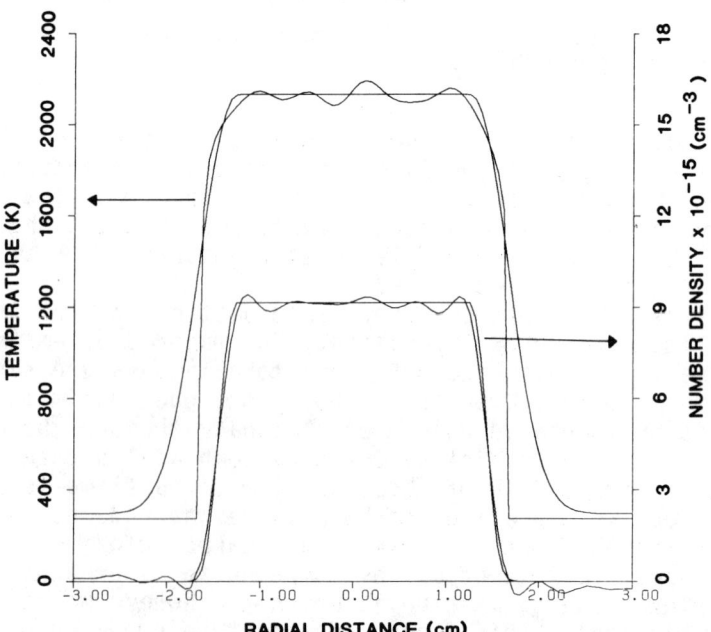

Fig. 6 Cross sectional comparison of reconstructed OH concentration and temperature with prescribed values - filtered.

added which acts to spread out the cells, due to its high thermal and mass diffusivities. With a judicious choice of proportions, a stable flame could be produced which appeared completely flat to the eye.

To perform the absorption measurements at the sodium doublet, a fine salt spray was introduced into the CO_2 flow. This was accomplished by passing the CO_2 through an orifice which had a small hollow tube placed at right angles to the exit. With a slight back pressure, a stream of salt solution was injected through the tube into the jet of CO_2 exiting from the orifice. By this action, the liquid jet was broken up to form a fine mist in a collection chamber located above the CO_2 jet. Some of this mist was continuously bled out and mixed with the other gaseous components. As the water droplets evaporated, microscopic particles of salt were left, which were carried by the gas flow to the burner. As a final precaution, a water trap was used to keep any liquid out of the burner. In this manner, a controllable supply of very fine salt powder could be introduced into the flame. The presence of water vapor in the supplied gas aided in further reducing the flame temperature, which was desirable. Fig. 8 shows an example of the measured transmission of light through a heavily absorbing flame. This measurement was made at line center of the 589.6-nm sodium line, at 0.05 in. (1.25 mm) from the honeycomb surface of the burner, which corresponds to a height slightly above the flame sheet. The resulting projection is plotted with the noise-free simulation projection for comparison in Fig. 9. As can be seen, the shape of the projections are quite similar. Differences near the wings of the projections are due to different gradients at the edges of the simulated and real flames.

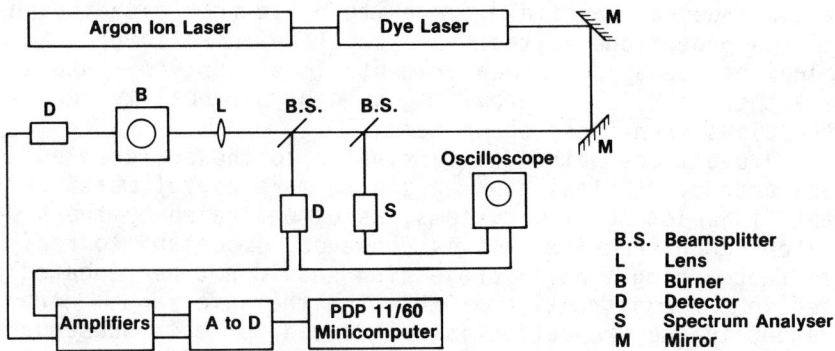

Fig. 7 Experimental configuration for single beam measurements.

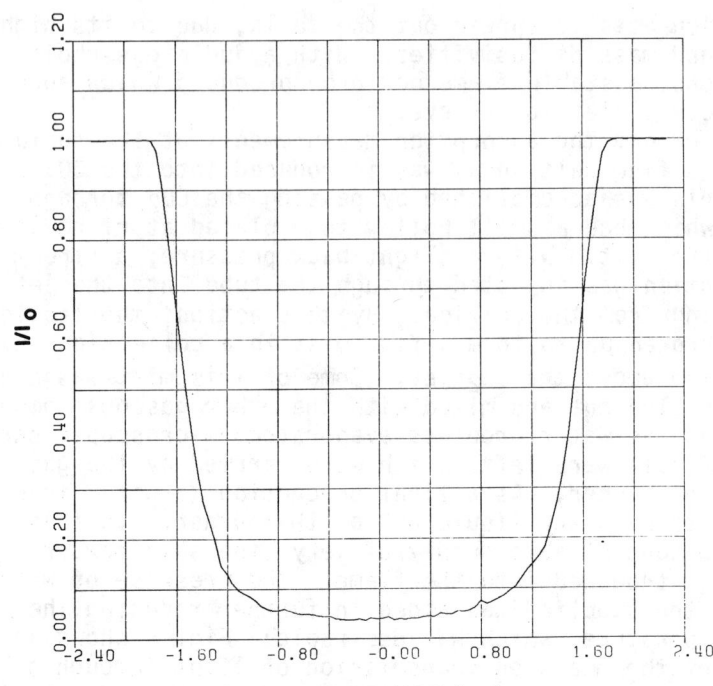

Fig. 8 Measurement of light transmitted through a seeded premixed flat flame.

It was from this kind of measurement that the level of noise used in the simulation was estimated. It should be noted that the source of noise in the measurement presented was not in the measurement system itself, but rather fluctuations in the supply of salt spray to the burner. This can be seen from the fact that the zero absorption readings at the edges of the field are quite noise-free even though the two photodiode voltages are still being ratioed. Therefore, tomographic measurements in combustion systems have the potential of providing even higher quality reconstructions than those shown here.

Projections with quality similar to those presented here are nevertheless good enough to make useful measurements in reacting flow systems, as demonstrated by the simulated reconstruction. It is, however, important to realize that a single noisy projection should not be repeatedly used in a reconstruction. Otherwise the same random error present in one projection is reinforced for each subsequent angle used in the calculation, leading to artificial circular structure in the reconstructed field. For example, us-

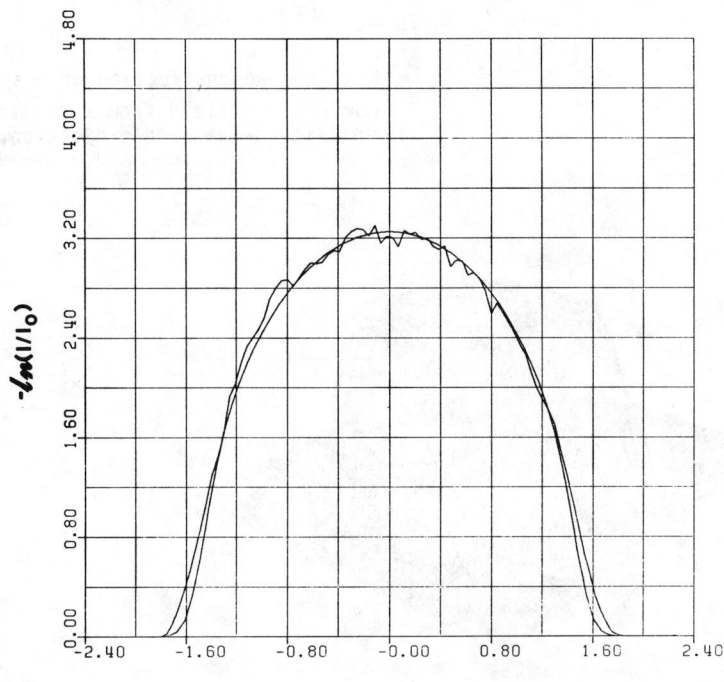

Fig. 9 Comparison of experimental and simulated flat flame projections.

ing the unfiltered experimental projection of Fig. 9 for all the angles in a reconstruction gives the field shown in Fig. 10, displaying the expected circular structure particularly at the center of the field where the error is repeated many times. This problem will not be present with the full multiangular experiment, since the errors will then tend to cancel.

Some preliminary experiments have also been carried out to obtain real-time measurements, with a typical temporal resolution of 10-50 μs. Thus, transient events occurring in a test field can be recorded and followed as a function of time. The repetition rate of these measurements is currently 2 kHz. This is accomplished by using a linear detector array to record the intensity across a sheet of laser light which passes through the test field. The signals from this array are then digitized with a high-speed data acquisition system and stored in a minicomputer.[30] The present work is limited to measurements at a single angle, which is suitable for an axisymmetric field. A typical projection obtained in real time is shown in Fig. 11.

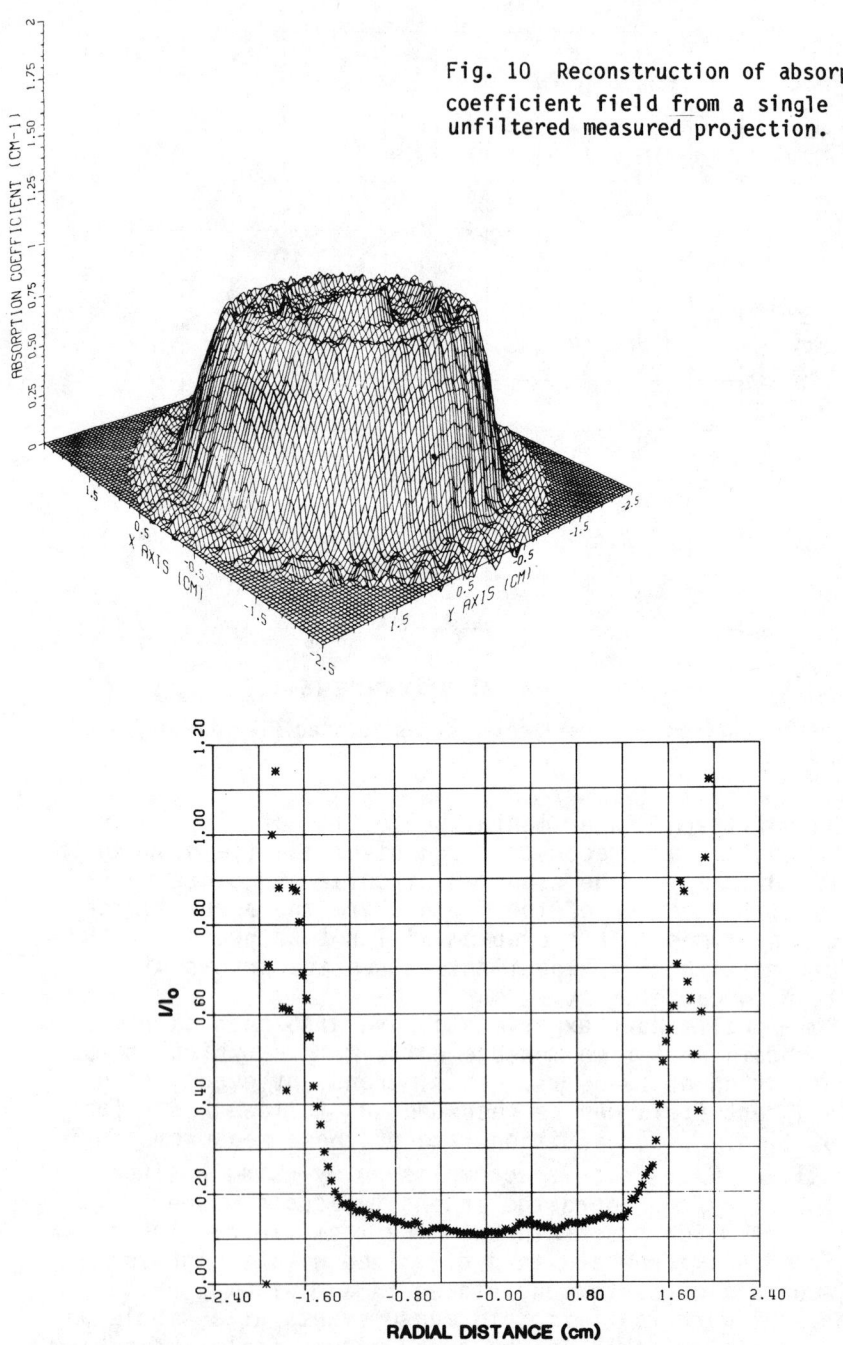

Fig. 10 Reconstruction of absorption coefficient field from a single unfiltered measured projection.

Fig. 11 High-speed measurement of light transmitted through a seeded premixed flat flame.

This projection exhibits some spherical aberration effects, which appear as the sheet is focussed onto the detector array. This problem will be rectified by using a fiber optic reducing bundle to collect the transmitted light and channel it to the detector without the use of lenses. An alternative approach, using a single beam and a high-speed rotating mirror scanner, will also be attempted.

Conclusions

Results have been presented on the application of laser tomography for real-time measurements of temperature and concentration in chemically reacting flows. Experimental data have been obtained in a sodium seeded premixed flat flame. Based on the experimental results, computer simulations have been carried out to evaluate the effect of noise on the temperature and concentration reconstructions. The simulations were carried out with random noise superimposed on the calculated projections, with an amplitude similar to that observed experimentally (4% of the signal). With the "typical" noisy projections, it has been demonstrated that the reconstructed values of concentration and temperature are within 3% of the true (prescribed) values inside the flame; this deviation can be reduced to less than 1.7% by using filtering techniques. It was found that the experimental noise was primarily due to fluctuations in the sodium seeding rate, rather than the optical system. Hence much higher quality data and more accurate reconstructions could be obtained when absorption measurements due to species naturally occurring in the flame are utilized.

Experimental measurements obtained with two different optical configurations have been presented. The first set of experiments was carried out with a single laser beam, which was translated through the flame. The second set of experiments provided real-time measurements, obtained with sheet of laser light and a linear array detector. These measurements were performed with a repetition rate of 2 kHz, and a temporal resolution of 10 μs.

The computer simulation results and the measurements performed to date give a clear indication thta laser tomography can be a valuable tool for diagnostics in chemically reacting flows. The technique is particularly useful for diagnostics in nonuniform flowfields, since no assumption of symmetry is required for the reconstruction process. The remaining task is to extend the system to a full multiangular geometry, in order to develop the laser tomography technique into a practical diagnostic tool.

References

[1] Regnier, P. R. and Taran, J. P., "On the Possibility of Measuring Gas Concentrations by Stimulated Anti-Stokes Scattering," Applied Physics Letters, Vol. 23, 1973, p. 240.

[2] Eckbreth, A. C., "Recent Advances in Laser Diagnostics for Temperature and Species Concentration in Combustion," Proceedings of the 18th Symposium on Combustion, The Combustion Institute, Pittsburgh, Pa., 1981, p. 1471.

[3] Lapp, M. and Penney, C. M., Laser Raman Gas Diagnostics, Plenum Press, New York, 1973.

[4] Smith, J. R., "Instantaneous Temperature and Density by Spontaneous Raman Scattering in a Piston Engine," AIAA Paper 80-1359, AIAA 15th Thermophysics Conference, Snowmass, Co., 1980.

[5] Morley, C., "The Mechanism of NO Formation from Nitrogen Compounds in Hydrogen Flames Studied by Laser Fluorescence," Proceedings of the 18th Symposium on Combustion, The Combustion Institute, Pittsburgh, Pa., 1981, p. 23.

[6] Eckbreth, A. C., Bonczyk, P. A. and Verdieck, J. F., "Combustion Diagnostics by Laser Raman and Fluorescence Techniques," Progress in Energy and Combustion Science, Vol. 5, 1979, p. 253.

[7] Dibble, R. W. and Hollenbach, R. E., "Laser Rayleigh Thermometry in Turbulent Flames," Proceedings of the 18th Symposium on Combustion, The Combustion Institute, Pittsburgh, Pa., 1981, p. 1489.

[8] Eckbreth, A. C. and Hall, R. J., "CARS Thermometry in a Sooting Flame," Combustion and Flame, Vol. 36, 1979, p. 87.

[9] Tourin, R. H., Spectroscopic Gas Temperature Measurements, Elsevier, New York, 1966.

[10] Brewer, L. E. and Limbaugh, C. C., "Infrared Bond Model Technique for Combustion Diagnostics," Applied Optics, Vol. 11, 1972, p. 1200.

[11] Chen, F. P. and Goulard, R., "Retrieval of Arbitrary Concentration and Temperature Fields by Multiangular Scanning Techniques," Journal of Quantitative Spectroscopy and Radiative Transfer, Vol. 16, 1976, p. 819.

[12] Blair, D. W., "An Analysis of Error Propagation in Abel Inversions of Spectral Emission-Absorption Data," Journal of Quantitative Spectroscopy and Radiative Transfer, Vol. 14, 1974, p. 325.

[13] Cormack, A. M., "Representation of a Function by its Line Integrals, with some Radiological Applications, II," *Journal of Applied Physics*, Vol. 35, No. 10, 1964, p. 2908.

[14] Hounsfield, G.N., "A Method of and Apparatus for Examination of a Body by Radiation such as X or Gamma Radiation," Patent #1283915, London Patent Office, 1972.

[15] Ramachandran, G. N. and Lakshminarayanan, A. V., "Three-Dimensional Reconstruction from Radiographs and Electron Micrographs: Application of Convolutions instead of Fourier Transforms," *Proceedings of National Academy of Sciences USA*, Vol. 68, 1971, p. 2236.

[16] Shepp, L. A. and Logan, B. F., "The Fourier Reconstruction of a Head Section," *IEEE Trans. Nuclear Science*, Vol. NS-21, 1974, p. 21.

[17] Emmerman, P. J., Goulard, R., Santoro, R. J. and Semerjian, H.G., "Multiangular Absorption Diagnostics of a Turbulent Argon-Methane Jet," *Journal of Energy*, Vol. 4, 1980, p. 70.

[18] Santoro, R. J., Semerjian, H. G., Emmerman P. J., and Goulard, R., "Optical Tomography for Flow Field Diagnostics," *International Journal of Heat and Mass Transfer*, Vol. 24, 1981, p. 1139.

[19] Tretter, S. A., *Introduction to Discrete-Time Signal Processing*, John Wiley, New York, 1976.

[20] Bracewell, R. N., *The Fourier Transform and Its Applications*, McGraw-Hill, New York, 1978.

[21] Penner, S. S., *Quantitative Molecular Spectroscopy and Gas Emissivities*, Addison-Wesley, Reading, Mass., 1959.

[22] Anketell, J. and Perry-Thorne, A., "Oscillator Strengths in the $2\Sigma^+ - 2\Pi$ Band System of OH by the Hook Method," *Proceedings of the Royal Society of London, Series A*, Vol. 301, 1967, p. 343.

[23] Dieke, G. H. and Crosswhite, H. M., "The Ultraviolet Bands of OH. Fundamental Data," *Journal of Quantitative Spectroscopy and Radiative Transfer*, Vol. 2, 1962, p. 97.

[24] Learner, R. C. M., "The Influence of Vibration-Rotation Interaction on Intensities in the Electronic Spectra of Diatomic Molecules. I. The Hydroxyl Radical," *Proceedings of the Royal Society of London, Series A*, Vol. 269, 1962, p. 311.

[25] Rouse, P. E. and Engleman, R., "Oscillator Strengths from Line Absorption in a High-Temperature Furnace - I. The (0,0) and (1,0) Bands of the $A^2\Sigma^+ - X^2\Pi_i$ Transition in OH and OD," *Journal of Quantitative Spectroscopy and Radiative Transfer*, Vol. 13, 1973, p. 1503.

[26] Semerjian, H. G., Santoro, R. J., Goulard, R. and Emmerman, P.J., "Optical Tomography for Diagnostics in Combusting Flows," *Fluid Mechanics of Combustion Systems*, edited by T. Morel, R. P. Lohman and J. M. Rackley, ASME, New York, 1981, p. 119.

[27] Semerjian, H. G., Ray, S. R. and Santoro, R. J., "Laser Tomography for Diagnostics in Reacting Flows," AIAA Paper 82-0854, 3rd Joint Thermophysics, Fluids, Plasma and Heat Transfer Conference, St. Louis, Mo., 1982.

[28] Schmidt, S. C. and Malte, P. C., "Spectroscopic Absorptance Measurements of OH in High-Intensity Continuous Hydrogen/Air and Methane/Air Combustion," *Journal of Quantitative Spectroscopy and Radiative Transfer*, Vol. 16, 1976, p. 963.

[29] Bulewicz, E. M., Padley, P. J. and Smith, R. E., "Spectroscopic Studies of C_2, CH and OH Radicals in Low Pressure Acetylene and Oxygen Flames," *Proceedings of the Royal Society of London, Series A*, Vol. 315, 1970, p. 129.

[30] Ray, S. R. and Semerjian, H. G., "Laser Tomography for Temperature and Concentration Measurement in Reacting Flows," Fall Technical Session, Western States Section, Combustion Institute, Livermore, Calif., 1982.

Flow Measurement in a Model Combustion Chamber

P. Magre,* J. Labbé,† and G. Collin‡
Office National d'Etudes et de Recherches Aérospatiales (ONERA)
Châtillon, France

Abstract

The flowfield is investigated in a model combustion chamber by means of a bicolor laser Doppler velocimeter. Both flows with and without combustion are analyzed. Results concerning the two components of time-averaged velocity, velocity fluctuations, and the corresponding correlations are presented. The vortex that stabilizes itself in the bottom of the combustion chamber, the mixing zone and the flame stabilization zone can be seen; the zones where turbulence is produced can also be recognized. Effect of heat release on the flowfield (time averaged and fluctuations) is also analyzed.

Introduction

Modeling of gas turbine combustion chambers requires the computation of three-dimensional turbulent flows with heat release. Such models, thoroughly described in several review paper (see for instance Lilley[1] or Jones and Whitelaw[2]) have to be validated by detailed comparison between theoretical predictions and experimental results.

The advent of optical diagnostic techniques is a great help for such experiments. These nonintrusive techniques have a good time and space resolution. They present many advantages if compared to conventional techniques. Measurements of concentration and temperature distributions by means of laser systems (see, for example, the review paper by Bechtel and Chraplyvy[3]) give more adequate results than the conventional probes and thermocouples.

Presented as Paper 83-1550 at the AIAA 18th Thermophysics Conference, Montreal, Canada, June 1-3, 1983. Copyright © 1983 American Institute of Aeronautics and Astronautics, Inc. All rights reserved.
*Ingénieur, Departement de l'Energétique.
†Chef de Groupe, Departement de l'Energétique.
‡Technicien, Departement de l'Energétique.

Laser velocimetry has also been a great help for the analysis of flowfields in reacting media. A lot of experiments compare reacting and nonreacting flows in basic two-dimensional configurations. This is the case of the step of Pitz and Daily[4], of the simple shaped flame holders of Fujii and Eguchi[5], or Whitelaw and Taylor[6], of some semi-industrial tests such as the swirlers of Fujii, Eguchi, and Gomi[7] or fully industrial tests such as the Baker, Hutchinson, Whitelaw furnace[8].

A basic research initiated several years ago at ONERA aims at the modeling of the primary zone of a gas turbine combustion chamber. The model is based on a simplified version of an angular sector of an actual annular combustion chamber. In the tests described the fuel is vaporized instead of being injected through swirlers. The flowfield is fully three-dimensional.

The model has all the characteristics of an actual combustion chamber: primary air-jets, air fuel injectors, turbulent recirculation zones for the mixing of combustion products with fresh air. The main advantage of the simplification of the model is a more easy description of the flow field. In parallel with the experimental research, numerical modeling of two- and three-dimensional flowfields are developed using time marching numerical techniques (Hirsinger and Tichtinsky[9]) with its adaptation to recirculating reacting flows in three-dimensional configuration (Tichtinsky[10]).

As a preliminary test, conventional visualization techniques in a water tunnel have been used (Hebrard and Magre[11]). But the necessity of the determination of the flowfield in the actual model combustion chamber and the perturbating effects due to heat release in the flow lead to the use of a two-color laser Doppler velocimeter (most of the previous tests used only one-directional LDV). The present paper describes the time-averaged velocity distribution, and the turbulence level in reacting and nonreacting flows.

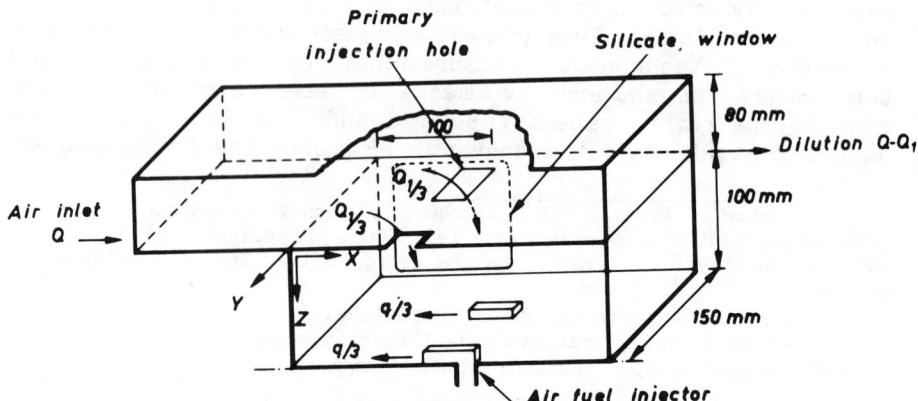

Fig. 1 Schematic view of the half combustion chamber.

Test Facility and Experimental Technique

Test Facility

The combustion chamber (Fig. 1) is a 100 x 300 x 610 mm rectangular channel. On its upper face three square air inlets (40 x 40 mm) divide the main flow of mass flow rate Q into two parts (see Table 1).

1) Q_1 enters the primary zone of the combustion chamber and gives three primary air jets;

2) $Q-Q_1$ is the dilution flow, nonrepresented in this model, centered on the primary zone.

The three fuel injectors are also square shaped (20 x 20 mm) and placed on the opposite wall of the chamber: the air fuel jets are directed towards the rear side of the chamber. These fuel injectors of air mass flow rate q and fuel addition are placed in the symmetry planes of the primary injection holes (planes Y = 0, Y = -100, and Y = 100 mm).

Two silicate windows on the lateral walls are incorporated for optical access of the laser beams into the combustion chamber.

Early Experiments and Test Conditions

Flow visualization (Fig. 2) has shown that there is a vortex situated in the rear of the combustion chamber[11]. From a detailed analysis of the various visualization results, a flow pattern has been derived (Fig. 3). Each primary jet of mass flow rate $Q_1/3$ is divided in three parts.

1) Q_{R1} forms the rear vortex, axis parallel to OY, i.e., normal to the combustion chamber axis; the airflow q/3 entering through the fuel injectors mixes with air flow Q_{R1}.

2) Q_{R2} forms a vortex of axis parallel to OX, i.e., parallel to the chamber axis.

Table 1 Test Conditions

	Q a)	Q_1 b)	q c)	T_1 d)	φ_1 e)	V_{NR}	V_R f)
	g/s	g/s	g/s	K		m/s	m/s
Nonreacting flow	970	270	51	320	0	43.3	...
Reacting flow	920	223	42	750	0.33	...	84.5

a) Q Main flow rate.
b) Q_1 Primary flow rate.
c) q Injection flow rate.
d) T_1 Inlet temperature.
e) φ_1 Equivalence ratio.
f) V_{NR}, V_R Reference velocity.

Fig. 2 Flow pattern recorded in a water tunnel modeling the combustion chamber.

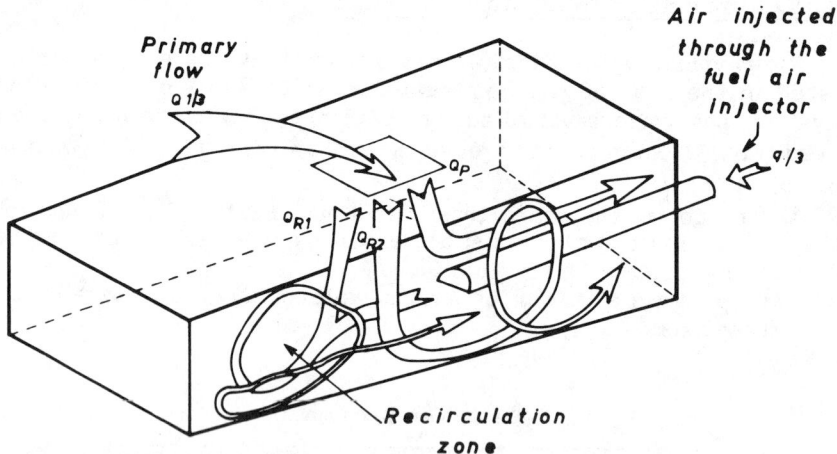

Fig. 3 Flow patterns derived from water tunnel visualization.

3) Q_P is directly evacuated from the chamber.

In parallel with the visualization techniques, residence time measurements techniques, by colorimetry or by particles tracking have also been developed and some results have been obtained. Figure 4 gives an example of the variation of the fraction $3Q_{R1}/Q_1$ of the primary flow that recirculates at the rear of the chamber. The effect of mean flow angle α_m, primary jet blockage factor B, ratio q/Q_1 have been investigated.

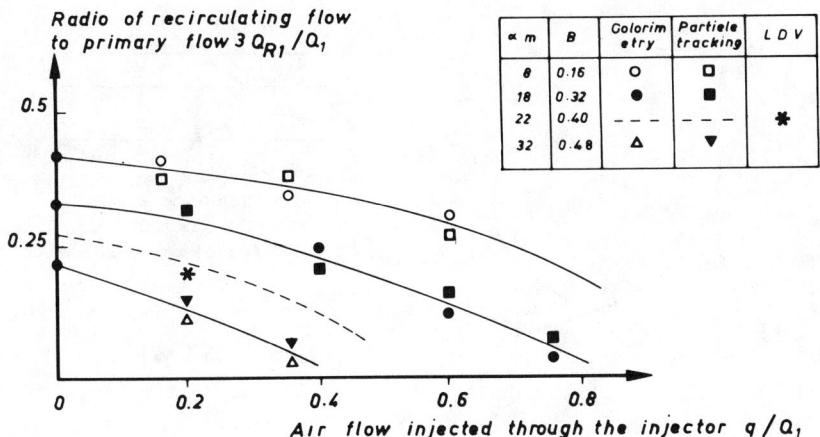

Fig. 4 Ratio of recirculating flow $3Q_{R1}/Q_1$ deduced from water tunnel visualization.

These purely aerodynamic investigation do not give any indication on the behavior of the combustion chamber when heat is released in the primary zone. Conventional gas sampling in various planes and their analysis by a gas chromatograph shows the nonuniformity in concentration of fresh and burnt species (Fig. 5). This pattern is due to the flow recirculation in the primary zone.

Therefore systematic investigation of the flowfield in the recirculation zone by means of LDV has been chosen. Velocity components u and w in the X and Z direction are determined. Test conditions in cold and in reacting flows are summarized in Table 1. For the reacting flow, the temperature in the primary zone can be roughly estimated at around 1200 K. The tests are restricted to the recirculation zone (Q_{R1}) at the rear of the chamber, and to the jets. The lateral vortex (Q_{R2}) is not described.

Laser Velocimeter

The laser Doppler velocimeter used was developed at ONERA (Fig. 6). It is equipped with a 15-W argon ion laser. It is a two-color (λ_1 = 514 nm and λ_2 = 488 nm) system, allowing the simultaneous determination of two velocity components in plane Y = const, perpendicular to the incident beams. Bragg cells are included on each optical path. They take into account the direction of the velocity vector even in high turbulent flows.

The forward scattered light technique is used, since it gives the best signal-to-noise ratio. Location of the measuring volume can be adjusted by remote control of a three-dimensional displacement system with an accuracy of ± 10 μm. Simultaneous signals scattered by the same particle on each optical path are filtered, are recorded

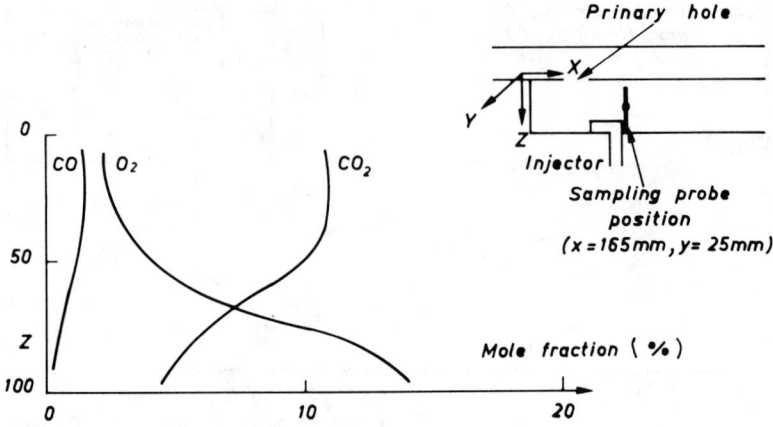

Fig. 5 Concentration profiles obtained by gas chromatography (fuel-air ratio corresponding to these tests φ = 0.58).

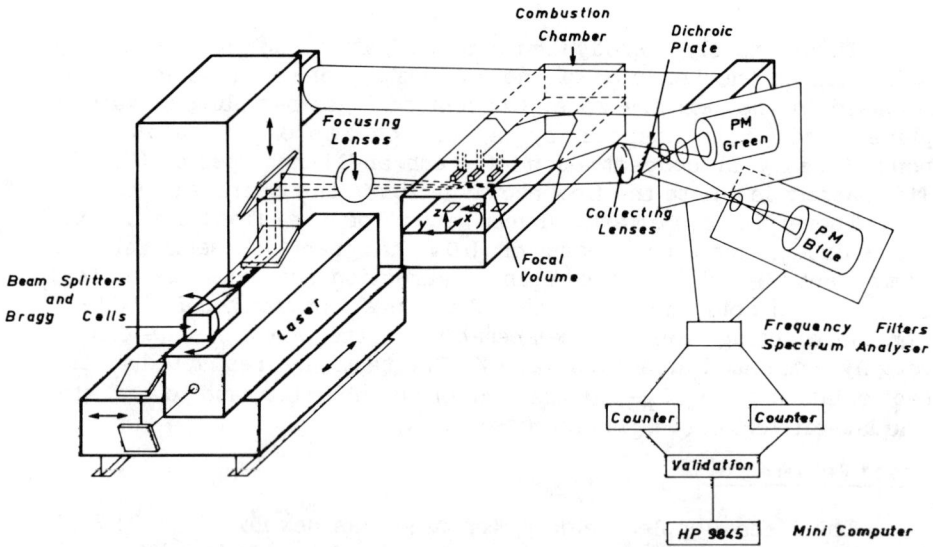

Fig. 6 Overall view of LDV and combustion chamber.

by two separate frequency counters. The signals are digitized, directed towards a mini-computer, and finally stored on floppy disks in a binary code.

For each point the local velocity is computed after sampling more than 1000 valid particles. The problem of bias towards high velocities in LDV (see McLaughlin and Tiederman[12]) is avoided if sampling of Doppler signals is made with some precautions. By taking a sampling frequency lower than the frequency of arrival of particles

(Durao, Laker, and Whitelaw[13]), it is possible to substantially reduce this bias. Therefore a particular attention is given to seeding in dead fluid zones, such as recirculation zones. The flow is seeded by injection of submicronic zircon particles (ZrO_2) far upstream in the main flow and in the injection pipes.

However, there were some difficulties proper to the test facility. The high noise level (approximately 100 dB) and the thermal radiation of the combustion required well-adapted shields on most of the optics and a thermal insulation of the combustion chamber. Zircon particle deposit on the windows was avoided by putting these windows at the outer end of chimneys outside of the combustion chamber. This solution restricts the tests to the vicinity of the primary center injection hole which is not perturbated by the cavities holding the windows. Finally, the change in flow direction inside the recirculation zone required an optimization of the fringe directions, which had to be set at 45 deg with respect to the mean flow direction. Therefore frequent resetting of the fringe pattern was necessary during mapping of the flowfield.

Experimental Results and Discussion

Due to the three-dimensional nature of the flow in the primary zone, several transverse planes were investigated:
1) plane A (Y = 0), symmetry plane of the primary air jet;
2) plane B (Y = 25 mm) at one quarter of the distance between two adjacent primary holes;
3) plane C (Y = 50 mm) at mid-distance between two adjacent primary holes.

In each of these planes mean velocity components $\bar{u}$ and $\bar{w}$ were measured as well as their fluctuations $\overline{u'^2}$, $\overline{w'^2}$ and the turbulent shear stress $\overline{u'w'}$. All these quantities were reduced to the mean velocity in the primary jet (we call it V_{NR} for nonreacting flows and V_R for reacting flows). These reduced quantities allow a better comparison of the flowfield with and without energy release.

Time-Averaged Velocities with and without Combustion

The velocity vectors plotted on Figs. 7a-7c show the flow pattern in planes A, B, and C. The primary jet enters the chamber with an average angle α_m of about 20 deg in respect to the vertical axis. This angle is practically the same with or without combustion. However, in the case of a reacting flow the jet extends in the X direction and retracts in the transverse Y direction (see Figs. 7a and 7b). The acceleration of the flow coming from the rear of the chamber explains this phenomenon which is limited to the upper half of the jet. As Z increases, i.e., far from the inlet of the jet, the jet tends to expand in the transverse direction in both cases, with and without combustion.

332   P. MAGRE, J. LABBE, AND G. COLLIN

The mixing region between the primary jet and the injector is greatly modified by combustion. Despite the scattered data for combustion measurements in this extremely high fluctuating region, we can make a few remarks concerning the behavior of the flow in the vicinity of the injector.

In the nonreacting flow case, the jet at the exit of the injector was well defined and directed, for the most part, towards the rear of the chamber (except for a small region $Z = 80$ mm, $90 < X < 110$ mm).

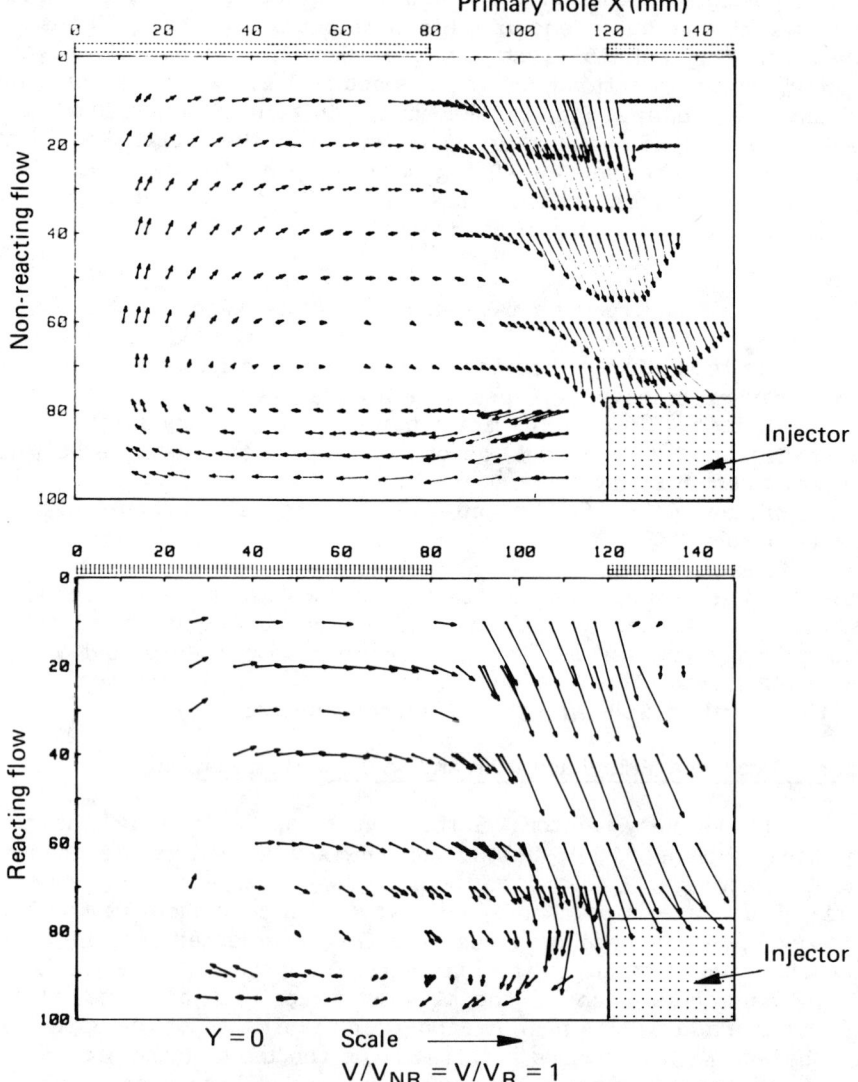

Fig. 7a Time-averaged velocity vectors in the symmetry plane A of primary hole.

In the case of reacting gases, this jet is more disturbed by the primary jet which is accelerated, the combustion being stabilized in the layer between the potential core and the recirculation zone. The blue luminous aspect of the flame and the increasing velocities in this layer support this hypothesis. The jet issuing from the injector first passes round the vortex in plane A, then expands towards the rear of

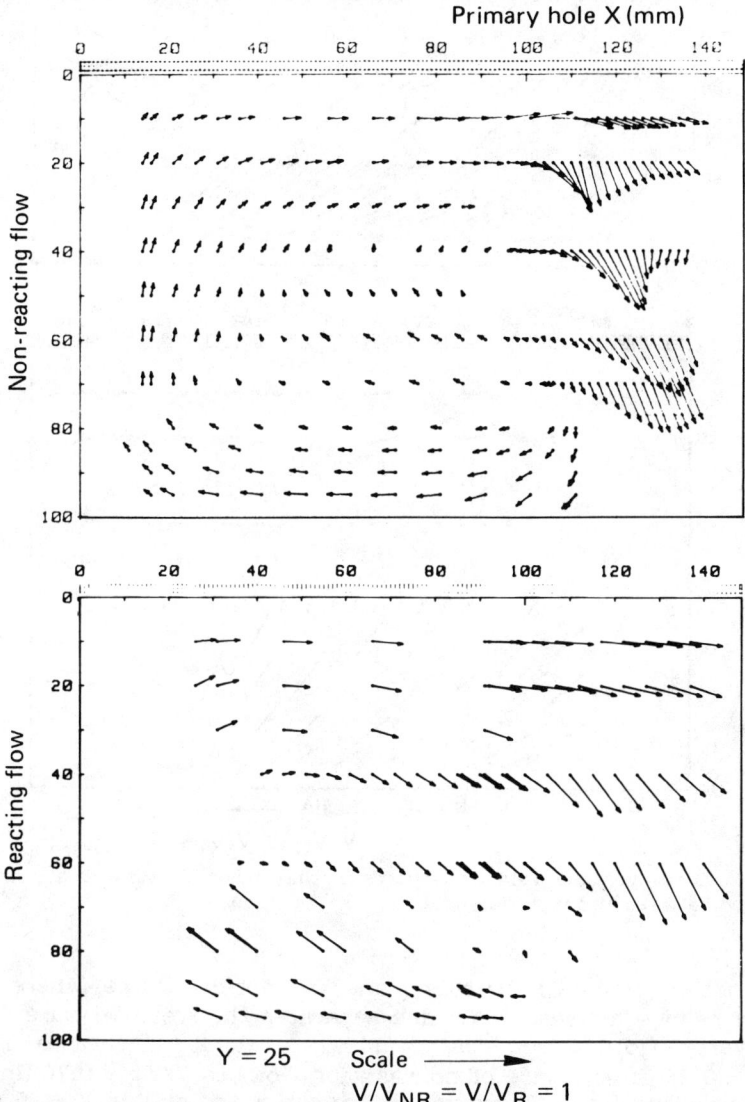

Fig. 7b Time-averaged velocity vectors in plane B situated at one quarter distance between two injection holes.

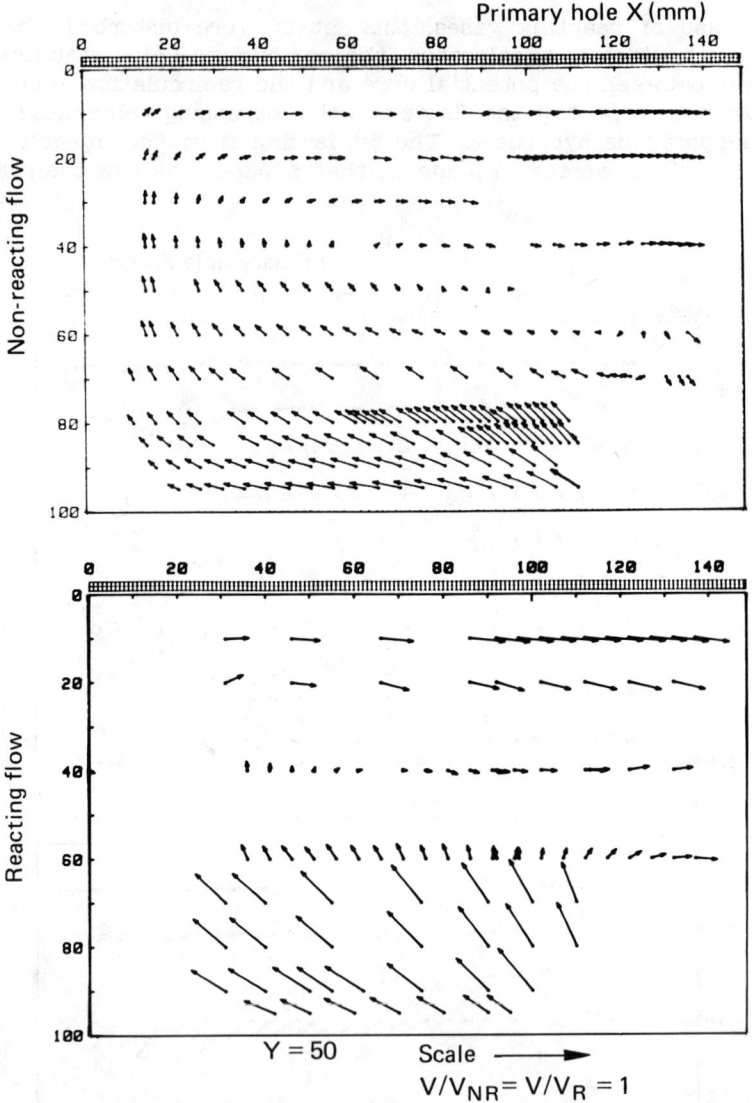

Fig. 7c Time-averaged velocity vectors in the plane C situated at mid-distance between two injection holes.

the chamber (plane B). In this plane we notice a line where the velocity nearly vanishes. This line separates the fresh mixture from the reverse flow where the maximum velocity increases from $V/V_{NR} = 0.35$ (in the case of nonreacting flow) to $V/V_R = 0.70$ (in the case of reacting flow). These values are estimated at line $Y = 25$ mm, $Z = 20$ mm. The strongly accelerated reverse flow is clearly shown on concentration profiles (Fig. 5). At abscissa lower than $Z = 50$ mm, a

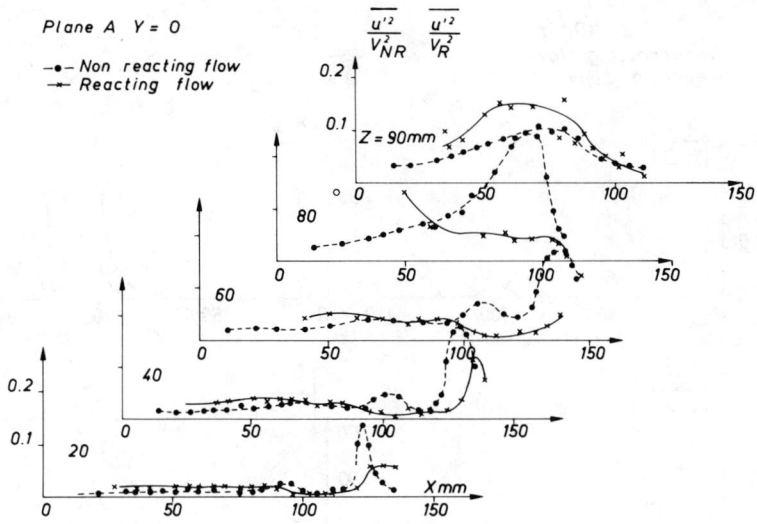

Fig. 8a Nondimensional velocity fluctuations $\overline{u'^2}/v^2$.

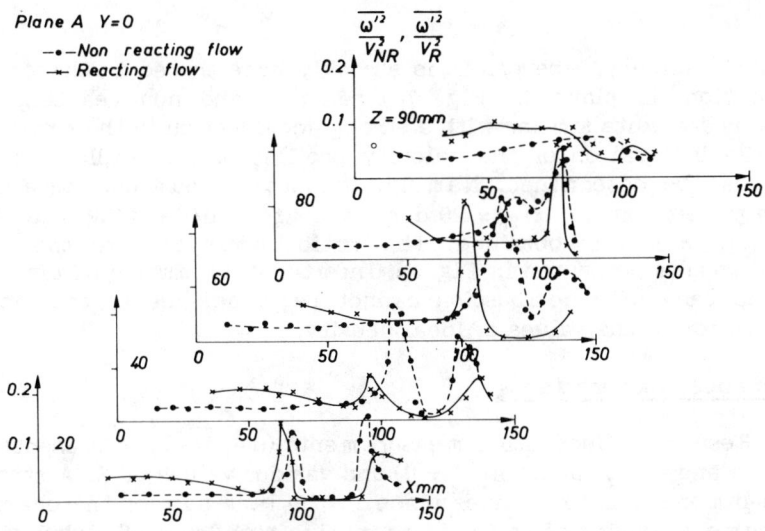

Fig. 8b Nondimensional velocity fluctuations $\overline{w'^2}/v^2$.

high concentration of carbon dioxide and a low concentration of oxygen indicate a nearly completed reaction for the gases entering the rear part of the combustion chamber.

Another difference between flow configuration with and without heat release is shown in plane B (Fig. 7b). Without combustion the jet issuing from the injector remained at practically

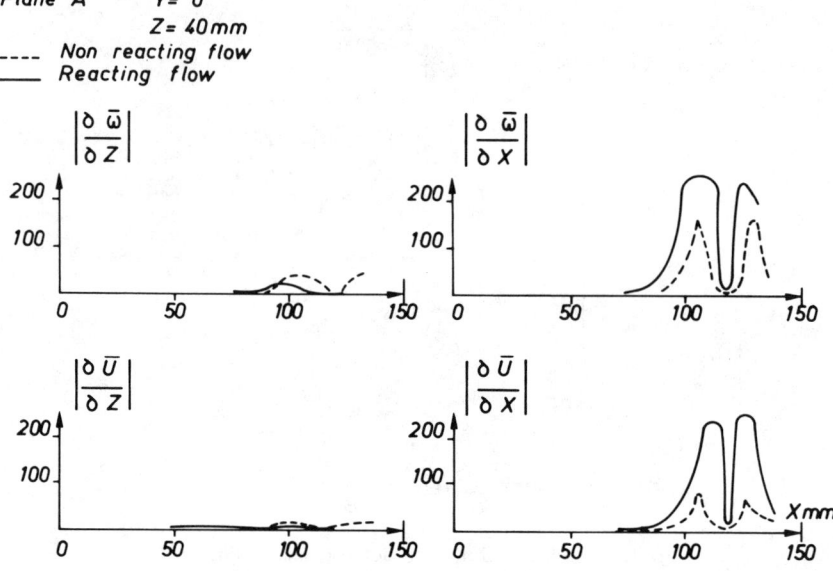

Fig. 9 Mean velocity gradients (arbitrary units).

constant velocity, whereas it is strongly accelerated in the case of combustion. In plane C (Fig. 7c) reacting and non reacting flow patterns are quite similar with a strong acceleration in the first case.

By integration of the velocity profiles, the overall mass flow rates can be determined. Namely, without combustion, for a mean primary jet angle $\alpha_m = 20$ deg and for $q/Q_1 \simeq 0.2$, a ratio $3 Q_{R1}/Q_1 = 0.19$ is obtained. This ratio corresponds to the water tunnel results presented in Fig. 4. Unfortunately similar computation for the case with combustion cannot be made, due to the lack of information of the values of local density.

Turbulence Measurements

Results of fluctuation measurements ($\overline{u'^2}$, $\overline{w'^2}$) are presented on Figs. 8a and 8b for plane A (Y = 0) and various values of Z. A strongly three-dimensional flow takes place in the rear part of the chamber. Impingment, mixing of various jets involve high level of turbulence in this region. So one is entitled to find several peaks corresponding to the limits of the various jets.

Figure 8a, representing the fluctuation $\overline{u'^2}$, shows this particular type of fluctuation diagram for the nonreacting flow with a large peak at one end of the jet and a similar one at the other. These peaks seem smaller with combustion. Figure 8b corresponds to fluctuation w'^2 and the two peaks are clearly visible.

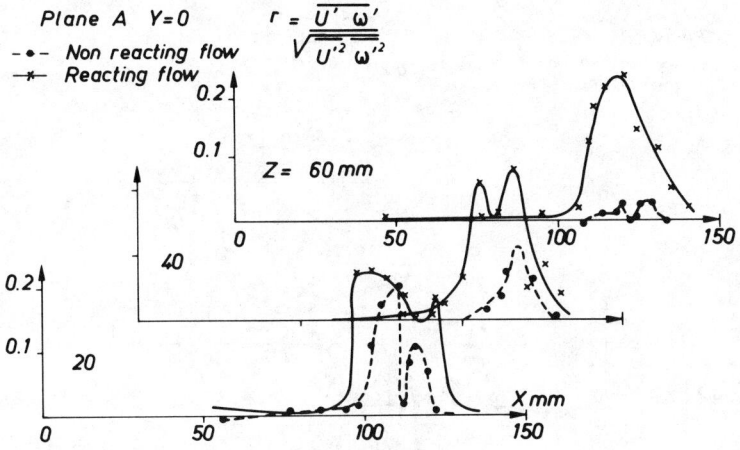

Fig. 10 Correlation factor.

Usually, in nonreacting flows, turbulence production is controlled by large time-averaged velocity gradients such as those of Fig. 9. The production terms of u'^2 and w'^2 can be written

production of u'^2

$$\alpha \quad -\left(\overline{u'^2}\frac{\partial \bar{u}}{\partial x} + \overline{u'v'}\frac{\partial \bar{u}}{\partial y} + \overline{u'w'}\frac{\partial \bar{u}}{\partial z}\right)$$

production of w'^2

$$\alpha \quad -\left(\overline{u'w'}\frac{\partial \bar{w}}{\partial x} + \overline{v'w'}\frac{\partial \bar{w}}{\partial y} + \overline{w'^2}\frac{\partial \bar{w}}{\partial z}\right)$$

It can be seen that the amplitude of these gradients increases when heat is released in the flow (mainly in the X direction). However the increase of the gradients due to the combustion is partially compensated for by a greater dissipation since combustion does not seem to have a serious effect on the value of $\overline{w'^2}$.

High fluctuation levels are expected at the impact of the primary jet with the jet issuing from the injector (plane Y = 0, Z = 80, $90 \leqslant X \leqslant 110$): as a matter of fact these high fluctuation levels (namely for $\overline{u'^2}$) seem to decrease with combustion (see Fig. 8a). However, as the jet proceeds towards the rear of the chamber both fluctuations $\overline{u'^2}$ and $\overline{w'^2}$ increase (Z = 90 mm).

Double peaks are also noticed on the correlation factor curve $r = \frac{\overline{u'w'}}{\sqrt{\overline{u'^2}\,\overline{w'^2}}}$ (Fig. 10) mainly for the primary jet. Maximum values of r of the order of 0.2 have been measured. r is of the order of 5.10^{-3} outside of the primary jet.

In plane B (at one quarter distance between two primary injection holes, Y = 25 mm), for Z = 20 mm the characteristic

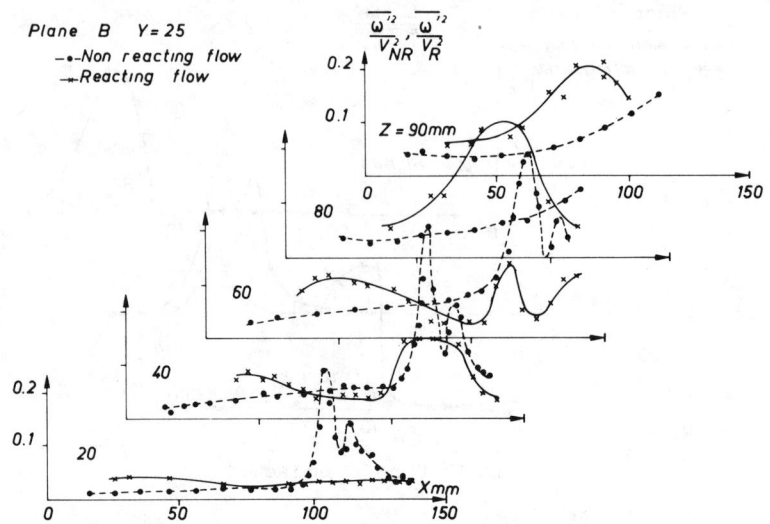

Fig. 11 Nondimensional velocity fluctuations $\overline{w'^2}/v^2$.

fluctuations of a mixing zone with two peaks are noticed (Fig. 11); these jets being less developed in the Y direction. For higher values of Z (Z = 80 mm and Z = 90 mm) combustion amplifies the fluctuations and their reduced values increases from $\overline{\omega'^2}/V_{NR}^2 = 0.05$ without combustion, to $\overline{\omega'^2}/V_R^2 = 0.2$ with combustion. This phenomenon seems to be related to high time-averaged velocity gradients (Fig. 12). Similar remarks have been made by Fujii, Eguchi, and Gomi[7] in their tests of a swirler, for which they found an amplification of turbulence in the high gradient regions. In our case, the amplification of turbulence with combustion does not seem restricted to the high gradient regions. In the core of the vortex (Y = 25 mm, Z = 60 mm, X = 50 mm), the nondimensional fluctuation increases from $\overline{\omega'^2}/V_{NR}^2 = 0.05$ without combustion, to $\overline{\omega'^2}/V_R^2 = 0.11$ with combustion (a peak value of 0.16 has been found for the point Y = 50 mm, Z = 40 mm, X = 80 mm). Again similar results have been found for a reacting flow by Whitelaw and Taylor[6] for a flame stabilized by a circular flame holder for which an increase of the fluctuations from 0.06 to 0.12 have been observed.

From the various results obtained it seems that turbulence is produced wherever there are high velocity gradients (limit of jets, flame stabilization zones). In these regions the correlation factor is high (r = 0.2). Then it seems that the turbulence kinetic energy is convected from these regions where turbulence is produced to the recirculation zones. There, turbulent fluctuations are still high but completely uncorrelated. Although this analysis gives a first description of what happens inside a combustion chamber, it is

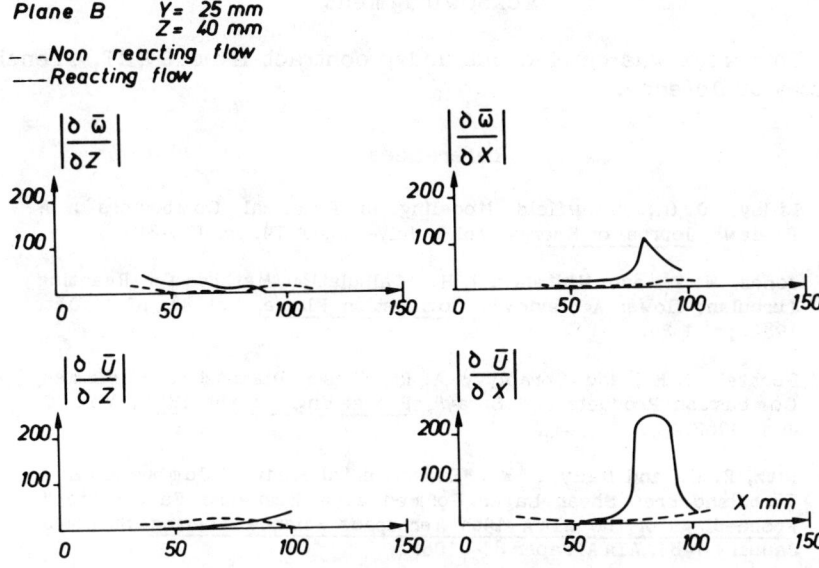

Fig. 12 Mean velocity gradient (arbitrary units).

insufficient for a detailed description of the phenomena. A better knowledge of the dissipation process in reacting flows is required.

Conclusion

A two-color laser Doppler velocimeter was used for the determination of the time-averaged velocity components $\bar{u}$ and $\bar{w}$ in the primary zone of a model combustion chamber. The corresponding values of velocity fluctuations $\overline{u'^2}$ and $\overline{w'^2}$ correlation $\overline{u'w'}$ were also recorded. Flows with and without combustion have been compared. The effect of combustion on the average flow field and turbulent fluctuations has been analyzed.

As it is, the whole volume of experimental results obtained will be of great help in the validation of the numerical models developed at ONERA. The information concerning kinetic energy of turbulence can be used for validation of $k - \varepsilon$ turbulence models in recirculating flows. The experimental histograms of fluctuating quantities can support statistical models based on probability density functions (p.d.f.). Using the CARS technique developed at ONERA by Attal, Péalat, and Taran[14] further analysis of the primary zone will lead to a better knowledge of local temperature and concentration of major species.

Acknowledgment

This work was carried out under contract from DRET, French Ministry of Defence.

References

1. Lilley, D. G., "Flowfield Modeling in Practical Combustors : A Review", Journal of Energy, Vol 3, July-August 79, pp. 193-210.

2. Jones, W. P., and Whitelaw, J. H. "Calculation Methods for Reacting Turbulent flows: A Review", Combustion Flame, Vol. 48, n° 1, Oct. 1982, pp. 1-26.

3. Bechtel, J. M., and Chraplyvy, A. R., "Laser Diagnostics of Flames, Combustion Products and Spray", Proceedings of the IEEE, Vol. 70, June 1982,.

4. Pitz, R. W., and Daily, J. W., "Experimental Study of Combustion in A Turbulent Free Shear Layer Formed at a Rearward Facing Step", Proceedings of the AIAA 19th Aerospace Science Meeting, St Louis, January 1981, AIAA Paper 81-0106.

5. Fujii, S., and Eguchi, K., "A Comparison of Cold and Reacting Flows Around a Bluff Body Flame Stabilizer", Transaction of ASME, Vol. 103, June 1981,

6. Whitelaw, J. H., and Taylor, A. M. K., "Velocity and Temperatures Measurements in the Premixed Flame Within and Axisymmetric Combustor", Proceedings of the 55th Propulsion and Energetics Panel AGARD, Bruxelles, May 1980.

7. Fujii, S., Eguchi, K., and Gomi, M., "Swirling Jets With and Without Combustion", AIAA Journal, Vol. 19, Nov. 1981, pp. 1438-1442.

8. Baker, R. J., Hutchinson, P., and Whitelaw, J. H., "Velocity Measurements in the Recirculation Region of an Industrial Burner Flame by Laser Anemometry with Light Frequency Shifting", Combustion and Flame, Vol. 23, Aug. 1974, pp. 57-71.

9. Hirsinger, F., and Tichtinsky, H., "Modélisation de Zones de Combustion en Régime Instationnaire", 54th Propulsion and Energetics Panel AGARD, Cologne, Oct. 1979.

10. Tichtinsky, H., "Three-Dimensional Flow Computation with Combustion and Recirculation", 6th International Symposium on Air Breathing Engines, Paris, June 1983.

11. Hebrard, P., and Magre, P., "Etude de l'Aérodynamique d'une Chambre de Combustion en vue d'une Modélisation Semi Empirique", Proceedings of the 54th Propulsion and Energetics Panel AGARD, Cologne, Oct. 1979.

12. McLaughlin, D. K., and Tiederman, W.G., "Biasing Correction for Individual Realization of Laser Anemometer Measurements in Turbulent Flows", The Physics of Fluids, Vol. 16, Dec. 1973, pp. 2082-2088.

13 Durao, D. F. G., Laker, J., and Whitelaw, J.H., "Biasing Correction for Individual Realization of Laser Anemometer Measurements in Turbulent Flows", <u>J. Phys. E. Sci. Instrum.</u>, Vol. 13, 1980, pp. 442-445.

14 Attal, B., Péalat, M., and Taran, J. P., "Diagnostics des Combustions par DRASC," Proceedings of the 55th Propulsion and Energetics Panel AGARD, Bruxelles, May 1980.

Author Index for Volume 92

Antcliff, R.R. 45
Benhachmi, D. 270
Bradley, R.P. 82
Collin, G. 325
Dobbins, R.A. 208
Eckbreth, A.C. 58
Emmerman, P.E. 270
Fontijn, A. 147
Goss, L.P. 24,82
Goulard, R. 270
Hall, R.J. 3, 58
Hirleman, E.D. 177
Jarrett, O. Jr. 45
Labbé, J. 325
Lewis, J.W.L. 132
MacDonald, B.G. 24
Magre, P. 325
McDaniel, J.C. 107

Powell, E.A. 238
Ray, S.R. 300
Reeves, C.M. 82
Roquemore, W.M. 82
Santavicca, D.A. 255
Santoro, R.J. 208
Selman, J.D. 132
Semerjian, H.G. 208, 300
Stufflebeam, J.H. 3
Stutrud, J.S. 82
Switzer, G.L. 24,82
Trump, D.D. 24, 82
Verdieck, J.F. 3, 58
Yahout, H. 270
Younes, N. 270
Zinn, B.T. 238
Zur Loye, A. 255

PROGRESS IN ASTRONAUTICS AND AERONAUTICS SERIES VOLUMES

VOLUME TITLE/EDITORS

*1. **Solid Propellant Rocket Research** (1960)
Martin Summerfield
Princeton University

*2. **Liquid Rockets and Propellants** (1960)
Loren E. Bollinger
The Ohio State University
Martin Goldsmith
The Rand Corporation
Alexis W. Lemmon Jr.
Battelle Memorial Institute

*3. **Energy Conversion for Space Power** (1961)
Nathan W. Snyder
Institute for Defense Analyses

*4. **Space Power Systems** (1961)
Nathan W. Snyder
Institute for Defense Analyses

*5. **Electrostatic Propulsion** (1961)
David B. Langmuir
Space Technology Laboratories, Inc.
Ernst Stuhlinger
NASA George C. Marshall Space Flight Center
J.M. Sellen Jr.
Space Technology Laboratories, Inc.

*6. **Detonation and Two-Phase Flow** (1962)
S.S. Penner
California Institute of Technology
F.A. Williams
Harvard University

*7. **Hypersonic Flow Research** (1962)
Frederick R. Riddell
AVCO Corporation

*8. **Guidance and Control** (1962)
Robert E. Roberson
Consultant
James S. Farrior
Lockheed Missiles and Space Company

*9. **Electric Propulsion Development** (1963)
Ernst Stuhlinger
NASA George C. Marshall Space Flight Center

*10. **Technology of Lunar Exploration** (1963)
Clifford I. Cummings and Harold R. Lawrence
Jet Propulsion Laboratory

*11. **Power Systems for Space Flight** (1963)
Morris A. Zipkin and Russell N. Edwards
General Electric Company

*12. **Ionization in High-Temperature Gases** (1963)
Kurt E. Shuler, Editor
National Bureau of Standards
John B. Fenn, Associate Editor
Princeton University

*13. **Guidance and Control—II** (1964)
Robert C. Langford
General Precision Inc.
Charles J. Mundo
Institute of Naval Studies

*14. **Celestial Mechanics and Astrodynamics** (1964)
Victor G. Szebehely
Yale University Observatory

*15. **Heterogeneous Combustion** (1964)
Hans G. Wolfhard
Institute for Defense Analyses
Irvin Glassman
Princeton University
Leon Green Jr.
Air Force Systems Command

*16. **Space Power Systems Engineering** (1966)
George C. Szego
Institute for Defense Analyses
J. Edward Taylor
TRW Inc.

*17. **Methods in Astrodynamics and Celestial Mechanics** (1966)
Raynor L. Duncombe
U.S. Naval Observatory
Victor G. Szebehely
Yale University Observatory

*18. **Thermophysics and Temperature Control of Spacecraft and Entry Vehicles** (1966)
Gerhard B. Heller
NASA George C. Marshall Space Flight Center

*19. **Communication Satellite Systems Technology** (1966)
Richard B. Marsten
Radio Corporation of America

---
*Out of print.

*20. **Thermophysics of Spacecraft and Planetary Bodies: Radiation Properties of Solids and the Electromagnetic Radiation Environment in Space** (1967)
Gerhard B. Heller
NASA George C. Marshall Space Flight Center

*21. **Thermal Design Principles of Spacecraft and Entry Bodies** (1969)
Jerry T. Bevans
TRW Systems

*22. **Stratospheric Circulation** (1969)
Willis L. Webb
Atmospheric Sciences Laboratory, White Sands, and University of Texas at El Paso

*23. **Thermophysics: Applications to Thermal Design of Spacecraft** (1970)
Jerry T. Bevans
TRW Systems

24. **Heat Transfer and Spacecraft Thermal Control** (1971)
John W. Lucas
Jet Propulsion Laboratory

25. **Communication Satellites for the 70's: Technology** (1971)
Nathaniel E. Feldman
The Rand Corporation
Charles M. Kelly
The Aerospace Corporation

26. **Communication Satellites for the 70's: Systems** (1971)
Nathaniel E. Feldman
The Rand Corporation
Charles M. Kelly
The Aerospace Corporation

27. **Thermospheric Circulation** (1972)
Willis L. Webb
Atmospheric Sciences Laboratory, White Sands, and University of Texas at El Paso

28. **Thermal Characteristics of the Moon** (1972)
John W. Lucas
Jet Propulsion Laboratory

29. **Fundamentals of Spacecraft Thermal Design** (1972)
John W. Lucas
Jet Propulsion Laboratory

30. **Solar Activity Observations and Predictions** (1972)
Patrick S. McIntosh and Murray Dryer
Environmental Research Laboratories, National Oceanic and Atmospheric Administration

31. **Thermal Control and Radiation** (1973)
Chang-Lin Tien
University of California at Berkeley

32. **Communications Satellite Systems** (1974)
P.L. Bargellini
COMSAT Laboratories

33. **Communications Satellite Technology** (1974)
P.L. Bargellini
COMSAT Laboratories

34. **Instrumentation for Airbreathing Propulsion** (1974)
Allen E. Fuhs
Naval Postgraduate School
Marshall Kingery
Arnold Engineering Development Center

35. **Thermophysics and Spacecraft Thermal Control** (1974)
Robert G. Hering
University of Iowa

36. **Thermal Pollution Analysis** (1975)
Joseph A. Schetz
Virginia Polytechnic Institute

37. **Aeroacoustics: Jet and Combustion Noise; Duct Acoustics** (1975)
Henry T. Nagamatsu, Editor
General Electric Research and Development Center
Jack V. O'Keefe, Associate Editor
The Boeing Company
Ira R. Schwartz, Associate Editor
NASA Ames Research Center

38. **Aeroacoustics: Fan, STOL, and Boundary Layer Noise; Sonic Boom; Aeroacoustic Instrumentation** (1975)
Henry T. Nagamatsu, Editor
General Electric Research and Development Center
Jack V. O'Keefe, Associate Editor
The Boeing Company
Ira R. Schwartz, Associate Editor
NASA Ames Research Center

39. **Heat Transfer with Thermal Control Applications** (1975)
M. Michael Yovanovich
University of Waterloo

SERIES LISTING

40. **Aerodynamics of Base Combustion** (1976)
S.N.B. Murthy, Editor
Purdue University
J.R. Osborn, Associate Editor
Purdue University
A.W. Barrows and J.R. Ward, Associate Editors
Ballistics Research Laboratories

41. **Communications Satellite Developments: Systems** (1976)
Gilbert E. LaVean
Defense Communications Agency
William G. Schmidt
CML Satellite Corporation

42. **Communications Satellite Developments: Technology** (1976)
William G. Schmidt
CML Satellite Corporation
Gilbert E. LaVean
Defense Communications Agency

43. **Aeroacoustics: Jet Noise, Combustion and Core Engine Noise** (1976)
Ira R. Schwartz, Editor
NASA Ames Research Center
Henry T. Nagamatsu, Associate Editor
General Electric Research and Development Center
Warren C. Strahle, Associate Editor
Georgia Institute of Technology

44. **Aeroacoustics: Fan Noise and Control; Duct Acoustics; Rotor Noise** (1976)
Ira R. Schwartz, Editor
NASA Ames Research Center
Henry T. Nagamatsu, Associate Editor
General Electric Research and Development Center
Warren C. Strahle, Associate Editor
Georgia Institute of Technology

45. **Aeroacoustics: STOL Noise; Airframe and Airfoil Noise** (1976)
Ira R. Schwartz, Editor
NASA Ames Research Center
Henry T. Nagamatsu, Associate Editor
General Electric Research and Development Center
Warren C. Strahle, Associate Editor
Georgia Institute of Technology

46. **Aeroacoustics: Acoustic Wave Propagation; Aircraft Noise Prediction; Aeroacoustic Instrumentation** (1976)
Ira R. Schwartz, Editor
NASA Ames Research Center
Henry T. Nagamatsu, Associate Editor
General Electric Research and Development Center
Warren C. Strahle, Associate Editor
Georgia Institute of Technology

47. **Spacecraft Charging by Magnetospheric Plasmas** (1976)
Alan Rosen
TRW Inc.

48. **Scientific Investigations on the Skylab Satellite** (1976)
Marion I. Kent and Ernst Stuhlinger
NASA George C. Marshall Space Flight Center
Shi-Tsan Wu
The University of Alabama

49. **Radiative Transfer and Thermal Control** (1976)
Allie M. Smith
ARO Inc.

50. **Exploration of the Outer Solar System** (1976)
Eugene W. Greenstadt
TRW Inc.
Murray Dryer
National Oceanic and Atmospheric Administration
Devrie S. Intriligator
University of Southern California

51. **Rarefied Gas Dynamics, Parts I and II (two volumes)** (1977)
J. Leith Potter
ARO Inc.

52. **Materials Sciences in Space with Application to Space Processing** (1977)
Leo Steg
General Electric Company

53. **Experimental Diagnostics in Gas Phase Combustion Systems** (1977)
Ben T. Zinn, Editor
Georgia Institute of Technology
Craig T. Bowman, Associate Editor
Stanford University
Daniel L. Hartley, Associate Editor
Sandia Laboratories
Edward W. Price, Associate Editor
Georgia Institute of Technology
James G. Skifstad, Associate Editor
Purdue University

54. **Satellite Communications: Future Systems** (1977)
David Jarett
TRW Inc.

55. **Satellite Communications: Advanced Technologies** (1977)
David Jarett
TRW Inc.

56. **Thermophysics of Spacecraft and Outer Planet Entry Probes** (1977)
Allie M. Smith
ARO Inc.

57. **Space-Based Manufacturing from Nonterrestrial Materials** (1977)
Gerard K. O'Neill, Editor
Princeton University
Brian O'Leary, Assistant Editor
Princeton University

58. **Turbulent Combustion** (1978)
Lawrence A. Kennedy
State University of New York at Buffalo

59. **Aerodynamic Heating and Thermal Protection Systems** (1978)
Leroy S. Fletcher
University of Virginia

60. **Heat Transfer and Thermal Control Systems** (1978)
Leroy S. Fletcher
University of Virginia

61. **Radiation Energy Conversion in Space** (1978)
Kenneth W. Billman
NASA Ames Research Center

62. **Alternative Hydrocarbon Fuels: Combustion and Chemical Kinetics** (1978)
Craig T. Bowman
Stanford University
Jorgen Birkeland
Department of Energy

63. **Experimental Diagnostics in Combustion of Solids** (1978)
Thomas L. Boggs
Naval Weapons Center
Ben T. Zinn
Georgia Institute of Technology

64. **Outer Planet Entry Heating and Thermal Protection** (1979)
Raymond Viskanta
Purdue University

65. **Thermophysics and Thermal Control** (1979)
Raymond Viskanta
Purdue University

66. **Interior Ballistics of Guns** (1979)
Herman Krier
University of Illinois at Urbana-Champaign
Martin Summerfield
New York University

67. **Remote Sensing of Earth from Space: Role of "Smart Sensors"** (1979)
Roger A. Breckenridge
NASA Langley Research Center

68. **Injection and Mixing in Turbulent Flow** (1980)
Joseph A. Schetz
Virginia Polytechnic Institute and State University

69. **Entry Heating and Thermal Protection** (1980)
Walter B. Olstad
NASA Headquarters

70. **Heat Transfer, Thermal Control, and Heat Pipes** (1980)
Walter B. Olstad
NASA Headquarters

71. **Space Systems and Their Interactions with Earth's Space Environment** (1980)
Henry B. Garrett and Charles P. Pike
Hanscom Air Force Base

72. **Viscous Flow Drag Reduction** (1980)
Gary R. Hough
Vought Advanced Technology Center

73. **Combustion Experiments in a Zero-Gravity Laboratory** (1981)
Thomas H. Cochran
NASA Lewis Research Center

74. **Rarefied Gas Dynamics, Parts I and II** (two volumes) (1981)
Sam S. Fisher
University of Virginia at Charlottesville

75. **Gasdynamics of Detonations and Explosions** (1981)
J.R. Bowen
University of Wisconsin at Madison
N. Manson
Université de Poitiers
A.K. Oppenheim
University of California at Berkeley
R.I. Soloukhin
Institute of Heat and Mass Transfer, BSSR Academy of Sciences

76. **Combustion in Reactive Systems** (1981)
J.R. Bowen
University of Wisconsin at Madison
N. Manson
Université de Poitiers
A.K. Oppenheim
University of California at Berkeley
R.I. Soloukhin
Institute of Heat and Mass Transfer, BSSR Academy of Sciences

77. **Aerothermodynamics and Planetary Entry** (1981)
A.L. Crosbie
University of Missouri-Rolla

78. **Heat Transfer and Thermal Control** (1981)
A.L. Crosbie
University of Missouri-Rolla

SERIES LISTING

79. **Electric Propulsion and Its Applications to Space Missions** (1981)
Robert C. Finke
NASA Lewis Research Center

80. **Aero-Optical Phenomena** (1982)
Keith G. Gilbert and Leonard J. Otten
Air Force Weapons Laboratory

81. **Transonic Aerodynamics** (1982)
David Nixon
Nielsen Engineering & Research, Inc.

82. **Thermophysics of Atmospheric Entry** (1982)
T.E. Horton
The University of Mississippi

83. **Spacecraft Radiative Transfer and Temperature Control** (1982)
T.E. Horton
The University of Mississippi

84. **Liquid-Metal Flows and Magnetohydrodynamics** (1983)
H. Branover
Ben-Gurion University of the Negev
P.S. Lykoudis
Purdue University
A. Yakhot
Ben-Gurion University of the Negev

85. **Entry Vehicle Heating and Thermal Protection Systems: Space Shuttle, Solar Starprobe, Jupiter Galileo Probe** (1983)
Paul E. Bauer
McDonnell Douglas Astronautics Company
Howard E. Collicott
The Boeing Company

86. **Spacecraft Thermal Control, Design, and Operation** (1983)
Howard E. Collicott
The Boeing Company
Paul E. Bauer
McDonnell Douglas Astronautics Company

87. **Shock Waves, Explosions, and Detonations** (1983)
J.R. Bowen
University of Washington
N. Manson
Université de Poitiers
A.K. Oppenheim
University of California at Berkeley
R.I. Soloukhin
Institute of Heat and Mass Transfer, BSSR Academy of Sciences

88. **Flames, Lasers, and Reactive Systems** (1983)
J.R. Bowen
University of Washington
N. Manson
Université de Poitiers
A.K. Oppenheim
University of California at Berkeley
R.I. Soloukhin
Institute of Heat and Mass Transfer, BSSR Academy of Sciences

89. **Orbit-Raising and Maneuvering Propulsion: Research Status and Needs** (1984)
Leonard H. Caveny
Air Force Office of Scientific Research

90. **Fundamentals of Solid-Propellant Combustion** (1984)
Kenneth K. Kuo
The Pennsylvania State University
Martin Summerfield
Princeton Combustion Research Laboratories, Inc.

91. **Spacecraft Contamination: Sources and Prevention** (1984)
J.A. Roux
The University of Mississippi
T.D. McCay
NASA Marshall Space Flight Center

92. **Combustion Diagnostics by Nonintrusive Methods** (1984)
T.D. McCay
NASA Marshall Space Flight Center
J.A. Roux
The University of Mississippi

(Other Volumes are planned.)